Optimization of Intelligent Transport Systems Planning Energy Efficiency and Environmental Responsibility

Optimization of Intelligent Transport Systems Planning Energy Efficiency and Environmental Responsibility

Guest Editors

Marcin Kłos
Grzegorz Sierpiński

MDPI

Basel • Beijing • Wuhan • Barcelona • Belgrade • Novi Sad • Cluj • Manchester

Guest Editors
Marcin Kłos
Department of Transport
Systems, Traffic Engineering
and Logistics
Silesian University of
Technology
Katowice
Poland

Grzegorz Sierpiński
Department Transport
Systems, Traffic Engineering
and Logistics
Silesian University of
Technology
Katowice
Poland

Editorial Office
MDPI AG
Grosspeteranlage 5
4052 Basel, Switzerland

This is a reprint of the Special Issue, published open access by the journal *Energies* (ISSN 1996-1073), freely accessible at: https://www.mdpi.com/journal/energies/special_issues/WKD0SVCPW4.

For citation purposes, cite each article independently as indicated on the article page online and as indicated below:

Lastname, A.A.; Lastname, B.B. Article Title. *Journal Name* **Year**, *Volume Number*, Page Range.

ISBN 978-3-7258-5167-6 (Hbk)
ISBN 978-3-7258-5168-3 (PDF)
https://doi.org/10.3390/books978-3-7258-5168-3

Contents

Marcin Jacek Kłos and Grzegorz Sierpiński
The Optimization of Intelligent Transport Systems: Planning, Energy Efficiency and Environmental Responsibility
Reprinted from: *Energies* **2025**, *18*, 4518, https://doi.org/10.3390/en18174518 . . . . **1**

Petar Sarajcev and Dino Lovric
Manifold Learning in Electric Power System Transient Stability Analysis
Reprinted from: *Energies* **2023**, *16*, 7810, https://doi.org/10.3390/en16237810 . . . . **6**

Miroslaw Smieszek, Vasyl Mateichyk and Jakub Mosciszewski
The Influence of Stops on the Selected Route of the City ITS on the Energy Efficiency of the Public Bus
Reprinted from: *Energies* **2024**, *17*, 4179, https://doi.org/10.3390/en17164179 . . . . **26**

Cong Ji, Elkhatib Kamal and Reza Ghorbani
Reliable Energy Optimization Strategy for Fuel Cell Hybrid Electric Vehicles Considering Fuel Cell and Battery Health
Reprinted from: *Energies* **2024**, *17*, 4686, https://doi.org/10.3390/en17184686 . . . . **52**

Wojciech Miechowicz, Marcin Kiciński, Izabela Miechowicz and Agnieszka Merkisz-Guranowska
The Attractiveness of Regional Transport as a Direction for Improving Transport Energy Efficiency
Reprinted from: *Energies* **2024**, *17*, 4844, https://doi.org/10.3390/en17194844 . . . . **78**

Krystian Woźniak, Beata Kurc, Łukasz Rymaniak, Natalia Szymlet, Piotr Pielecha and Jakub Sobczak
High-Capacity Energy Storage Devices Designed for Use in Railway Applications
Reprinted from: *Energies* **2024**, *17*, 5904, https://doi.org/10.3390/en17235904 . . . . **99**

Henryk Bąkowski, Iga Przytuła, Wioletta Cebulska, Damian Hadryś and Janusz Ćwiek
The Impact of Mechanical Failure of 18650 Batteries on the Safety of Electric Transport Operations
Reprinted from: *Energies* **2024**, *17*, 5980, https://doi.org/10.3390/en17235980 . . . . **114**

Piotr Sawicki, Hanna Sawicka, Marek Karkula and Krzysztof Zajda
Combined Rough Sets and Rule-Based Expert System to Support Environmentally Oriented Sandwich Pallet Loading Problem
Reprinted from: *Energies* **2025**, *18*, 268, https://doi.org/10.3390/en18020268 . . . . **125**

Thanapong Champahom, Chinnakrit Banyong, Thananya Janhuaton, Chamroeun Se, Fareeda Watcharamaisakul, Vatanavongs Ratanavaraha and Sajjakaj Jomnonkwao
Deep Learning vs. Gradient Boosting: Optimizing Transport Energy Forecasts in Thailand Through LSTM and XGBoost
Reprinted from: *Energies* **2025**, *18*, 1685, https://doi.org/10.3390/en18071685 . . . . **173**

energies MDPI

Editorial

The Optimization of Intelligent Transport Systems: Planning, Energy Efficiency and Environmental Responsibility

Marcin Jacek Kłos * and Grzegorz Sierpiński

Department of Transport Systems, Traffic Engineering, and Logistics, Faculty of Transport and Aviation Engineering, Silesian University of Technology, Krasinskiego 8 Street, 40-019 Katowice, Poland; grzegorz.sierpinski@polsl.pl
* Correspondence: marcin.j.klos@polsl.pl

1. Introduction

Modern urban development processes have created new demands on mobility systems, shaped by growing spatial constraints, evolving land-use patterns, and a marked increase in individual travel behaviors [1]. These dynamics have been accompanied by a surge in private vehicle ownership, often at the expense of public and active transport modes [2]. The consequences, including congestion, environmental degradation, rising energy consumption, and deteriorating urban health indicators, have reinforced the urgent need to rethink how transport systems are planned and managed.

In this Special Issue, the focus was placed on the optimization of intelligent transport systems (ITS) within the broader context of energy efficiency and environmental responsibility. Contributors addressed the inherent tensions between limited urban space and the growing demand for mobility, offering multidimensional perspectives on how ITS and smart planning tools can support more sustainable urban futures [3]. The accepted articles reflect the increasing relevance of approaches that combine technological innovation with spatial and social awareness [4].

Key themes covered in this issue include the optimization of public transport operations, spatial and infrastructural planning for electromobility, the integration of information technologies into transport systems, and the development of transit-oriented strategies aligned with the principles of sustainable mobility. Several contributions also explored the role of autonomous vehicles, MaaS (Mobility-as-a-Service) platforms, and novel driving systems in enhancing system-wide efficiency and reducing environmental impact.

Collectively, the articles in this Special Issue demonstrate the critical importance of holistic and data-driven planning approaches in modern transport systems. They also underscore the potential of ITS as a catalyst for systemic change, enabling not only better service quality but also measurable improvements in energy use and environmental outcomes. The findings presented here are of interest not only to researchers but also to decision-makers tasked with shaping the future of urban mobility.

2. A Short Review of the Contributions in This Special Issue

Brief summaries of the six selected papers belonging to this Special Issue of *Energies*, titled "The Optimization of Intelligent Transport Systems: Planning, Energy Efficiency and Environmental Responsibility", are included in the next subsections.

2.1. Manifold Learning in Electric Power System Transient Stability Analysis [Contribution 1]

Sarajcev and Lovrić propose an innovative application of manifold learning methods for the transient stability analysis (TSA) of electric power systems. Using data derived from

IEEE 39-bus system simulations and wide-area PMU monitoring, they compare multiple dimensionality reduction techniques including PCA, kernel PCA, isomap, MDS, LLE, and t-SNE. Their findings show that kernel PCA, when combined with a support vector machine classifier, offers robust performance for identifying system disturbances, even in a fully unsupervised learning setting. The study contributes both methodological advancements and practical insights into the integration of machine learning with real-time power grid stability diagnostics.

2.2. *The Influence of Stops on the Selected Route of the City ITS on the Energy Efficiency of the Public Bus [Contribution 2]*

This study investigates how bus stops, both scheduled and traffic-induced, affect energy consumption in urban public transport, using data from Rzeszów, Poland. Based on GPS tracking and fuel usage records from a diesel-powered city bus operating on route 13, the analysis quantifies energy losses resulting from stopping, idling, and subsequent acceleration. The findings indicate that approximately 26.2% of total fuel consumption is attributable to stops at bus stops, while 42.5% is due to stops caused by traffic conditions. The study highlights that implementing on-demand stops and improving traffic flow management through intelligent transport systems (ITS) could significantly enhance the energy efficiency of urban bus operations.

2.3. *Reliable Energy Optimization Strategy for Fuel Cell Hybrid Electric Vehicles Considering Fuel Cell and Battery Health [Contribution 3]*

This article presents a novel hierarchical energy management strategy (HSEMS) for fuel cell hybrid electric vehicles (FCHEVs) aimed at reducing hydrogen consumption while enhancing energy efficiency and component durability. The upper-level controller employs fuzzy fault-tolerant mechanisms to ensure reliable operation and preserve the health of the battery and proton-exchange membrane fuel cell (PEMFC) under fault conditions. The lower-level controller uses dynamic programming and Pontryagin's minimum principle for optimal real-time power distribution between the fuel cell and battery, integrating a proportional–integral controller for precise tracking. Simulations using TruckMaker/MATLAB demonstrate the proposed strategy's superiority over equivalent consumption minimization and fuzzy logic methods, yielding up to 9.8% hydrogen savings while effectively managing faults. The strategy's predictive capability is enhanced through adaptive fuzzy and neuro-fuzzy SOC estimators, contributing to extended component lifespan and robust performance under varying driving cycles.

2.4. *The Attractiveness of Regional Transport as a Direction for Improving Transport Energy Efficiency [Contribution 4]*

This article examines the role of regional public transport frequency in improving energy efficiency by encouraging a modal shift from private to public transport. Based on empirical observations from 27 regional bus lines across 19 Polish districts, the authors develop novel indicators (AttK and AttYK) to assess transport attractiveness for different demographic groups. Their statistical analysis confirms a strong positive correlation (Spearman's Rs = 0.807, $p = 0.001$) between the number of daily connections and attractiveness for adults and seniors, suggesting that at least four daily pairs of connections are necessary to make regional transport a viable alternative to private cars. The study estimates that each additional connection can attract 1.5 new adult passengers and reduce energy consumption by 0.33–0.69 kWh per kilometer, contributing to the EU's transport energy efficiency goals. The findings highlight the importance of service frequency in combating transport exclusion and achieving sustainability in non-urban areas.

2.5. High-Capacity Energy Storage Devices Designed for Use in Railway Applications [Contribution 5]

This article investigates the potential of high-capacity supercapacitors, particularly those using starch-derived carbon materials, for application in railway transport to recover energy from electrodynamic braking. Through experimental testing on a diesel–electric multiple unit, the authors demonstrate that approximately 28.5% of energy can be recuperated during braking, irrespective of the drive type. The paper highlights the synthesis and performance of starch-based carbon electrodes, which achieved capacitances of up to 130 F/g, offering a sustainable and cost-effective alternative to conventional carbon materials. It discusses the electrochemical behavior, challenges of energy density and temperature sensitivity, and the need for effective voltage balancing in railway systems. The study further estimates that storing 15 kWh of recovered energy would require about 245 kg of supercapacitors, emphasizing both the feasibility and the technical considerations for large-scale implementation. Ultimately, the paper contributes to advancing environmentally responsible energy solutions for modern rail transport through supercapacitor integration.

2.6. The Impact of Mechanical Failure of 18,650 Batteries on the Safety of Electric Transport Operations [Contribution 6]

This study investigates the impact of mechanical damage on the safety of 18,650 lithium-ion batteries used in electric transport systems. Through finite element method (FEM) simulations and crash tests at various speeds, the authors demonstrate that collisions can lead to structural deformation, internal short circuits, and increased internal resistance, significantly impairing battery performance and potentially causing thermal runaway. Despite severe damage at high speeds (e.g., 200 km/h), thermal imaging confirmed that no internal short circuits occurred during charging, though capacity and reusability were compromised. The study underscores the importance of robust battery design, effective thermal management, and real-time monitoring systems to mitigate risks and improve safety and reliability in electromobility applications.

2.7. Combined Rough Sets and Rule-Based Expert System to Support Environmentally Oriented Sandwich Pallet Loading Problem [Contribution 7]

This article presents an integrated methodology that addresses the environmentally responsible optimization of the Sandwich Pallet Loading Problem (SPLP) within the logistics of fast-moving consumer goods. The authors combine Dominance-Based Rough Set Theory (DRST) and a rule-based expert system to minimize the number of stacked load units (LU), thereby reducing transport-related energy consumption. Key attributes such as product weight, height, and fragility are used to classify LUs. DRST handles imprecise data, while the expert system applies decision rules to determine optimal LU arrangements. The methodology is validated using real-world data from a major distribution company, achieving a 40% reduction in energy consumption per customer order. The paper demonstrates how machine learning and optimization can jointly support sustainable logistics operations by minimizing shipments, reducing resource use, and enhancing decision-making consistency.

2.8. Deep Learning vs. Gradient Boosting: Optimizing Transport Energy Forecasts in Thailand Through LSTM and XGBoost [Contribution 8]

This article investigates the accuracy of transport energy consumption forecasting in Thailand by comparing two advanced machine learning methods: Long Short-Term Memory (LSTM) neural networks and the XGBoost algorithm. Using data from 1993 to 2022, including vehicle registrations by size, vehicle kilometers traveled (VKT), GDP, and population, the study evaluates each model's predictive performance. Results show that XGBoost significantly outperforms LSTM across all statistical measures, achieving an R^2 of 0.9508 versus 0.2005. Feature importance analysis highlights that medium vehicle registrations and truck VKT are dominant predictors of energy use, together accounting for over 57% of model influence, while demographic and economic variables contribute about 15%. The research not only demonstrates the superior forecasting capability of XGBoost, but also offers practical insights for policymakers seeking to enhance transport energy efficiency through evidence-based fleet and infrastructure planning.

3. Conclusions and Future Works

This Special Issue provides a comprehensive overview of current advances in the optimization of intelligent transport systems (ITS) with a strong emphasis on energy efficiency, environmental responsibility, and transport safety. The contributions demonstrate the versatility of methods ranging from machine learning and rough set theory to empirical transport energy analyses and safety testing of vehicle components applied to both urban and regional contexts. Together, these studies illustrate how integrating intelligent technologies, operational optimization, and energy-aware planning can foster more sustainable and resilient mobility systems.

The results confirm that ITS optimization requires not only technological innovation but also a context-sensitive approach that considers the physical, environmental, and social constraints of transport networks. Whether in public transport, logistics, electromobility, or infrastructure forecasting, the alignment of digital solutions with broader sustainability goals is essential for realizing systemic improvements.

Looking ahead, several avenues for future research emerge. These include the following:

- Integrating ITS planning with real-time big data analytics to improve adaptive transport management;
- Investigating the long-term environmental and economic impacts of electromobility infrastructure and alternative energy storage technologies (e.g., supercapacitors, hydrogen cells);
- Exploring hybrid modeling frameworks that combine machine learning with rule-based and simulation models for transport system forecasting and risk assessment;
- Extending multimodal energy efficiency studies to include active transport and last-mile logistics;
- Assessing the social acceptability and behavioral implications of ITS and Mobility-as-a-Service (MaaS) platforms in diverse urban and regional settings.

These directions offer promising opportunities for advancing transport research and supporting policy frameworks aimed at decarbonization, accessibility, and resilience in the mobility sector.

Funding: This research received no external funding.

Conflicts of Interest: The author declares no conflicts of interest.

List of Contributions:

1. Sarajcev, P.; Lovric, D. Manifold Learning in Electric Power System Transient Stability Analysis. *Energies* **2023**, *16*, 7810. https://doi.org/10.3390/en16237810.
2. Smieszek, M.; Mateichyk, V.; Mosciszewski, J. The Influence of Stops on the Selected Route of the City ITS on the Energy Efficiency of the Public Bus. *Energies* **2024**, *17*, 4179. https://doi.org/10.3390/en17164179.
3. Ji, C.; Kamal, E.; Ghorbani, R. Reliable Energy Optimization Strategy for Fuel Cell Hybrid Electric Vehicles Considering Fuel Cell and Battery Health. *Energies* **2024**, *17*, 4686. https://doi.org/10.3390/en17184686.

4. Miechowicz, W.; Kiciński, M.; Miechowicz, I.; Merkisz-Guranowska, A. The Attractiveness of Regional Transport as a Direction for Improving Transport Energy Efficiency. *Energies* **2024**, *17*, 4844. https://doi.org/10.3390/en17194844.
5. Woźniak, K.; Kurc, B.; Rymaniak, Ł.; Szymlet, N.; Pielecha, P.; Sobczak, J. High-Capacity Energy Storage Devices Designed for Use in Railway Applications. *Energies* **2024**, *17*, 5904. https://doi.org/10.3390/en17235904.
6. Bąkowski, H.; Przytuła, I.; Cebulska, W.; Hadryś, D.; Ćwiek, J. The Impact of Mechanical Failure of 18650 Batteries on the Safety of Electric Transport Operations. *Energies* **2024**, *17*, 5980. https://doi.org/10.3390/en17235980.
7. Sawicki, P.; Sawicka, H.; Karkula, M.; Zajda, K. Combined Rough Sets and Rule-Based Expert System to Support Environmentally Oriented Sandwich Pallet Loading Problem. *Energies* **2025**, *18*, 268. https://doi.org/10.3390/en18020268.
8. Champahom, T.; Banyong, C.; Janhuaton, T.; Se, C.; Watcharamaisakul, F.; Ratanavaraha, V.; Jomnonkwao, S. Deep Learning vs. Gradient Boosting: Optimizing Transport Energy Forecasts in Thailand Through LSTM and XGBoost. *Energies* **2025**, *18*, 1685. https://doi.org/10.3390/en18071685.

References

1. Kłos, M.J.; Sierpiński, G. Siting of Electric Vehicle Charging Stations Method Addressing Area Potential and Increasing Their Accessibility. *J. Transp. Geogr.* **2023**, *109*, 103601. [CrossRef]
2. Drabicki, A.; Kucharski, R.; Szarata, A. Modelling the Public Transport Capacity Constraints' Impact on Passenger Path Choices in Transit Assignment Models. *Arch. Transp.* **2017**, *43*, 7–28. [CrossRef]
3. Claire, P.; Dupont-Kieffer, A.; Palmier, P. Potential Accessibility to the Workplace by Public Transit and Its Social Distribution in Lille, France: A Scenario-Based Equity Appraisal. *Transp. Policy* **2022**, *125*, 256–266. [CrossRef]
4. Shanmukhappa, T.; Ho, I.W.H.; Tse, C.K. Spatial Analysis of Bus Transport Networks Using Network Theory. *Phys. A Stat. Mech. Its Appl.* **2018**, *502*, 295–314. [CrossRef]

energies

Article

Manifold Learning in Electric Power System Transient Stability Analysis

Petar Sarajcev * and Dino Lovric

Department of Electric Power Systems, Faculty of Electrical Engineering, Mechanical Engineering and Naval Architecture, University of Split, R. Boskovica 32, HR-21000 Split, Croatia; dino.lovric@fesb.hr
* Correspondence: petar.sarajcev@fesb.hr

Abstract: This paper examines the use of manifold learning in the context of electric power system transient stability analysis. Since wide-area monitoring systems (WAMSs) introduced a big data paradigm into the power system operation, manifold learning can be seen as a means of condensing these high-dimensional data into an appropriate low-dimensional representation (i.e., embedding) which preserves as much information as possible. In this paper, we consider several embedding methods (principal component analysis (PCA) and its variants, singular value decomposition, isomap and spectral embedding, locally linear embedding (LLE) and its variants, multidimensional scaling (MDS), and others) and apply them to the dataset derived from the IEEE New England 39-bus power system transient simulations. We found that PCA with a radial basis function kernel is well suited to this type of power system data (where features are instances of three-phase phasor values). We also found that the LLE (including its variants) did not produce a good embedding with this particular kind of data. Furthermore, we found that a support vector machine, trained on top of the embedding produced by several different methods was able to detect power system disturbances from WAMS data.

Keywords: electric power system; bulk power system; transient stability analysis; rotor angle stability; wide-area monitoring system; machine learning; manifold learning

1. Introduction

The electric power system features the most complex transport system to date. It is presently undergoing a very rapid and intensive transformation, as it is transitioning toward the renewable and sustainable future. The most important aspect of that transformation is the accelerating change in the mix of generating resources, which reduces the proportion of conventional power plants (with large synchronous generators) while increasing the share of inverter-based resources (IBRs), i.e., wind and solar power plants. This change in the generation mix represents a fundamental shift in the operational characteristics of the power system as such, with potential (significant) reliability implications for the electric grid [1]. Namely, generating resources need to provide on an ongoing basis, among others, voltage control, frequency support, and ramping capability in order to balance and maintain the grid [2]. However, the vast majority of IBRs today do not participate in this support. At the same time, generating resources need to provide system stability during transient events, brought on by faults (i.e., short-circuits), severe weather, and other internal and external disturbances. On this front, new resource mix will significantly (and irreversibly) reduce the power system inertia, thereby exacerbating the power system transient stability (TSA) problems of tomorrow. The response of the power grids with low system inertia to disturbances necessitates new approaches for dealing with the TSA issues [3]. The electric grid is also under increasing strain from the growing share of electric vehicles, which increase in the overall grid load and electricity demands. These combined stresses on the electric power system are without precedent and pose considerable technical challenges, which need to be successfully tackled if the electrification of the transport sector, and transition to the renewable energy, is to be realized.

These aforementioned developments are accompanied by still more innovations in the power system sector, one of which is the introduction of wide-area monitoring (WAM) aspects to the electric grid, based on extensive communication infrastructure and distributed phasor measurement units (PMUs). WAMSs bring a "big data" paradigm into the operational portfolio of the transmission system operators and provide novel avenues for tackling various power system problems, from load flow, state estimation, relay protection, to transient analyses [4,5]. Furthermore, electric power systems are integrating an increasing number of power electronic components, FACTS, AC/DC devices, thereby changing the response characteristics of the system to various disturbances [6]. All these different aspects set new challenges and, at the same time, provide opportunities for novel technical solutions. Namely, some well-known traditional solutions—particularly in the domain of transient stability analysis—have acknowledged weaknesses that may compromise the security of the bulk power system under increased stress emanating from the accelerated pace of its ongoing transformation.

The transient stability analysis of the electric power system is currently the focus of intense research [7], as it is increasingly becoming obvious that traditional solutions are not future-proof and are becoming increasingly maladjusted to rapid changes in modern systems. Neither time-domain simulations, direct methods (equal-area and Lyapunov methods), nor different hybrid schemes are able to deliver solutions that would satisfy the reliability requirements of modern power systems [8]. Hence, a new breed of methods is slowly emerging, based on artificial intelligence and WAMS data troves.

Machine learning (ML) and its subset of deep learning have been thus far applied to the TSA problem, with promising results on benchmark test cases. For example, Tan et al. used a deep imbalanced learning framework for the TSA of a power system [9]. Liu et al. introduced transient stability assessment models based on ensemble learning with network topology [10], distance metric learning [11] and kernel regression. Su et al. proposed a probabilistic stacked denoising autoencoder for a power system transient stability prediction with wind farms [12]. Autoencoder is a special kind of artificial neural network (ANN) that can be trained using an unsupervised learning approach. Tian et al. introduced a transient stability boundary, constructed using the broad learning approach [13]. Ren and Xu recommended a method for the TSA based on generative adversarial networks, where two ANNs contested with each other in an unsupervised manner [14]. Also, Ren et al. employed transfer learning for the transient stability analysis [15]. Transfer learning is the ability to reuse ANNs pretrained on one (general) task and repurpose them (with fine-tuning) for different but related (specialized) tasks. Li et al. suggested an intelligent TSA framework with continual learning ability [16]. An ensemble of decision trees was employed in [17–19] to tackle the TSA problem. Support vector machine (SVM) is a popular ML approach, used by many authors for tackling the TSA problem, e.g., [20–22]. In addition, Bashiri et al. introduced a learning framework for the TSA that relied on a twin convolutional SVM [23].

Inherent in the deep learning approach is the fact that the classical feature engineering step of the ML model building process is absent and replaced by an automatic learning of the features by the deep model itself during training [24]. This removes expert knowledge and instead relies on deep ANN architectures to learn useful features. However, a considerable preprocessing of the data (signals) is still needed for any deep learning model, even without classical feature engineering. Another important aspect of the TSA problem, which any machine learning model needs to tackle, is the fact that data space is multidimensional, often with many thousands of features or more. This poses a challenge to training ML models, and some kind of dimensionality reduction technique or embedding process is typically employed as part of the model training. For example, Chen in [25] applied an indirect principal component analysis (PCA) as a dimensionality reduction technique for a power system transient stability assessment. Bellizio et al. proposed a causality-based feature selection [26]. Gnanavignesh and Shenoy introduced a relaxation-based spatial-domain decomposition method [27]. Arvizu et al. used a diffusion maps approach for the dimensionality reduction in transient simulations [28]. Still, other signal feature extraction and dimension reduction techniques have been proposed, e.g., [29,30].

This paper is concerned with examining the use of manifold learning (as part of the wider ML landscape) in assessing the power system TSA, based on WAMS data and power system simulations. Manifold learning is an approach to the nonlinear dimensionality reduction of the data through its low-dimensional embedding that preserves maximum information. The paper features both supervised and unsupervised learning with different embedding methods in order to demonstrate their feasibility in the domain of power system transient stability analysis with WAMS-PMU data. Namely, our overview of the published peer-reviewed research did not show comprehensive comparisons of different embedding methods for the power system TSA with PMU data. Hence, we believe that this paper fills that need and brings new insights in the domain of applying manifold learning to the power system TSA problem. The following methods are examined [31]: (1) principal component analysis, (2) kernel PCA (kPCA), (3) truncated singular value decomposition (tSVD), (4) isomap embedding, (5) locally linear embedding (LLE) and its variants, (6) spectral embedding, (7) multidimensional scaling (MDS), and (8) *t*-distributed stochastic neighbor embedding (*t*-SNE). These methods are applied on the WAMS-PMU data derived from the IEEE New England 39-bus test-case benchmark power system [32]. The model hyperparameters are optimized using cross-validated metrics and a custom-tailored simulated-annealing stochastic optimization algorithm. When supervised learning is applied, it is based on a support vector machine classifier.

Our research hypothesis is that a low-dimensional embedding, derived from the WAMS-PMU data (features), preserves information and can detect power system TSA events in the supervised and unsupervised settings. However, it is shown that not all embedding methods produce equally good results on the TSA data, and that some are better suited than others. We find that a combination of kPCA with SVM classifier provides good overall results in the domain of supervised learning. The same can be said about the isomap embedding and MDS when used with the SVM. However, unlike kPCA, embeddings produced by isomap and MDS are not divisible into separate clusters, which makes them less favorable. We also find that LLE and its variants cannot produce useful embeddings with these data. Furthermore, when the unsupervised learning is concerned, we find that only kPCA is able to generate separable clusters. Finally, we find that the SVM classifier trained on top of several of the tested embeddings is able to detect power system disturbances from the WAMS-PMU data.

The paper is organized in the following manner. Section 2 first introduces in Section 2.1 a dataset which emanates from the IEEE New England test power system simulations and brings a description of the steps in its processing pipeline. Next, in Section 2.2, it exposes a simulated annealing stochastic optimization algorithm and its application to hyperparameter tuning. Section 2.3 introduces unsupervised and supervised learning for the power system transient stability assessment, with different embedding methods. The results of the models' applications on the IEEE New England dataset are provided in Section 3, which is followed by a discussion in Section 4 and a conclusion is Section 5.

2. Materials and Methods

The materials part introduces the dataset and describes its processing steps: statistical sampling, features engineering, shuffling, splitting, and normalizing. The methods part first describes the simulated annealing algorithm and then briefly introduces embedding models. Next, it provides details on the training process (i.e., supervised and unsupervised learning), including the hyperparameter optimization and model scoring.

2.1. Power System Stability Dataset

The power system stability dataset was produced from the extensive set of time-domain simulations on the IEEE New England 39-bus benchmark test power system, with varying degrees of loading and three different short-circuit types scattered throughout the network. Further details on the benchmark power system, time-domain simulations, and dataset construction can be found in [33]. The full 3.8 GB dataset of PMU signals was deposited on Zenodo with a CC BY license [34]. These raw signals were further processed according to the pipeline graphically depicted in Figure 1. It consisted of following steps: a statistical processing of data cases (from systematic into the stochastic domain), extracting features from the time-domain signals, splitting the data into training and test sets, and scaling them. The processing of data cases by random sampling without replacement followed a probability distribution of short-circuit events found in the actual power systems [33]. The splitting strategy used a stratified shuffle split in order to preserve the class imbalance between stable (majority) and unstable (minority) cases. The scaling standardized (training and test) the datasets by removing the mean and scaling them to a unit variance.

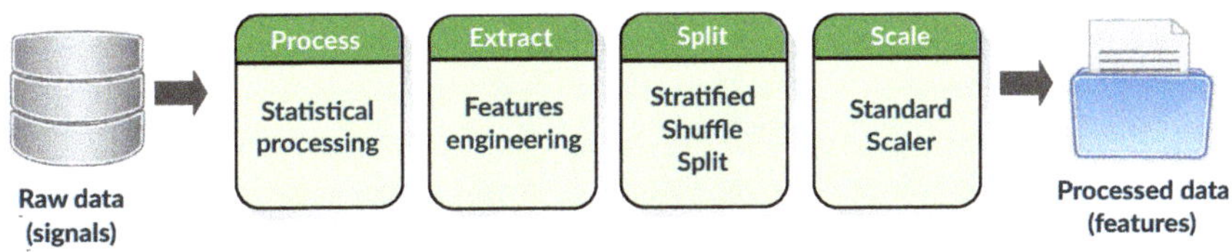

Figure 1. Data processing pipeline transforming the raw time-domain PMU signals into standardized training and test feature sets.

Table 1 presents selected features extracted from the time-domain PMU signals. Each time-domain signal was represented by only two points in time, based on prefault and postfault values of the measured quantity. It needs to be stated here that based on the fault location within the network, a coordination between generator out-of-step protection and distance relay protection of the incident transmission lines should determine the time instants for the feature selection. The labeling of data cases as stable or unstable was carried out by means of the power system transient stability index (TSI) [8,33]. The statistically processed dataset with 350+ features was also deposited on Zenodo with a CC BY license [35].

Table 1. Features derived from the time-domain PMU signals.

Feature Description	Number
Rotor speed	10
Rotor angle deviation	10
Rotor mechanical angle	10
Stator voltage	10
Stator d-component current	10
Stator q-component current	10
Power load angle	2×10 [1]
Generator active power	2×10 [1]
Generator reactive power	2×10 [1]
Bus voltage magnitude	$3 \times 2 \times 39$ [2]

[1] prefault and postfault values. [2] prefault and postfault values in all three phases.

It should be mentioned that we used expert knowledge in selecting these features, and that we had already removed some redundant features (e.g., one of the three phases, all phase angle values, etc.). Notwithstanding that, the number of features was still quite large for this small power system. More importantly, it grew considerably with the increase in number of buses and generators in the system and could easily become difficult to manage and process (i.e., curse of dimensionality). Hence, the main obstacle to the successful training of ML models was controlling and/or reducing this large number of features without significant loss of information.

2.2. Simulated Annealing Algorithm

Simulated annealing is a general-purpose stochastic optimization algorithm that usually consists of four main parts [36]: (1) an objective function that is being optimized, (2) a perturbation method that is used for generating candidate solutions, (3) an acceptance criterion, and (4) a temperature schedule (i.e., cooling). Our improvements on the classical algorithm constituted using the burn-in period, a multidimensional random walk based on Student's t-distribution as a perturbation method, and an early stopping criterion. This combination of burn-in with early stopping balanced exploration and exploitation while reducing the possibility of overfitting.

The objective function in our case constructed the model for which the hyperparameters were being optimized $f(\mathbf{x})$ and returned a cross-validated metric by which it was scored.

We employed an n-dimensional random walk as a perturbation method, where n is the model's number of hyperparameters x_i, $i = 1, \ldots, n$. For that purpose, we drew uncorrelated random samples from the multivariate statistical distribution in the following manner. During the initial phase of the algorithm's exploration of the search space (i.e., during a select number of initial iterations $i \leq k$), random samples were drawn from the multivariate Student's t-distribution as follows, $for\ i = 1, \ldots, k$:

$$\mathbf{z}_i \sim MVN(\mathbf{0}, \mathbf{\Sigma}), \tag{1}$$

$$u_i \sim \chi^2(\nu), \tag{2}$$

$$\mathbf{w}_i = \sqrt{\frac{\nu}{u_i}} \cdot \mathbf{z}_i, \tag{3}$$

with $\mathbf{z}$ being the random 1D vector drawn from the multivariate normal distribution, where $\mathbf{\Sigma} = diag(\sigma_1^2, \ldots, \sigma_n^2)$ is the diagonal matrix of covariances for the hyperparameters (as uncorrelated random variables); u is the random variable from the uniform χ^2 distribution with ν degrees of freedom, and $\mathbf{w}$ is finally the random 1D vector from the multivariate Student's t-distribution with ν degrees of freedom. This phase of the algorithm (where $i \leq k$) is called the burn-in period, and it emphasizes the exploration of the search space (by using low values of the ν parameter). After the burn-in period, the samples for the subsequent steps ($i > k$) are drawn directly from the multivariate normal distribution, but (generally) with a lower variance, as follows:

$$\mathbf{w}_i \sim MVN(\mathbf{0}, \mathbf{\Sigma} \cdot f_s);\ for\ i > k, \tag{4}$$

where $0 < f_s \leq 1$ is the factor by which the initial standard deviations (that form the diagonal covariance matrix) are scaled. If there is a statistical correlation between hyperparameters, that information can be readily introduced through the covariance matrix itself.

Hence, for the ith iteration, a random walk takes the steps $\mathbf{x_{new}}_i = \mathbf{x_{old}}_i + \mathbf{w}_i$, where $\mathbf{w}$ comes from Equation (3) if $i \leq k$ or from Equation (4) if $i > k$; $\mathbf{x}$ is the 1D vector holding the (either new or old) coordinates of hyperparameter values inside the multidimensional search space. It can be seen that the burn-in phase (first k iterations of the algorithm) take longer steps (from the Student's t-distribution which has more mass in the tail than a normal distribution) and thus is able to explore more of the search space. After the burn-in period, the step size is reduced and, in addition, the starting position of vector x at the beginning of iteration $k + 1$ is taken from the best coordinates found during the burn-in period of the previous k iterations. This starts the exploitation phase. Furthermore, bounds can be imposed on hyperparameters within the search space, such that each variable x from the 1D vector $\mathbf{x}$ can have upper and lower bounds, i.e., $x \in [x_l, x_u]$, where x_l is the lower bound and x_u is the upper bound. If, during the random walk, a step takes any variable x outside its imposed bounds, that step is reversed as follows: $x_{new} = x_{old} - abs(w)/2$ if

$x_{new} > x_u$ and $x_{new} = x_{old} + abs(w)/2$ if $x_{new} < x_l$. In other words, if the step brings a variable above the upper bound, it is reversed to the left; alternatively, if the step brings it below the lower bound, it is reversed to the right.

The acceptance criterion is based on the Metropolis algorithm [31], where the difference between each two successive objective function's evaluations is computed as $\Delta E = f(\mathbf{x_{new}}) - f(\mathbf{x_{old}})$. If $\Delta E \leq 0$, the new solution is readily accepted. On the other hand, the case when $\Delta E > 0$ is treated as probabilistic. First, a random number is drawn from theuniform distribution $r \sim \mathcal{U}(0,1)$. Then, a probability that the new solution is accepted is derived from the *Boltzman* probability distribution as follows:

$$p(\Delta E) = (2\pi T)^{-n/2} \cdot \exp(\frac{-|\Delta E|^2}{2T}), \tag{5}$$

where T is the current temperature. It can be seen that this probability is related to the temperature, and that it decreases as the system state cools down. Now, a new solution is accepted only if $r < p(\Delta E)$; otherwise, it is rejected. As can be seen, this last step is stochastic and allows us to accept inferior solutions at the beginning, when the temperature is still high, which helps to explore the search space. Later, when the temperature is decreased, the acceptance of inferior solutions is far less probable.

The temperature schedule (i.e., cooling) followed the exponential law, where the temperature for the ith iteration can be determined as follows:

$$T_i = T_0 \cdot \exp(-i/\kappa), \tag{6}$$

where T_0 is the initial temperature and κ is a constant that determines the cooling time. A total number of iterations is determined directly from the cooling schedule, i.e., the algorithm proceeds until the system has cooled down to a certain predetermined temperature.

After each iteration step, the accepted solution is compared to the best solution that has been found thus far and takes its place when having a lower value (i.e., is closer to the optimum). Furthermore, when $|\Delta E| < \epsilon$ holds, where ϵ is an arbitrary small number, the algorithm does not converge. This is monitored and if there is no progress after a certain number of iterations (i.e., waiting period), the algorithm is terminated under the early stopping criterion.

The complete simulated annealing algorithm, described in pseudocode, is depicted in Algorithm 1.

Algorithm 1 Simulated annealing algorithm with burn-in and early stopping.

input: k, $\mathbf{x_0}$, $[x_l, x_u]_{i=1,n}$, $\mathbf{\Sigma}$, f_s, T_0, κ, ϵ, θ, stop
$\mathbf{x} = \mathbf{x_0}$ ▷ initial values
$E = f(\mathbf{x_0})$
$\mathbf{x}_{best} = \mathbf{x}$
$\mathbf{E}_{best} = \mathbf{E}$
i = 0; wait = 0
while $T \geq \theta$ **do**
  if $i \leq k$ **then** ▷ burn-in phase
    $\mathbf{z} \sim MVN(\mathbf{0}, \mathbf{\Sigma})$
    $u \sim \chi^2(\nu)$
    $\mathbf{w} = \sqrt{\frac{\nu}{u}} \cdot \mathbf{z}$
  else
    $\mathbf{w} \sim MVN(\mathbf{0}, \mathbf{\Sigma} \cdot f_s)$
  $\mathbf{x_{new}} \leftarrow \mathbf{x} + \mathbf{w}$ ▷ random walk
  for $j = 1, n$ **do**
    if $x_{new} < x_l$ **then** ▷ below the lower bound
      $x_{new} \leftarrow x + abs(w)/2$
    else if $x_{new} > x_u$ **then** ▷ above the upper bound
      $x_{new} \leftarrow x - abs(w)/2$
  $E_{new} \leftarrow f(\mathbf{x_{new}})$
  $\Delta E = E_{new} - E$
  if $\Delta E \leq 0$ **then**
    $\mathbf{x} = \mathbf{x_{new}}$
    $E = E_{new}$
  else ▷ Metropolis acceptance
    $r \sim \mathcal{U}(0,1)$
    $p(\Delta E) = (2\pi T)^{-n/2} \cdot \exp(\frac{-|\Delta E|^2}{2T})$
    if $r < p(\Delta E)$ **then**
      $\mathbf{x} = \mathbf{x_{new}}$
      $E = E_{new}$
  T $\leftarrow T_0 \cdot \exp(-T/\kappa)$
  if $E < E_{best}$ **then**
    $\mathbf{x}_{best} = \mathbf{x_{new}}$
    $\mathbf{E}_{best} = \mathbf{E_{new}}$
  if $abs(\Delta E) \leq \epsilon$ **then**
    wait += 1
  if wait > stop **then**
    break
  i += 1

2.3. Embedding Models

The following embedding methods, as already mentioned, are briefly introduced hereafter: PCA, kPCA, tSVD, isomap embedding, LLE and its variants (modified and Hessian), spectral embedding, MDS, and *t*-SNE. Interested readers are at this point advised to consult [31] for more information and additional mathematical background on some of these different methods.

2.3.1. PCA, kPCA, and tSVD

Principal component analysis is a well-known method for decomposing a multivariate dataset into a set of orthogonal components (i.e., principal components). It employs a linear projection and relies on the SVD algorithm. Kernel PCA is a direct extension of PCA, which introduces a nonlinear dimensionality reduction through the use of kernels (i.e., the kernel trick); see [31] for more information. Different kernels can be applied, but we examined only the radial basis function (RBF) kernel. Truncated SVD is another method closely related to PCA. It uses, as the name suggests, a singular value decomposition of the feature matrix and retains only its largest singular values, thereby reducing the dimensionality of the feature space [31]. If k is the dimensionality of the projection, then a truncated SVD applied to feature matrix X produces a low-rank approximation $X \approx X_k = U_k \Sigma_k V_k^\top$, where Σ_k holds the top k singular values (from the main diagonal), while U_k and V_k hold, respectively, the left- and right-singular vectors.

2.3.2. Isomap Embedding

Isomap embedding can be seen as an extension of the kernel PCA. It seeks a lower-dimensional embedding that maintains geodesic distances between all data points, through the following three-step process [37]: (1) construct a neighborhood graph from distances between data points, (2) compute the shortest paths on the neighborhood graph, and (3) construct a low-dimensional embedding from the partial eigenvalue decomposition. The nearest search is based on the ball-tree algorithm [38]. The embedding is encoded in the eigenvectors corresponding to the largest eigenvalues of the isomap kernel.

2.3.3. Locally Linear Embedding

Locally linear embedding is yet another extension of PCA which seeks a lower-dimensional projection of the data that preserves distances within local neighborhoods. This method can be viewed as an application of a series of local PCAs, which are then globally compared to find the best nonlinear embedding [38]. This method also comprises a tree-step process as follows [39]: (1) assign a neighborhood to each data point based on local distances, (2) compute weights W_{ij} that best linearly reconstruct the data points from their neighbors, and (3) compute the low-dimensional embedding from these weights W_{ij}, by finding the smallest eigenmodes of the sparse symmetric matrix $M_{ij} = \delta_{ij} - W_{ij} - W_{ji} + \sum_k W_{ki} W_{kj}$, where $\delta_{ij} = 1$ if $i = j$ and 0 otherwise. Although it relies only on linear algebra, since the data points are reconstructed only from their neighbors, this method can still produce highly nonlinear embeddings. If one were to use multiple weight vectors in each neighborhood (instead of a single one), that would result in the so-called modified LLE method. Furthermore, if the locally linear structure were to be recovered by means of the Hessian-based quadratic form, that would result in the so-called Hessian eigenmapping. This is a variant of LLE also known as Hessian-based LLE; see [40] for more information.

2.3.4. Spectral Embedding

Spectral embedding with Laplacian eigenmaps is yet another nonlinear embedding method which is closely related to the locally linear embedding discussed above. It preserves local information and implicitly emphasizes natural clusters in the data. This method constructs a low-dimensional representation of the data using a spectral decomposition of the graph Laplacian. It also consists of three individual steps, as follows [41]: (1) construct the adjacency graph from nearest neighbors, (2) weight the graph edges by forming the matrix W, where $W_{ij} = 1$ if vertices i and j are connected by an edge and $W_{ij} = 0$ otherwise, and (3) compute eigenvalues and eigenvectors for the generalized eigenvalue problem of the form $Lf = \lambda Df$, where D is a diagonal weight matrix $D_{ii} = \sum_j W_{ji}$ and L is the Laplacian matrix $L = D - W$.

2.3.5. Multidimensional Scaling

The central concept in multidimensional scaling is the dissimilarity data δ_{ij} expressed through the distances in geometric spaces. The MDS algorithm minimizes the so-called *stress* function for the configuration of points $x_1, \ldots, x_n$ in t-dimensional space ($t < n$), which is described by the following root-mean-square residual:

$$S = \sqrt{\frac{\sum (d_{ij} - \hat{d}_{ij})^2}{\sum d_{ij}}}, \tag{7}$$

where d_{ij} are distances between points $x_1, \ldots, x_n$, and $\hat{d}_{ij}$ are those values that minimize S, subject to the constraint that $\hat{d}_{ij}$ have the same rank order as the dissimilarity measures δ_{ij}. The stress function is seen as a measure of how well the low-dimensional representation matches the data. The minimization of the multiparameter stress function can be tackled by means of the steepest decent or some other suitable optimization algorithm; see [42] for more information.

2.3.6. *t*-Distributed Stochastic Neighbor Embedding

The *t*-SNE method is based on converting the affinities of data points into probabilities, where the affinities in the original space are represented by Gaussian joint probabilities and the affinities in the embedded space are represented by the Student's *t*-distributions. The *t*-SNE method then minimizes the Kullback–Leibler (KL) divergence between a joint probability distribution, P, in the high-dimensional space and a joint probability distribution, Q, in the low-dimensional space [43]:

$$KL(P||Q) = \sum_i \sum_j p_{ij} \log \frac{p_{ij}}{q_{ij}}, \tag{8}$$

with $p_{ij} = p_{ji}, q_{ij} = q_{ji}, \forall i, j$, and $p_{ii} = q_{ii} = 0$. The pairwise similarities, for the low-dimensional map q_{ij}, are given by:

$$q_{ij} = \frac{\exp\left(-\|y_i - y_j\|^2\right)}{\sum_{k \neq l} \exp\left(-\|y_k - y_l\|^2\right)}, \tag{9}$$

and for a high-dimensional map, it is stated that $p_{ij} = (p_{i|j} + p_{j|i})/2n$, where $p_{i|j}$ is the conditional probability [43]

$$p_{i|j} = \frac{\exp\left(-\|x_i - x_j\|^2/2\sigma^2\right)}{\sum_{k \neq l} \exp\left(-\|x_k - x_l\|^2/2\sigma^2\right)}, \tag{10}$$

scaled by the variance σ of the Gaussian that is centered on the data point x.

The KL divergence is minimized by means of the steepest descent algorithm [43]. The *t*-SNE focuses on the local structure of the data and tends to extract clustered local groups more than the other previously mentioned algorithms. This ability to group samples based on the local structure might be beneficial in some circumstances.

2.4. *Model Training with Supervised and Unsupervised Learning*

Producing a low-dimensional embedding from the high-dimensional data entails training the associated model by means of unsupervised learning. However, some models (such as the PCA) can be trained using supervised learning as well; see [31] for more information. In addition, a classifier can be trained, using supervised learning, on top of the embedding produced by any of the above-introduced (dimensionality reduction) models. This essentially establishes a pipeline which consists of the embedding model (as a transformer) and a classifier. We used a support vector machine (SVM) with an RBF kernel as a classifier. The entire process of model training (both supervised and unsupervised) is graphically represented in Figure 2.

As can be seen from Figure 2, the unsupervised learning loop uses only data features (matrix X), while the supervised learning loops additionally use targets (vector y) that are fed to the classifier. In addition, supervised learning can optimize the entire pipeline (outer loop), or only the classifier (inner loop). The hyperparameter optimization (both of the embedding method and of the SVM classifier) was carried out using the same simulated annealing algorithm described in Section 2.2. Only the objective function was different, as is explained next.

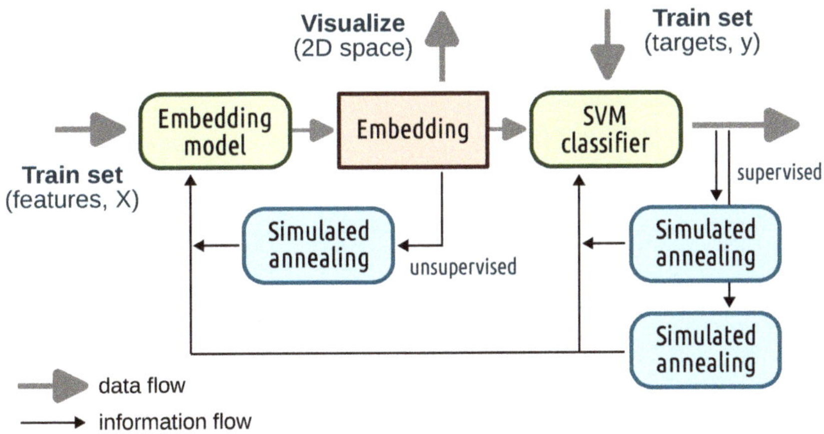

Figure 2. Training models using supervised and unsupervised learning.

2.4.1. Unsupervised Learning

Unsupervised learning used only the features of the dataset and no targets when optimizing the hyperparameters of the embedding models. We demonstrated this approach using two different embedding models: (a) kPCA and (b) *t*-SNE.

(a) The kPCA model has only one hyperparameter, the γ value of the underlying RBF kernel. The objective function here is the mean squared error (MSE). Namely, kPCA first constructs a low-dimensional embedding and then reconstructs the original high-dimensional data by the inverse transform. The MSE error between the original data and their reconstruction was minimized using the simulated annealing algorithm described in Section 2.2.

(b) The *t*-SNE model has one important hyperparameter, a perplexity (P) parameter [38]. The objective function here is based on the Kullback–Leibler divergence and can be described by the following relation [44]:

$$S(P) = 2 \cdot KL + \log(n) \cdot \frac{P}{n}, \tag{11}$$

where KL is the Kullback–Leibler divergence, P is the perplexity, and n is the number of samples. This function was minimized using the simulated annealing algorithm described in Section 2.2.

Unsupervised learning can be extended a step further by applying clustering on top of the low-dimensional embedding; see for example [45]. We used K-means clustering for separating the low-dimensional space of samples into stable and unstable cases [38]. Since we knew a priori that there are only two classes (stable and unstable), we assigned the K-means clustering with (hopefully) finding these clusters withing the embedding space.

2.4.2. Supervised Learning

Supervised learning used an SVM binary classifier with a radial basis function (RBF) kernel and class-weight balancing for classifying cases into stable and unstable ones. An SVM constructs a decision function of the form:

$$f(x) = \sum_{i \in SV} y_i \alpha_i \cdot K(x_i, x) + b, \tag{12}$$

where the summation of sample pairs (x_i, y_i) is over the space of the support vectors (SV), while $K(\cdot)$ is the kernel:

$$K(x, x') = \exp(-\gamma \|x - x'\|^2). \tag{13}$$

Unknown coefficients α_i and b are to be determined by solving the following optimization problem [31]:

$$\begin{aligned} &\min_{w,b,\zeta} \frac{1}{2}||w||^2 + C\sum_{i=1}^{n} \zeta_i, \\ &\text{s.t.} \begin{cases} y_i(w^T\phi(x_i) + b) \geq 1 - \zeta_i, \\ \zeta_i \geq 0, \quad i = 1, \ldots, n, \end{cases} \end{aligned} \tag{14}$$

where C is the penalty that acts as an inverse regularization parameter, while ζ_i is a slack variable, and n is again the number of samples. The penalty (C) and the RBF kernel coefficient (γ) are two crucial hyperparameters of the SVM that need to be optimized.

The supervised loop can be applied to the SVM classifier itself only or to the entire pipeline (see Figure 2). In case that it is applied to the classifier itself, there are two hyperparameters that need to be optimized. These are penalty $C \in (0.1 - 1000)$ and $\gamma \in (0.0001 - 1)$. In this case, the classifier "sits" on top of the embedding. The objective function here was the negative value of the cross-validated F_1-metric as a classification score. On the other hand, if the supervised loop is applied to the entire pipeline, it optimizes the hyperparameters of the embedding model in addition to those of the SVM, and it does that at the same time. We demonstrated this approach on a pipeline which consisted of kPCA as the embedding model (with an RBF kernel) and an SVM classifier (also with its own RBF kernel). In this case there were a total of three hyperparameters (one for kPCA and two for the SVM). Kernel PCA has its own γ parameter for the RBF kernel, that is independent of and unrelated to that of the SVM classifier. All three hyperparameters were optimized (by means of the simulated annealing) at the same time. The objective function was again the negative value of the cross-validated F_1-metric.

It is important to note that the hyperparameter search space during optimization (both supervised and unsupervised) is based on the common logarithm space (i.e., decadic logarithm). Namely, since the parameters can span large values, for example, $C \in (0.1 - 1000)$, in log-space, this transforms to $(10^{-1} - 10^3)$, which means searching within the interval $(-1,3)$. Since this is a much narrower band (when seen in log-space), it significantly improves the convergence speed.

2.4.3. Scoring Models

The SVM binary classifier, trained with supervised learning on top of the embedding, was scored on the test set. The scoring metric is reported as the mean value with a standard deviation ($\mu \pm \sigma$), obtained from a 3-fold cross validation. The process of scoring the model is graphically visualized in Figure 3. It can be seen that the embedding model (as a transformer) and the SVM classifier constituted a pipeline, which was fed the test dataset along with the optimal hyperparameters obtained from the training phase. The classifier produced predictions, which were compared to the ground-truth values (i.e., test-set targets) and a certain metric was reported as the score (i.e., measure of test accuracy). Since the dataset was class-imbalanced, we used the F_1-score as the harmonic mean of the classifier's precision and recall, i.e., $F_1 = 2TP/(2TP + FP + FN)$, where TP is a true positive, FP is a false positive, and FN is a false negative outcome (from the confusion matrix of the classifier).

It can be seen that both the objective function and the scoring method were based on the same metric, which was not necessary but streamlined the comparison. This scoring process enabled us to compare different embedding models in terms of quality. Namely, the SVM classifier had a higher score if the embedding on which it was trained preserved more of the informational content.

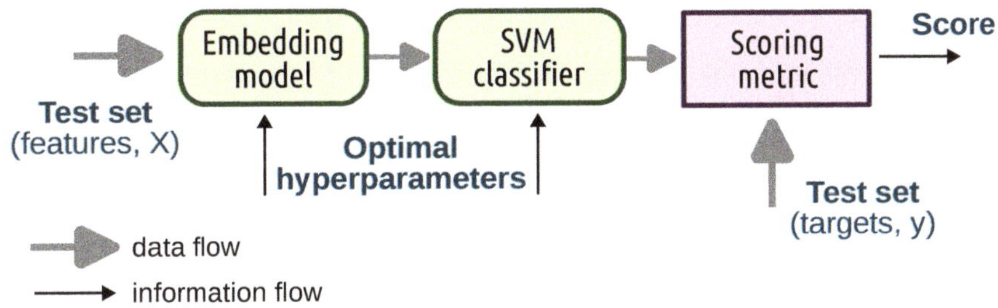

Figure 3. Scoring models on the test set.

3. Results

The previously discussed embedding models were applied to the IEEE New England 39-bus power system dataset (Section 2.1). This section presents the results and other findings which were derived from that application. In order to make the task more challenging for this small power system, each embedding model was required to transform the original high-dimensional space (with 350+ features) into a two-dimensional (2D) space, while preserving as much information as possible. Each embedding model was individually trained using the unsupervised learning approach (Section 2.4). A good embedding should feature only two (distinct) separable clusters, one (larger) for the stable and another (smaller) for the unstable cases. In order to further compare the quality of embeddings, the SVM classifier was individually trained (using the supervised training approach) on top of each of the embeddings. A classifier with a higher score indicated that the embedding model was able to preserve more of the information content and hence produced a better low-dimensional representation.

First, in order to demonstrate the convergence of our modified simulated annealing algorithm for hyperparameter tuning, as an example, Figure 4 presents the algorithm's convergence when optimizing the hyperparameters of the kPCA model using the unsupervised learning approach. It can be seen that the algorithm attained convergence before reaching 200 iterations.

The models produce embedding in a 2D space, which could be easily visualized and further inspected. Hence, Figure 5 demonstrates an example of a "good" embedding (from kPCA on the left side) and a "bad" embedding (from spectral embedding on the right side). It can be clearly seen that kPCA produced two distinct clusters, while spectral embedding was not able to provide a meaningful separation between stable and unstable cases. The color in Figure 5 was superimposed on the data points from the labels (after the fact) in order to visually emphasize stable and unstable cases (this information, however, was not available during the unsupervised training of these models).

Moreover, we found that from all the different embedding models that we tested on this dataset, only kPCA produced exactly two separable clusters. Furthermore, as can be seen from Figure 5 (left side), these clusters (roughly) coincided with stable and unstable classes. Other embedding models, such as the isomap embedding and MDS, preserved information that could be valuable to the classifier which could be trained on top of them but did not produce separable clusters (we found this to be the case with PCA and tSVD as well), or produced too many meaningless clusters (which was the case with *t*-SNE). For example, Figure 6 presents the embedding produced by *t*-SNE from unsupervised learning (unlabeled on the left and labeled on the right), where the color was added after the fact for visual emphasis. At the same time, Figure 7 presents the embedding produced by MDS, without labels (on the left) and with the addition of labels after the fact (on the right). It is clear from Figure 7 that this low-dimensional embedding produced a single (somewhat loose) cluster of data points; similar (but more dense) single clusters resulted from applying PCA and tSVD as well. However, if labels were present in the data, then these embeddings would still preserve enough information for training the classifier on top of them (with supervised learning or even semisupervised learning).

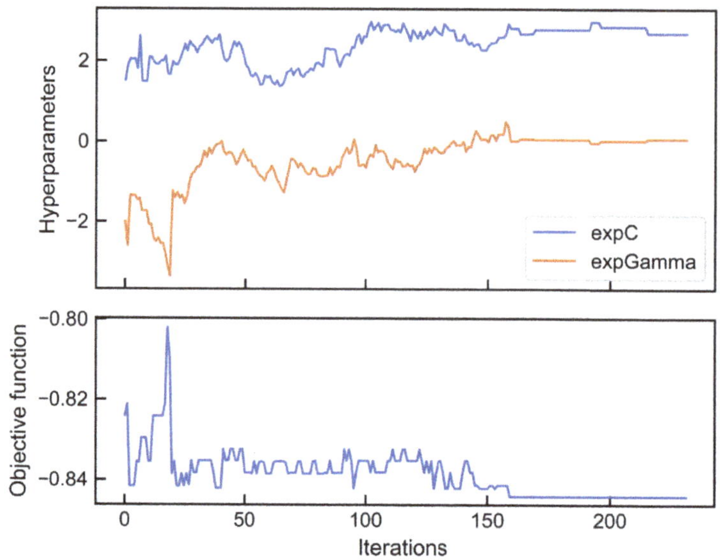

Figure 4. Convergence of the simulated annealing algorithm for optimizing hyperparameters of the kPCA model using the unsupervised learning approach.

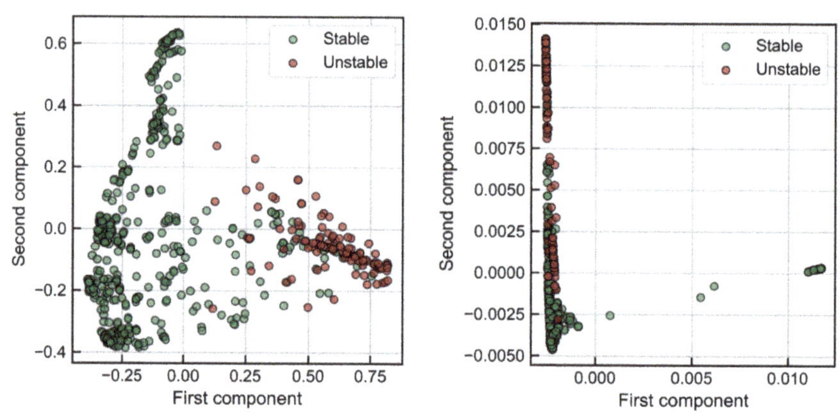

Figure 5. Embedding produced by kPCA (**left**) and spectral embedding (**right**), from the unsupervised learning approach (color added after the fact for visual emphasis).

In addition, when the kPCA model was trained together with the SVM classifier using the pipeline approach (i.e., using the outer optimization loop from Figure 2), we found that this prioritized the classifier's accuracy (as would be expected), and that the resulting embedding occasionally did not feature two clearly distinct clusters. An example of this outcome is depicted in Figure 8, where we show labeled data on the left and unlabeled data on the right. Of course, labels must be known a priori in order to apply this (supervised learning) approach.

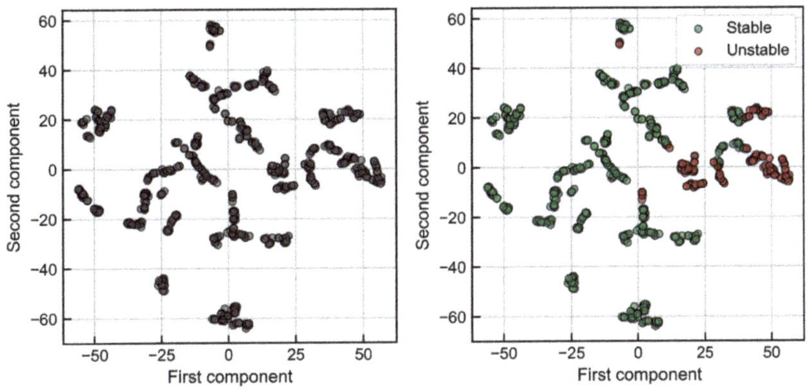

Figure 6. Embedding produced by *t*-SNE from the unsupervised learning approach without labels (**left**) and with the addition of labels after the fact (**right**).

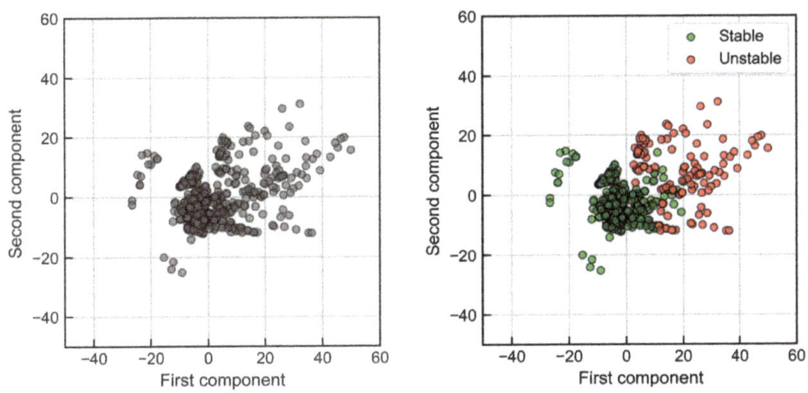

Figure 7. Embedding produced by MDS, from the unsupervised learning approach, without labels (**left**) and with the addition of labels after the fact (**right**).

Furthermore, interestingly, we found that locally linear embedding (LLE), including all of the variants that we tested (i.e., modified and Hessian), produced a 2D embedding that had very low informational content. Hence, classifiers trained on top of these embeddings had low accuracy, as shown later. Moreover, we found that the embedding produced by LLE (and its variants) was very sensitive to the number of neighbors used for producing it. In order to demonstrate these effects, Figure 9 provides the embedding computed by LLE with 10 (left) and 20 (right) neighbors. This high sensitivity was not found with other embedding methods and makes LLE less useful for this particular application.

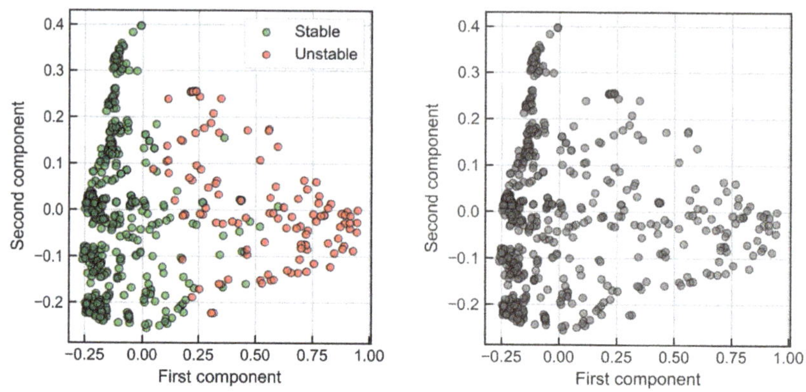

Figure 8. Example of the embedding produced by the pipeline composed of kPCA and SVM with supervised training; labeled data (**left**) and unlabeled data (**right**).

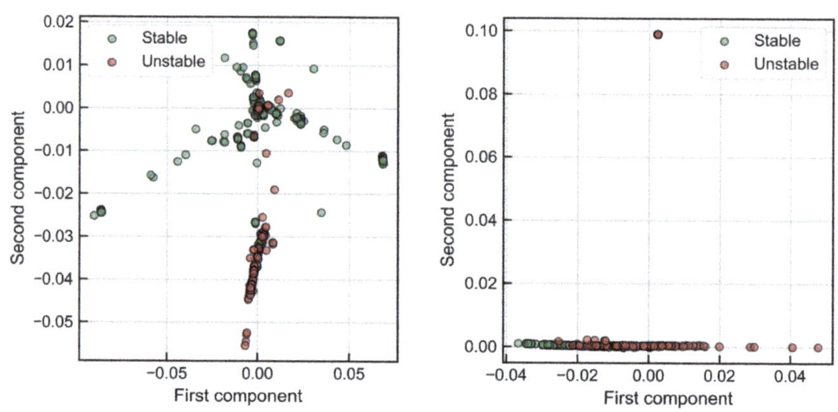

Figure 9. Embedding produced by locally linear embedding with 10 (**left**) and 20 (**right**) neighbors.

Finally, an SVM classifier was independently trained by means of supervised learning (i.e., using the inner optimization loop from Figure 2) on top of each embedding, and its classification metric (i.e., the threefold cross-validated F_1-score) is reported, along with its optimal hyperparameters (C, γ), in Table 2. The last column shows the number of iterations of the simulating annealing algorithm, along with a Y/N indicator that reports if early stopping was triggered. It should be mentioned that the SVM classifier was not calibrated.

In a fully unsupervised learning approach, one would take the unlabeled data and train the embedding model (which hopefully would produce only two clusters), which could then be discovered by the appropriate clustering algorithm (as another unsupervised learning method). We demonstrate this approach here on the embedding produced by the kPCA method trained using the unsupervised learning; see Figure 4 (left). The K-means clustering was applied on top of that embedding. The produced clusters are graphically presented in Figure 10 (left), where the color indicates the cluster membership of each data point. If this is compared to Figure 4 (left), it can be seen that many of the points from the clusters also belong to the actual classes, which means that a completely unsupervised learning approach (kPCA plus K-means) is able to detect power system disturbances from the WAMS-PMU data. Furthermore, Figure 10 (on the right) presents a decision boundary from the SVM classifier trained on top that embedding with a supervised learning approach.

An overlap between cluster and classifier boundaries can be discerned from Figure 10 by comparing the left- and right-hand sides, corroborating the fact that clustering on top of the kPCA embedding can separate power system TSA cases and identify unstable ones.

Table 2. The SVM classifier's parameters and scores after being applied to each of the embeddings.

Embedding Model	SVM (C, γ)		F_1-Score $(\mu \pm \sigma)$	No. Iters.
PCA	1.93	0.0064	0.8521 ± 0.0121	75 (Y)
kPCA	98.5	0.2610	0.8456 ± 0.0223	167 (Y)
tSVD	0.43	0.0118	0.8517 ± 0.0058	43 (Y)
t-SNE	4.51	0.0496	0.9316 ± 0.0301	232 (N)
MDS	14.95	0.0443	0.9281 ± 0.0133	67 (Y)
kPCA (pipeline)	398.1	6.5463	0.9114 ± 0.0226	232 (N)
LLE (default)	67.1	8.8491	0.7307 ± 0.0301	64 (Y)
LLE (modified)	147.7	1.2313	0.7352 ± 0.0297	39 (Y)
Hessian eigen.	212.6	0.0184	0.8213 ± 0.0406	37 (Y)
Spectral embed.	405.5	0.0733	0.7541 ± 0.0346	99 (Y)
Isomap embed.	5.14	0.0036	0.8984 ± 0.0391	65 (Y)

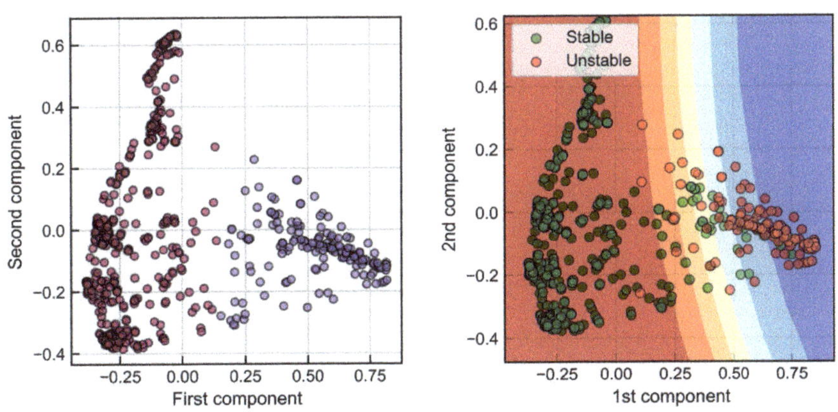

Figure 10. K-means clusters (**left**) and SVM decision boundary (**right**) on top of the embedding produced by kPCA with unsupervised learning.

4. Discussion

We compared our proposed simulated annealing approach to a hyperparameter optimization with a random search approach [38] and found that it produced better results with the same scoring method, while (generally) using fewer iterations. We further found that the random search was less consistent since it exhibited a larger deviation of the cross-validated scores. This is not completely unexpected, since a random search simply draws samples from (in our case exponential) distributions and has no overarching guiding principle in finding the objective function's optimum value. On the other hand, our simulated annealing balances exploration with exploitation through the use of burn-in and cooling schedule and avoids overfitting by means of the early stopping. Also, early stopping enables the simulated annealing to stop searching before all iterations have been exhausted, which is not possible with a random search [38]. However, the random search algorithm, unlike simulated annealing, can be run in parallel on multicore architectures, which helps reduce its execution time.

As far as different embedding methods are concerned, we found that the isomap embedding and MDS provided good bases for supervised classification (which can be seen from Table 2) but, unfortunately, could not produce separable clusters. Hence, they are not well suited for a fully unsupervised learning approach with this kind of data. The same can be said about the PCA and tSVD methods, which also could not produce separable clusters. Furthermore, although an SVM trained on top of the *t*-SNE embedding had a high score (as evident from Table 2), it suffered from a similar deficiency, which emanated from the fact that that embedding contained a lot of small and artificial clusters (see Figure 6). Hence, the *t*-SNE method could not be combined with clustering in a fully unsupervised learning approach, in contrast to what was proposed for some other datasets, e.g., see [45]. This finding is characteristic for the studied dataset and may be due to its peculiar information content (time instants of phase voltages, currents, active and reactive powers, etc.). We also found that only kPCA, when applied to this dataset, produced an embedding that could be utilized in a fully unsupervised manner, as evidenced from Figure 10 (left side).

It is important to mention that there are several sources of randomness that need to be considered here: in the dataset itself (statistical sampling), in the way it was split into training and test sets, in the embedding process (finding nearest neighbors, etc.), and in the simulated annealing (which rests on a random walk). All of these aspects contribute to the variability in the classifier scores between different runs. However, we found that there was a persistence where some embedding methods, when combined with the SVM, would reliably (and consistently) produce better results. These were the PCA, kPCA, and MDS methods (even their 2D embeddings were consistent between runs, despite different data splits). This is a reassuring finding. Also, the tSVD often produced embedding (almost) identical to that of the PCA, which is not surprising considering their similar backgrounds. On the other hand, we found that LLE (and its variants) was quite brittle, sensitive to inputs, and produced embeddings with a low informational content. Finally, although *t*-SNE often produced embeddings that happened to be fertile ground for the SVM training, it was not consistent between runs and could not produce meaningful (nor separable) clusters.

Also, it can be stated that the classifier scores improved across the board when more than two dimensions were retained for the low-dimensional embedding, which was something to be expected. For example, PCA with only seven principal components was able to explain 90% of the variance in the dataset. Classifiers trained on embeddings with more than two dimensions generally produce F_1-scores above the 0.9 level. However, if the goal is to use unsupervised (or semisupervised) learning, then two- and three-dimensional embeddings are particularly appealing, since results of applied clustering can be easily visualized, inspected, and explained (i.e., retaining the properties of the white-box model). We showed that kPCA with K-means clustering produced satisfactory results on the IEEE New England 39-bus dataset. We also found that contrary to the findings on other datasets [45], *t*-SNE was not able to produce meaningful clusters. Neither could the other embedding models that we tested.

In addition, we also tested an LLE variant that implemented a local tangent space alignment algorithm [38]. This approach proved no better than any other LLE variant. Furthermore, it can be mentioned in passing that we also examined a stacked autoencoder for embedding this same dataset and found that it could not consistently produce separable clusters [33]. However, we did not carry out exhaustive tests of different autoencoder architectures and hence, our conclusions are limited by these facts. Finally, there are still other embedding methods, but due to the limited space, they could not be tested here.

5. Conclusions

This paper introduced manifold learning to the electric power system transient stability analysis, based on the dataset derived from extensive transient simulations of the IEEE New England 39-bus benchmark power system. A majority of published papers that applied machine learning approaches to the power system TSA problem were based on supervised learning, with very few exceptions. This paper, however, explored supervised and unsupervised learning for the power system TSA problem. It compared and contrasted different embedding methods. The main contributions of this paper to the state of the art can be summarized through the following findings: (1) kPCA is able to produce a two-dimensional embedding with clearly separable clusters of stable and unstable cases; (2) K-means clustering can be linked with kPCA for a fully unsupervised learning approach; (3) several embedding methods (e.g., multidimensional scaling, isomap embedding) produce a low-dimensional embedding which preserves information without finding clearly separable clusters; (4) an SVM classifier can attain a high accuracy when trained on top of these low-dimensional embeddings; (5) *t*-SNE embedding could not produce meaningful clusters; and (6) simulated annealing can be used for hyperparameter optimization and provides better results than a random search approach.

It should be mentioned that some of these conclusions may be restricted to the examined dataset, although some findings reported here (e.g., the SVM classifier performance) have been corroborated by other researchers. Certainly, further research is needed to independently verify and extrapolate these findings on other power system datasets. Furthermore, additional research is needed to stress-test these findings in an environment that is not a benchmark test power system. This would include introducing various levels of noise into the dataset, along with missing values, measurement errors, etc., that would corrupt the data in different ways. Also, the resilience of different embeddings to the data drift could be examined. These are some future research directions that we can foresee at the present time.

Author Contributions: Conceptualization, P.S. and D.L.; methodology, P.S.; software, P.S. and D.L.; validation, P.S. and D.L.; investigation, D.L.; resources, P.S. and D.L.; data curation, P.S.; writing—original draft preparation, P.S.; writing—review and editing, D.L.; visualization, P.S. All authors have read and agreed to the published version of the manuscript.

Funding: This research received no external funding.

Institutional Review Board Statement: Not applicable.

Informed Consent Statement: Not applicable.

Data Availability Statement: The IEEE New England 39-bust test power system transient dataset with 350+ features has been deposited on Zenodo under the CC BY license, accessed on 1 October 2023 (www.zenodo.org/record/7350829), https://doi.org/10.5281/zenodo.7350828.

Conflicts of Interest: The authors declare no conflict of interest.

Abbreviations

The following abbreviations are used in this manuscript:

IBR	Inverter-based resource
TSA	Transient stability analysis
TSI	Transient stability index
ML	Machine learning
KL	Kullback–Leibler divergence
MVN	Multivariate normal distribution
ANN	Artificial neural network
PMU	Phasor measurement unit
WAMS	Wide-area monitoring system
PCA	Principal component analysis
kPCA	Kernel principal component analysis
tSVD	Truncated singular value decomposition
LLE	Locally linear embedding
MDS	Multidimensional scaling
t-SNE	t-Distributed stochastic neighbor embedding
SVM	Support vector machine
RBF	Radial basis-function kernel
MSE	Mean squared error
FACTS	Flexible AC transmission system

References

1. NERC. *Reliability Guideline: Performance, Modeling, and Simulations of BPS-Connected Battery Energy Storage Systems and Hybrid Power Plants*; North American Electric Reliability Corporation: Atlanta, GA, USA, 2021.
2. Eto, J.H.; Undrill, J.; Roberts, C.; Mackin, P.; Ellis, J. *Frequency Control Requirements for Reliable Interconnection Frequency Response*; Lawrence Berkeley National Laboratory: Berkeley CA, USA, 2018. Available online: https://eta-publications.lbl.gov/sites/default/files/frequency_control_requirements_lbnl-2001103.pdf (accessed on 1 October 2023).
3. Zhang, Y.; Liu, W.; Wang, F.; Zhang, Y.; Li, Y. Reactive Power Control Method for Enhancing the Transient Stability Total Transfer Capability of Transmission Lines for a System with Large-Scale Renewable Energy Sources. *Energies* **2020**, *13*, 3154. [CrossRef]
4. Phadke, A.G.; Bi, T. Phasor measurement units, WAMS, and their applications in protection and control of power systems. *J. Mod. Power Syst. Clean Energy* **2018**, *6*, 619–629. [CrossRef]
5. Oladeji, I.; Zamora, R.; Lie, T.T. An Online Security Prediction and Control Framework for Modern Power Grids. *Energies* **2021**, *14*, 6639. [CrossRef]
6. Li, Y.; Xia, Y.; Ma, J.; Peng, Y. Transient Stability Analysis for a Multi-VSC Parallel System Based on the CFND Method. *Energies* **2023**, *16*, 7243. [CrossRef]
7. Zhang, S.; Zhu, Z.; Li, Y. A Critical Review of Data-Driven Transient Stability Assessment of Power Systems: Principles, Prospects and Challenges. *Energies* **2021**, *14*, 7238. [CrossRef]
8. Alimi, O.A.; Ouahada, K.; Abu-Mahfouz, A.M. A Review of Machine Learning Approaches to Power System Security and Stability. *IEEE Access* **2020**, *8*, 113512–113531. [CrossRef]
9. Tan, B.; Yang, J.; Tang, Y.; Jiang, S.; Xie, P.; Yuan, W. A Deep Imbalanced Learning Framework for Transient Stability Assessment of Power System. *IEEE Access* **2019**, *7*, 81759–81769. [CrossRef]
10. Liu, X.; Zhang, X.; Chen, L.; Xu, F.; Feng, C. Data-driven Transient Stability Assessment Model Considering Network Topology Changes via Mahalanobis Kernel Regression and Ensemble Learning. *J. Mod. Power Syst. Clean Energy* **2020**, *8*, 1080–1091. [CrossRef]
11. Liu, X.; Min, Y.; Chen, L.; Zhang, X.; Feng, C. Data-driven Transient Stability Assessment Based on Kernel Regression and Distance Metric Learning. *J. Mod. Power Syst. Clean Energy* **2021**, *9*, 27–36. [CrossRef]
12. Su, T.; Liu, Y.; Zhao, J.; Liu, J. Probabilistic Stacked Denoising Autoencoder for Power System Transient Stability Prediction with Wind Farms. *IEEE Trans. Power Syst.* **2021**, *36*, 3786–3789. [CrossRef]
13. Tian, Y.; Wang, K.; Oluic, M.; Ghandhari, M.; Xu, J.; Li, G. Construction of Multi-State Transient Stability Boundary Based on Broad Learning. *IEEE Trans. Power Syst.* **2021**, *36*, 2906–2917. [CrossRef]
14. Ren, C.; Xu, Y. A Fully Data-Driven Method Based on Generative Adversarial Networks for Power System Dynamic Security Assessment with Missing Data. *IEEE Trans. Power Syst.* **2019**, *34*, 5044–5052. [CrossRef]
15. Ren, J.; Chen, J.; Li, B.; Zhao, M.; Shi, H.; You, H. A Method for Power System Transient Stability Assessment Based on Transfer Learning. In Proceedings of the 2020 IEEE Power Energy Society General Meeting (PESGM), Montreal, QUE, Canada, 2–6 August 2020; pp. 1–5. [CrossRef]
16. Li, X.; Yang, Z.; Guo, P.; Cheng, J. An Intelligent Transient Stability Assessment Framework with Continual Learning Ability. *IEEE Trans. Ind. Inform.* **2021**, *17*, 8131–8141. [CrossRef]
17. Mukherjee, R.; De, A. Development of an Ensemble Decision Tree-Based Power System Dynamic Security State Predictor. *IEEE Syst. J.* **2020**, *14*, 3836–3843. [CrossRef]
18. Behdadnia, T.; Yaslan, Y.; Genc, I. A new method of decision tree based transient stability assessment using hybrid simulation for real-time PMU measurements. *IET Gener. Transm. Distrib.* **2021**, *15*, 678–693. [CrossRef]
19. Mi, D.; Wang, T.; Gao, M.; Li, C.; Wang, Z. Online transient stability margin prediction of power systems with wind farms using ensemble regression trees. *Int. Trans. Electr. Energy Syst.* **2021**, *31*, e13057. [CrossRef]
20. Shao, D.; Xu, Y.; Ma, S.; Jing, Y.; Sun, W.; Zhou, T.; Yang, J. Transient Stability Assessment Method for Power System Based on SVM with Adaptive Parameters Adjustment. In Proceedings of the 2021 IEEE 4th International Electrical and Energy Conference, CIEEC 2021, Wuhan, China, 28–30 May 2021. [CrossRef]
21. Zhou, Y.; Wu, J.; Yu, Z.; Ji, L.; Hao, L. A Hierarchical Method for Transient Stability Prediction of Power Systems Using the Confidence of a SVM-Based Ensemble Classifier. *Energies* **2016**, *9*, 778. [CrossRef]
22. Hu, W.; Lu, Z.; Wu, S.; Zhang, W.; Dong, Y.; Yu, R.; Liu, B. Real-time transient stability assessment in power system based on improved SVM. *J. Mod. Power Syst. Clean Energy* **2019**, *7*, 26–37. [CrossRef]
23. Bashiri Mosavi, A.; Amiri, A.; Hosseini, H. A Learning Framework for Size and Type Independent Transient Stability Prediction of Power System Using Twin Convolutional Support Vector Machine. *IEEE Access* **2018**, *6*, 69937–69947. [CrossRef]
24. Zhu, Q.; Chen, J.; Zhu, L.; Shi, D.; Bai, X.; Duan, X.; Liu, Y. A Deep End-to-End Model for Transient Stability Assessment with PMU Data. *IEEE Access* **2018**, *6*, 65474–65487. [CrossRef]
25. Chen, K. Indirect PCA Dimensionality Reduction Based Machine Learning Algorithms for Power System Transient Stability Assessment. In Proceedings of the 2019 IEEE Innovative Smart Grid Technologies—Asia (ISGT Asia), Chengdu, China, 21–24 May 2019; pp. 4175–4179. [CrossRef]
26. Bellizio, F.; Cremer, J.; Sun, M.; Strbac, G. A causality based feature selection approach for data-driven dynamic security assessment. *Electr. Power Syst. Res.* **2021**, *201*, 107537. [CrossRef]

27. Gnanavignesh, R.; Shenoy, U. An improved relaxation based spatial domain decomposition method for parallel transient stability simulation. *Int. J. Electr. Power Energy Syst.* **2022**, *135*, 107561. [CrossRef]
28. Arvizu, C.M.C.; Messina, A.R. Dimensionality Reduction in Transient Simulations: A Diffusion Maps Approach. *IEEE Trans. Power Deliv.* **2016**, *31*, 2379–2389. [CrossRef]
29. Biswal, M.; Hao, Y.; Chen, P.; Brahma, S.; Cao, H.; De Leon, P. Signal features for classification of power system disturbances using PMU data. In Proceedings of the 19th Power Systems Computation Conference, PSCC 2016, Genoa, Italy, 19–23 June 2016. [CrossRef]
30. Hang, F.; Huang, S.; Chen, Y.; Mei, S. Power system transient stability assessment based on dimension reduction and cost-sensitive ensemble learning. In Proceedings of the 2017 IEEE Conference on Energy Internet and Energy System Integration, EI2 2017-Proceedings, Beijing, China, 26–28 November 2017; pp. 1–6. [CrossRef]
31. Murphy, K.P. *Machine Learning: A Probabilistic Perspective*, 4th ed.; MIT Press: Cambridge, MA, USA, 2013.
32. Moeini, A.; Kamwa, I.; Brunelle, P.; Sybille, G. Open data IEEE test systems implemented in SimPowerSystems for education and research in power grid dynamics and control. In Proceedings of the Universities Power Engineering Conference, Stoke-on-Trent, UK, 1–4 September 2015; pp. 3–8. [CrossRef]
33. Sarajcev, P.; Kunac, A.; Petrovic, G.; Despalatovic, M. Power system transient stability assessment using stacked autoencoder and voting ensemble. *Energies* **2021**, *14*, 3148. [CrossRef]
34. Sarajcev, P.; Kunac, A.; Petrovic, G.; Despalatovic, M. Power System Transient Stability Assessment Simulations Dataset—IEEE New England 39-Bus Test Case. 2021. Available online: https://zenodo.org/records/4521886 (accessed on 1 October 2023). [CrossRef]
35. Sarajcev, P.; Kunac, A.; Petrovic, G.; Despalatovic, M. IEEE New England 39-Bus Test Case: Dataset for the Transient Stability Assessment. 2021. Available online: https://zenodo.org/records/7350829 (accessed on 1 October 2023). [CrossRef]
36. Kirkpatrick, S.; Gelatt, C.D., Jr.; Vecchi, M.P. Optimization by Simulated Annealing. *Science* **1983**, *220*, 671–680. [CrossRef] [PubMed]
37. Tenenbaum, J.B.; de Silva, V.; Langford, J.C. A Global Geometric Framework for Nonlinear Dimensionality Reduction. *Science* **2000**, *290*, 2319–2323. [CrossRef] [PubMed]
38. Pedregosa, F.; Varoquaux, G.; Gramfort, A.; Michel, V.; Thirion, B.; Grisel, O.; Blondel, M.; Prettenhofer, P.; Weiss, R.; Dubourg, V.; et al. Scikit-learn: Machine Learning in Python. *J. Mach. Learn. Res.* **2011**, *12*, 2825–2830.
39. Roweis, S.T.; Saul, L.K. Nonlinear Dimensionality Reduction by Locally Linear Embedding. *Science* **2000**, *290*, 2323–2326. [CrossRef]
40. Donoho, D.L.; Grimes, C. Hessian eigenmaps: Locally linear embedding techniques for high-dimensional data. *Proc. Natl. Acad. Sci. USA* **2003**, *100*, 5591–5596. [CrossRef]
41. Belkin, M.; Niyogi, P. Laplacian Eigenmaps for Dimensionality Reduction and Data Representation. *Neural Comput.* **2003**, *15*, 1373–1396. [CrossRef]
42. Kruskal, J.B. Nonmetric multidimensional scaling: A numerical method. *Psychometrika* **1964**, *29*, 115–129. [CrossRef]
43. van der Maaten, L.; Hinton, G. Visualizing Data using t-SNE. *J. Mach. Learn. Res.* **2008**, *9*, 2579–2605.
44. Cao, Y.; Wang, L. Automatic Selection of t-SNE Perplexity. *arXiv* **2017**, arXiv:1708.03229.
45. Liu, L.P.; Wang, Q.; Dong, M.N.; Zhang, Z.Y.; Li, Y.; Wang, Z.Z.; Wang, S.Q. Application of K-Means ++ Algorithm Based on t-SNE Dimension Reduction in Transformer District Clustering. In Proceedings of the 2020 Asia Energy and Electrical Engineering Symposium (AEEES), Chengdu, China, 28–31 May 2020; pp. 74–78. [CrossRef]

energies

MDPI

Article

The Influence of Stops on the Selected Route of the City ITS on the Energy Efficiency of the Public Bus

Miroslaw Smieszek , Vasyl Mateichyk * and Jakub Mosciszewski

Department of Technical Systems Engineering, Rzeszow University of Technology, al. Powstancow Warszawy 8, 35-959 Rzeszow, Poland; msmieszk@prz.edu.pl (M.S.); j.mosciszews@prz.edu.pl (J.M.)
* Correspondence: vmate@prz.edu.pl

Abstract: Public transport is an important part of sustainable economic development, sustainable cities, and communities. Reducing energy consumption in public transport can be achieved through better organisation of the transport system, changes in infrastructure, the use of new energy-efficient means of transport, and other ways to achieve intelligent mobility. The operation of a city bus involves frequent stops. These stops are due to the need to exchange passengers at bus stops and traffic conditions. Each stop and the subsequent acceleration process require additional energy consumption. In this paper, an analysis of bus operation within the Rzeszów ITS on a selected route is carried out to determine the energy consumption in these special modes. First, the number and duration of stops were determined based on data recorded during the bus operation using the tracker. Then, taking into account the idle fuel consumption and the energy consumption required to reach a set speed, the total energy consumption associated with the stops was determined. The results obtained on the selected route indicate a significant share of energy associated with stops at bus stops and outside bus stops in total fuel consumption. These shares are about 26.2% and about 42.5%, respectively. The opportunity to improve the energy efficiency of the city bus on the route due to the reduction of stops at bus stops by introducing on-demand stops as one of the elements of ITS has been evaluated. The number of stops related to traffic conditions can be reduced by further improving traffic management and measures to modify urban infrastructure.

Keywords: transport system; public bus; traffic condition; stops; energy efficiency

1. Introduction

The energy consumption of public buses and, in the case of conventional vehicles, their fuel consumption depends on many factors. Research on the impact of these factors has been conducted by researchers from various countries. The literature contains information on the analysis of the impact of traffic conditions taking into account many factors, as well as selected isolated factors. Some of the works [1–4] take into account various driving profiles reflecting specific traffic conditions. Many studies were conducted in real operating conditions [5–16]. A number of works analyse the impact of selected factors on fuel consumption. Many authors dealt with driving technique and drivers' skills [1–3,6–10]. The influence of the number of passengers on fuel consumption [6,13], the influence of the length of sections [17], and the accelerations and decelerations obtained [6,18] were also considered. Several studies also analysed the impact of road inclination on fuel consumption [5,6,19]. One paper dealt with the impact of engine oil viscosity on fuel consumption [20]. In [14], the significant impact of air conditioning, outside temperature, average speed, congestion index, and average acceleration on the energy consumption of an electric bus was demonstrated. In [15], a significant impact of external temperature on the energy consumption of an electric bus was demonstrated. The work [21] showed that the number of bus stops is one of the three most significant factors influencing the energy consumption of an electric bus. However, this work does not indicate the exact increase in energy consumption because it involves creating a predictive model and does not consider the exact impact.

The energy demand in urban traffic characterised by high variability of traffic parameters is much higher than in steady-state traffic [4,11]. The range and dynamics of changes in these parameters are largely dependent on external traffic conditions and the time of day. These changes include maximum speed, average speed, and number and duration of stops. Urban traffic is characterised by repeated phases of accelerated and decelerated motion. During the deceleration motion resulting from braking, the kinetic energy acquired from the propulsion engine is dissipated. In vehicles without a recuperation system, this energy is lost. The work of friction forces in the brakes is converted into heat dissipated into the atmosphere.

The stoppage of the bus and the associated need to dissipate kinetic energy are the results of two factors. The first is traffic conditions. In this case, the factor forcing the bus to stop can be traffic lights, pedestrian crossings and exits from a subordinated road. The second factor results from the need to stop at a bus stop. The number of stops at bus stops can be regulated to some extent by public transport authorities. Reducing the overall number of stops and thus increasing traffic flow should contribute to reducing energy demand.

There are several systems for organising public transport in Europe. These systems may require stopping at all bus stops or only when requested by passengers on the bus or at a bus stop. There is also an intermediate system, where only part of the bus stops are on-demand. These bus stops are most often characterised by low passenger exchange.

The aim of this paper is to determine the energy loss associated with the stopping process at bus stops and outside bus stops. The second group is associated with stops forced by external traffic conditions. The subject of impact of stops on fuel consumption is not widely researched and this paper aims to increase the knowledge of this subject.

The article consists of the following parts:

- The first part provides an introduction;
- The second part presents the energy balance of the bus;
- In the third part, the examined case is described and a statistical analysis of the vehicle's motion profile is made;
- The fourth part presents the simulation of energy consumption;
- The fifth part summarizes the paper.

2. Energy Balance of the Bus

A system of forces shown in Figure 1 [22,23] acts on a moving vehicle. In the figure presented, the road is assumed to be horizontal, hence the absence of the resistance force of the hill F_w.

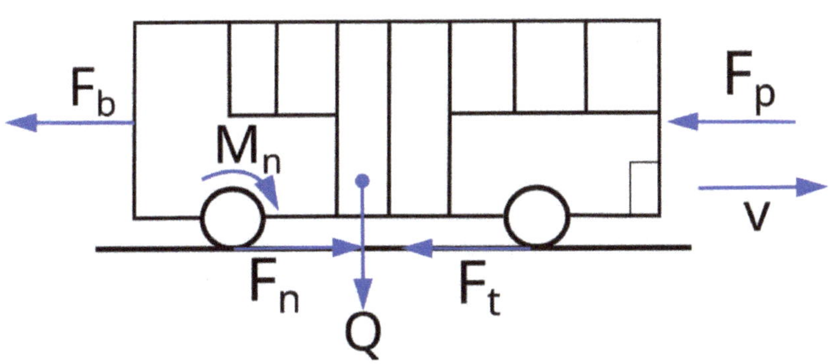

Figure 1. System of forces acting on an accelerating vehicle moving on a horizontal surface (F_b—inertia force, F_p—air resistance force, F_t—rolling resistance force, F_n—driving force, Q—vehicle weight, M_n—driving torque, v—vehicle speed).

Due to the speeds reached by the bus (below 80 km/h), the rolling resistance force Ft is assumed to be a constant value and not dependent on speed v. The other forces acting on the vehicle, apart from the gravity force Q, depend on the traffic conditions, which include speed and acceleration.

Figure 2 shows a fragment of a simplified driving cycle of the city bus. The acceleration, steady-state movement, and braking phases, as well as the stop time, can be distinguished in Figure 2.

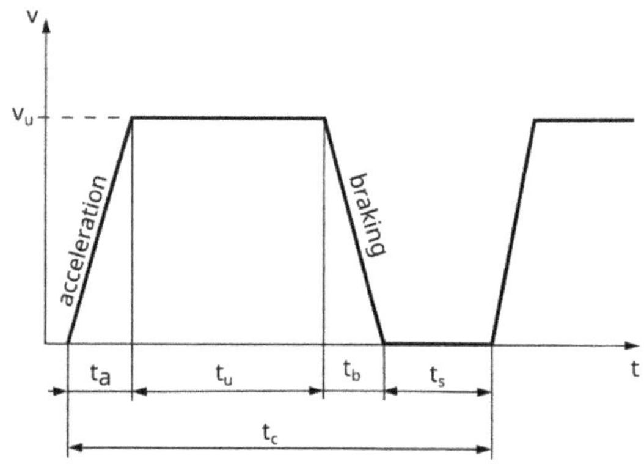

Figure 2. Theoretical driving cycle between stops (t_a—acceleration time, t_u—steady-state motion time, t_b—braking time, t_s—stop time, t_c—cycle time).

According to the principle of dynamics, the sum of all forces acting on the vehicle equals 0. For forces acting horizontally, it can be represented by the following relation:

$$F_n - F_t - F_p - F_b - F_w = 0 \quad (1)$$

In the further considerations presented in the article, it is assumed that the road inclination angle α = 0. This assumption results from the fact that the vast majority of the bus stops of the analysed line are located on flat terrain, characterised by an altitude of 200 to 205 metres above sea level. Then, F_w = 0 and Equation (1) takes the form

$$F_n - F_t - F_p - F_b = 0 \quad (2)$$

The individual parts of the equation can be described by the following relationships:

$$F_t = m * f_t * g \quad (3)$$

$$F_p = \frac{\rho * A * c_x * v^2}{2} \quad (4)$$

where

c_x—air resistance coefficient;
A—vehicle frontal area, [m^2];
ρ—air density, [kg/m^3];
v—vehicle speed, [m/s];
f_t—rolling resistance coefficient;
m—vehicle weight, [kg];
g—acceleration due to gravity, [m/s^2].

During acceleration, the driving force overcomes the movement resistances F_t and F_p and the inertia force F_b. At the final stage of the acceleration process, the vehicle has the kinetic energy described by the following relation:

$$E_k = \frac{m * v^2}{2} \tag{5}$$

During the steady-state motion phase, the kinetic energy of the vehicle does not change. The energy supplied by the engine covers the energy losses due to the drag and rolling forces described by the relationships:

$$W_p = F_p * s_u = \frac{\rho * A * c_x * v_u^2 * s_u}{2} \tag{6}$$

$$W_t = F_t * s_u = m * g * f_t * s_u \tag{7}$$

where

s_u—length of a road section of a steady-state motion, [m];
c_x—air resistance coefficient;
A—vehicle frontal area, [m^2];
ρ—air density, [kg/m^3];
v_u—speed of steady-state motion, [m/s];
f_t—rolling resistance coefficient;
m—vehicle weight, [kg];
g—acceleration due to gravity, [m/s^2].

In the case of braking, the kinetic energy of the vehicle E_k is converted into the work of braking forces W_b, the work of rolling resistance forces W_t, and the work of air resistance forces W_p. This is described by the following dependencies:

$$E_k = W_b + W_t + W_p \tag{8}$$

$$W_t = F_t * s_b = m * g * f_t * s_b \tag{9}$$

$$s_b = v_b * t_b - \frac{a_b * t_b^2}{2} \tag{10}$$

where

s_b—length of braking distance, [m];
a_b—deceleration, [m/s^2];
t_b—braking time, [s];
v_b—the speed at which braking is initiated, [m/s].

The force of air resistance depends on the speed, which during the braking process changes from v_u to 0 on the distance s_b. The energy lost as a result of air resistance is presented as

$$W_p = \int_0^{s_b} F_p * ds = \int_0^{s_b} \frac{\rho * A * c_x * v(s)^2}{2} ds \tag{11}$$

According to Equation (8), the energy lost as a result of the braking forces is equal to the kinetic energy of the vehicle at the start of the braking process minus the works W_t and W_p. The ratio of the work of rolling resistance forces to the total kinetic energy according to Equation (12) is a value independent of the vehicle speed at the start of the braking. This ratio is influenced only by the rolling resistance coefficient and the deceleration during braking.

$$\frac{W_t}{E_k} = \frac{g * f_t}{a} \tag{12}$$

The ratio of the work of air resistance forces to the total kinetic energy is largely dependent on the vehicle's speed at the start of braking.

Figure 3 shows the ratio of these two components to the total kinetic energy at the start of the braking.

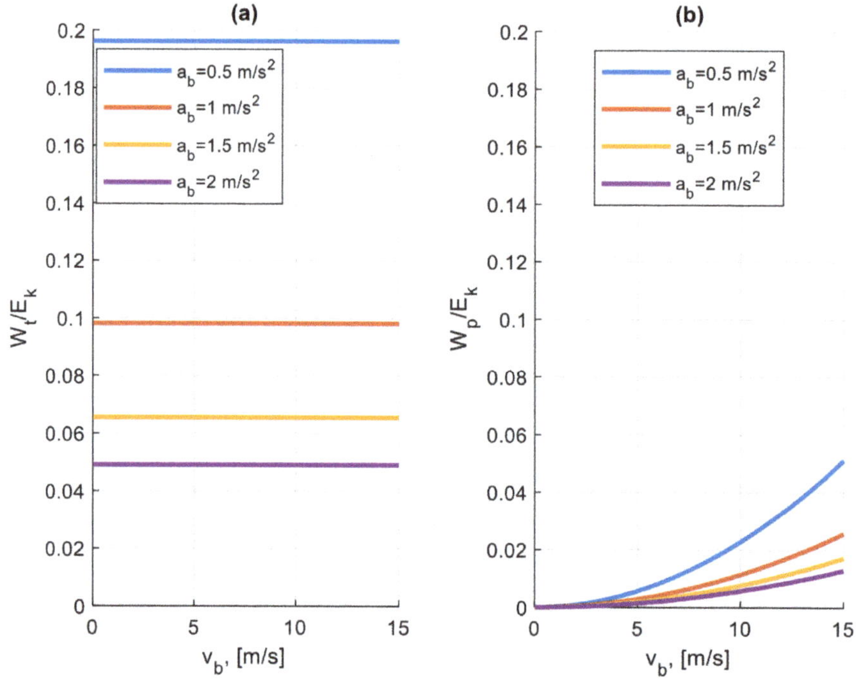

Figure 3. The ratio of the work of the motion resistance forces to the kinetic energy E_k at the start of braking for different values of deceleration a_b (v_b—the speed at which braking is initiated), (**a**) the ratio of the work W_t of the rolling resistance forces to the kinetic energy E_k, (**b**) the ratio of the work W_p of the air resistance forces to the kinetic energy E_k.

It can be seen that for the same values of braking deceleration, the work of the rolling resistance forces is many times greater than that of the air resistance forces. In both cases, the magnitude of these forces is strongly influenced by braking deceleration. As the value of deceleration increases, the braking distance shortens and the ratio of these two quantities to the kinetic energy decreases.

3. Case Description and Analysis of Parameters Related to Vehicle Movement

3.1. Intelligent Transport System in Rzeszów

There is an Intelligent Transport System in the city of Rzeszów. This system includes the following subsystems: Area Traffic Control System, Public Transport Management System, Passenger Information System, Electronic Toll Collection System, Dynamic Vehicle Weighing System, and ICT Platform integrating the previously mentioned subsystems. The Area Traffic Control System consists of a traffic light control system that ensures smooth traffic flow and minimises waiting times at intersections, a right-of-way system for public transport that assigns priority to public transport and emergency vehicles, and a driver information system that uses variable message signs to provide quick information on traffic impediments, changes in traffic organisation and recommended detours. The basis of the signalling system are induction loops embedded in the asphalt and a camera system, which allows the signalling system to adjust the traffic intensity. In addition, at intersections with traffic lights, there is also automatic detection of pedestrian and bicycle traffic, thanks to which there is no need to press a button to change the lights. Assigning priority to public transport and emergency vehicles involves transmitting information from the vehicle to the traffic light controller, checking whether it is possible to pass without having to stop. Depending on the situation, there may be an extension of the phase of the traffic lights allowing passage, or a shortening of the phases preceding the next phase of the lights allowing passage. The Public Transport Management System enables identification of the location of vehicles, counting the number of passengers and controlling the quality standards of passenger service. This allows the current needs of public transport in the city to be responded to. Thanks to this system, passengers are informed in real time about bus delays. The Passenger Information System aims to provide efficient information to public transport passengers. It includes information boards at bus stops and a web-based system that provides more information than the information boards. The Electronic Fare Collection System is an electronic public transport ticketing system. Under this system, it is possible to buy time-based or single-trip tickets, the price of which depends on the number of bus stops travelled. The Dynamic Vehicle Weighing System is designed to detect overloaded vehicles and to record information about detected overloads. This system consists of sensors embedded in the asphalt that monitor axle loads and a central system that collects data [24–26].

3.2. Characteristics of the Considered Route

The analysis and research carried out in this article concerns a selected public transport route in Rzeszów. It is one of the main public transport routes that practically runs through the entire city. All subsystems of the city ITS are actively involved in this route. The recorded data are from the days 16 to 18 March 2022, 21 to 25 March 2022, 29 March 2022, and 4 April 2022, and these are working days. Data are from one diesel-powered bus, a Mercedes-Benz O530 operating on city route number 13. A diagram of the route number 13 is shown in Figure 4.

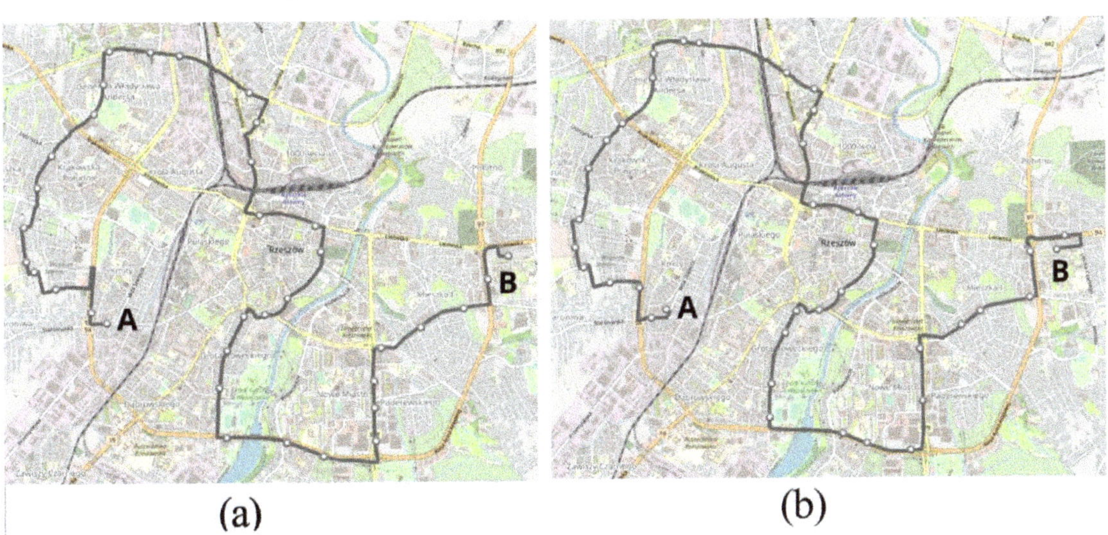

Figure 4. Diagram of the route no. 13: (**a**) direction A to B; (**b**) direction B to A.

Due to the road layout with separated traffic directions streets, both routes are not the same. There are 35 bus stops on the route in the A to B direction and 38 bus stops in the B to A direction. In the A to B direction, the route is 16.376 km long and in the B to A direction 16.805 km long. Depending on the time of day and the direction of traffic, the total number of stops on a given course varies. The highest number of stops occurs during the morning and afternoon rush hours. This is shown in Figure 5.

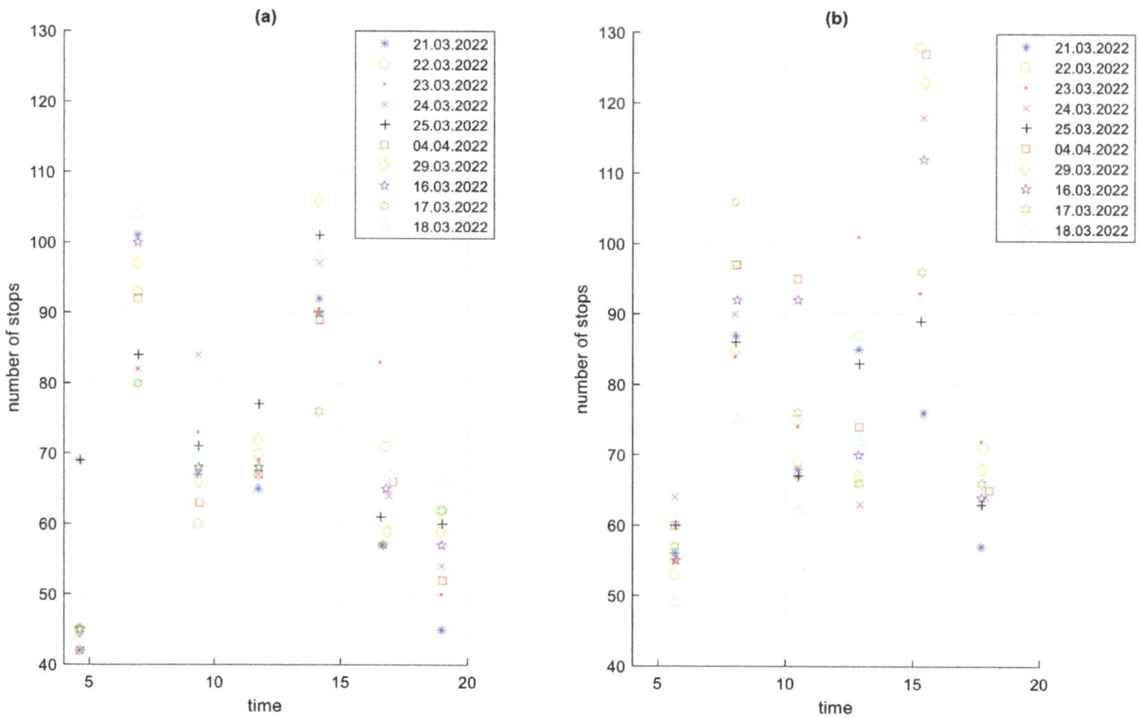

Figure 5. Total number of stops on courses in the analysed period depending on the time of the start of the course: (**a**) direction A to B; (**b**) direction B to A.

The data used in the study come from a GPS data recorder installed in the vehicle, which records data at a frequency of 1 Hz. A description of the system and its characteristics is presented in papers [27–29].

Fuel consumption data were provided by MPK Rzeszów (Municipal Transport Company, Rzeszów, Poland). The bus was refuelled at the end of each day after work and the amount of fuel refuelled was recorded.

3.3. Characteristics of the Traffic Profile

All the routes under consideration are operated by the same bus running the same daily schedule. In direction A to B, the bus makes seven trips per day and in direction B to A six trips per day. Figure 6a shows an example of the speed profile during the day on 25 March 2022. The speed of the bus in urban traffic very rarely reaches values greater than 15 m/s, which corresponds to a value of 54 km/h.

As can be seen in Figure 6b, there are four phases in the movement of a public bus: the acceleration phase, the braking phase, the steady speed phase, and the stopping phase. The stopping phase results from the need to stop at a bus stop and from the current traffic conditions, e.g., stopping in front of traffic lights.

The varying traffic conditions and the number of passengers transported on a particular course affect the duration of the course and the average course speed. A chart of course durations in the studied period is shown in Figure 7.

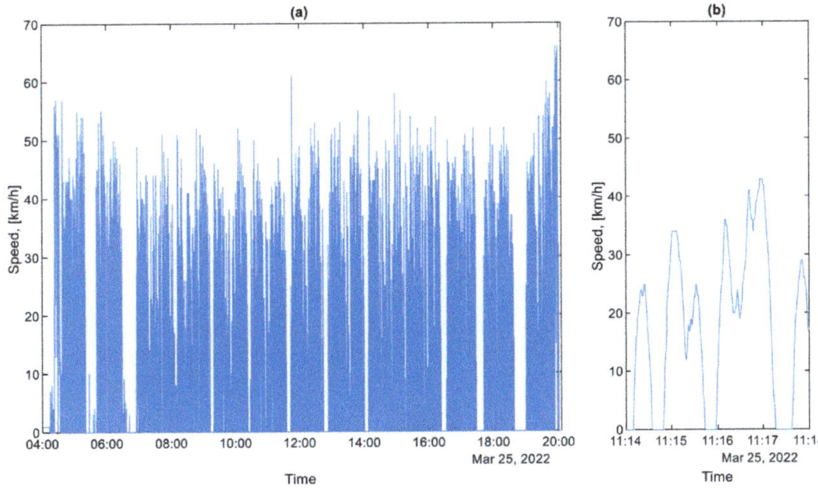

Figure 6. Bus speed profile for a selected working day: (**a**) from the entire day; (**b**) from the selected section.

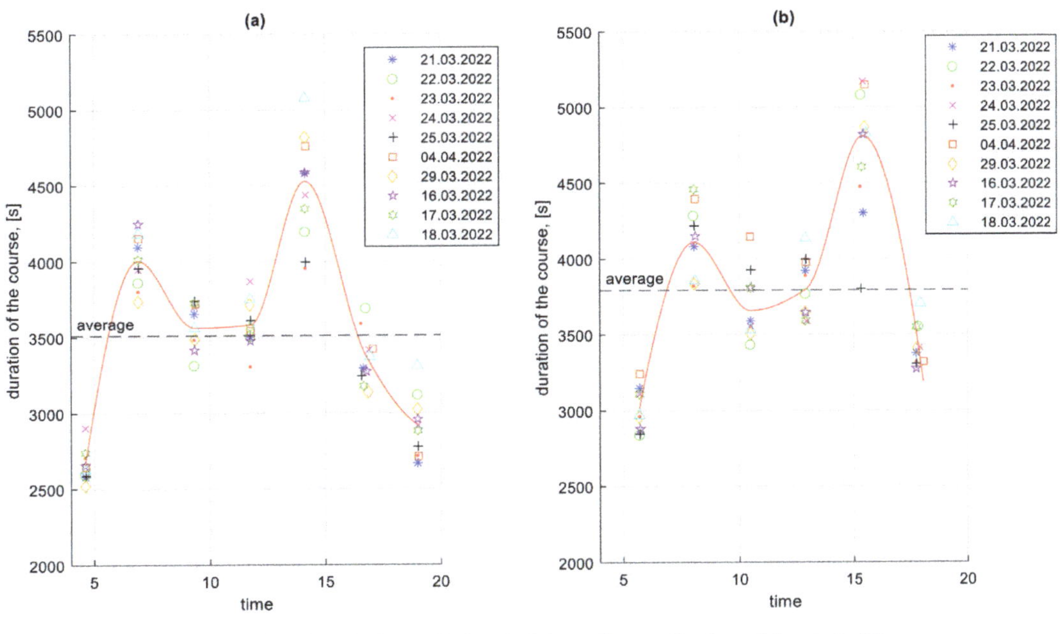

Figure 7. Course durations in the studied period depending on the time of the start of the course: (**a**) in the direction A to B; (**b**) in the B to A direction.

To determine the energy lost in the braking process, it is necessary to know the vehicle speed v_b at the start of the braking. In most cases, this speed does not correspond to the speed at the end of the acceleration process v_a, nor to the maximum speed reached by the bus between bus stops, as illustrated in Figure 8.

As each braking process is characterised by different initial conditions and intensity, the determination of the energy lost as a result of this process requires a case-by-case consideration.

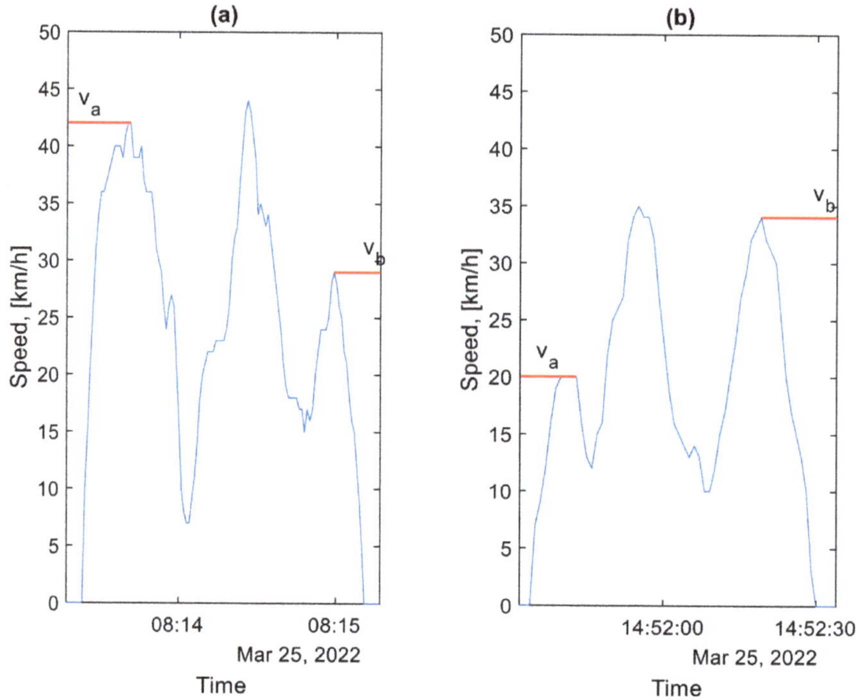

Figure 8. Graph of bus speed from two selected parts of the route between bus stops (v_a—acceleration speed after the stop, v_b—braking speed before the stop): (**a**)—morning rush hour, (**b**)—afternoon rush hour.

3.4. Analysis of Braking Process Parameters

As part of the study, all braking speeds v_b were determined from the recorded data. The results were divided into two groups. The first is related to the braking process before the bus stop. The second group is related to stopping forced by traffic conditions. The distribution of these speeds is shown in Figures 9 and 10. Figure 9 concerns the braking speed before the bus stop. Figure 10 concerns braking speeds forced by traffic conditions.

The distributions shown in Figures 9 and 10 differ. In the case of Figure 10, the distribution has a flattened nature. This is due to traffic conditions in traffic jams. Vehicles involved in this traffic move at low speeds and stop frequently. In the case of Figure 9, most of the braking processes are carried out from a much higher speed. Figure 11 shows the average braking speeds before a bus stop on the course as a function of the average speed on the course. Figure 12 shows the average braking speeds outside the bus stops on the course as a function of the average speed on the course.

In the case of Figure 11, there is little change in the average braking speed before a bus stop on the course from the average speed on the course. As Figure 9 shows, the majority of braking speeds before the bus stop are clustered between values of 6 m/s and 12 m/s. In Figure 12, the significant effect of the average speed on the course is seen on the average braking speed outside the bus stops on the course. When moving in a traffic jam, the vehicle moves at low speeds, hence the low braking speeds. At higher average speeds on a given course, the braking process resulting from, for example, pedestrians at a crossing or traffic lights takes place at a higher speed, close to the braking speed before the bus stop.

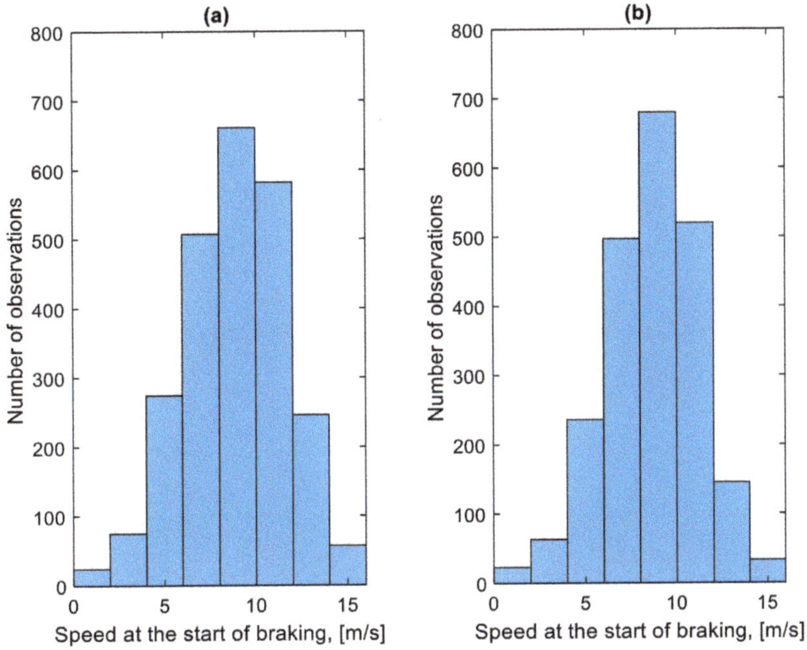

Figure 9. Distribution of braking speeds before the bus stop in the studied period: (**a**) in the direction A to B; (**b**) in the B to A direction.

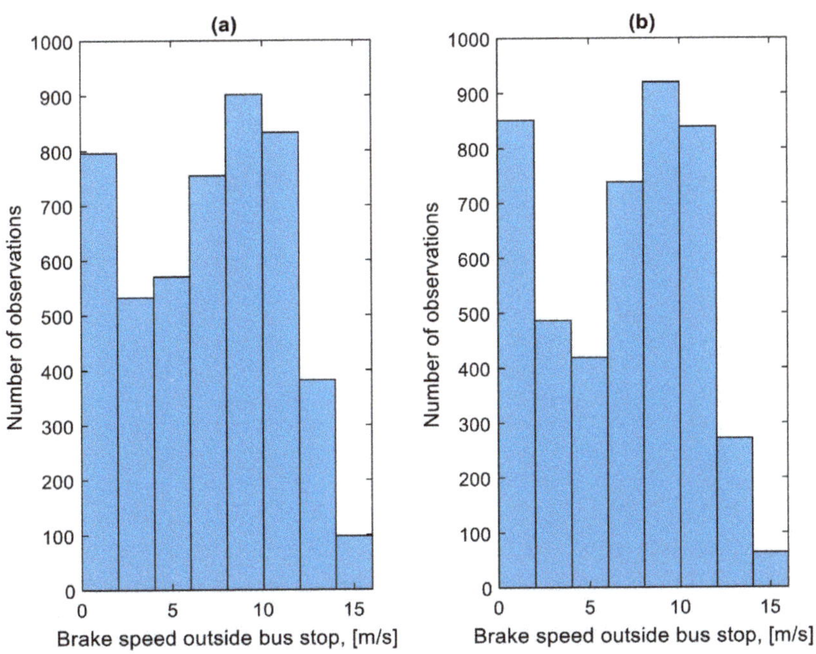

Figure 10. Distribution of braking speeds forced by traffic conditions in the studied period: (**a**) in the direction A to B; (**b**) in the B to A direction.

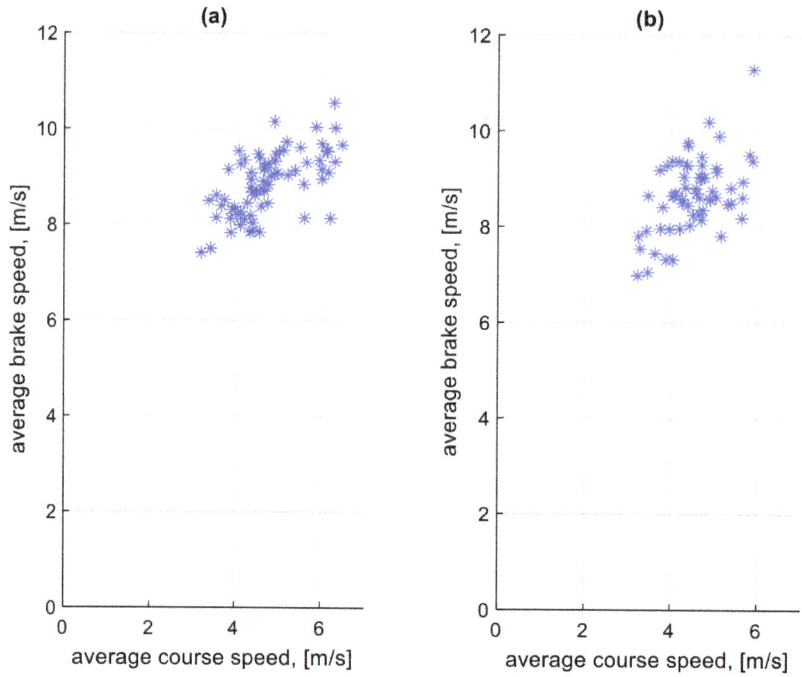

Figure 11. Average braking speed before bus stop on a course depending on the average speed on a given course: (**a**) in the direction A to B; (**b**) in the B to A direction.

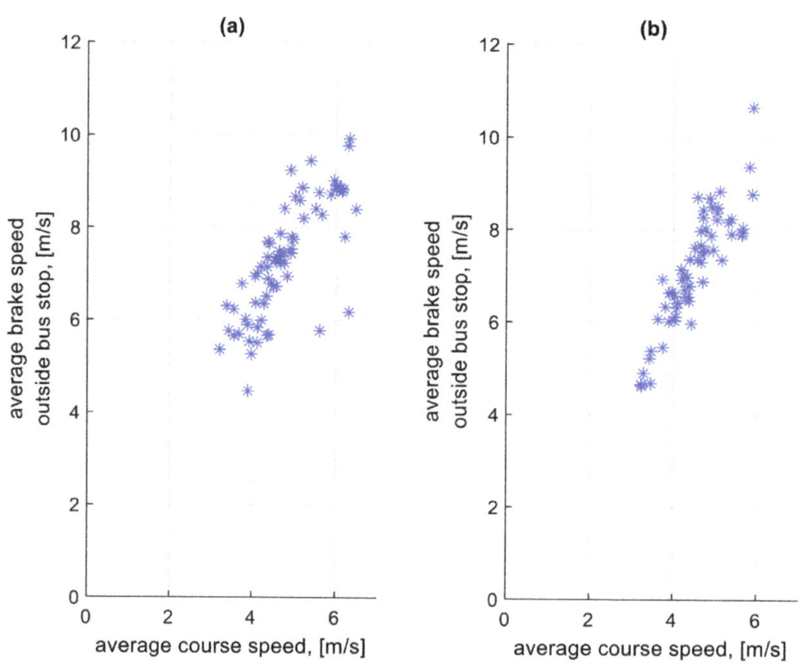

Figure 12. Average braking speed outside bus stops on a course depending on the average speed on a given course: (**a**) in the direction A to B; (**b**) in the B to A direction.

Figure 13 shows the distribution of decelerations during the braking process before bus stops during the period studied. Figure 14 shows the distribution of decelerations during the braking process outside the bus stops during the period studied. In the case of the decelerations of the braking process shown in Figures 13 and 14, no significant differences can be seen between the distributions of these decelerations. Most of the decelerations achieved do not exceed 1.25 m/s^2. This value does not cause discomfort to passengers and does not pose any threat.

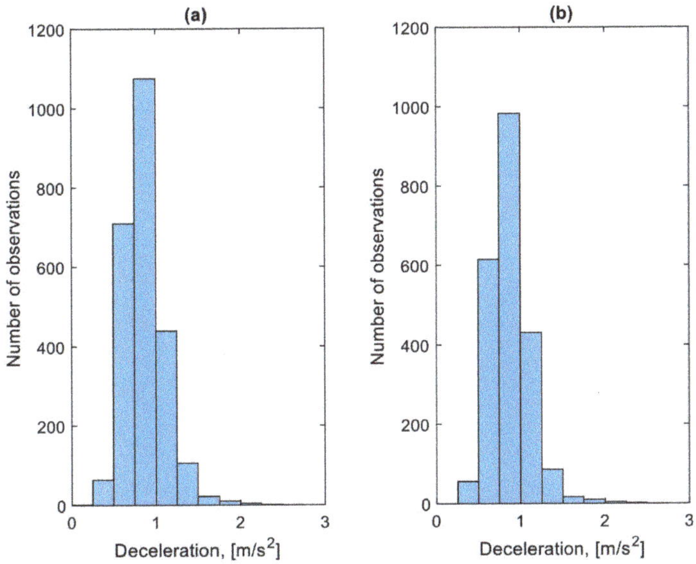

Figure 13. Distribution of decelerations during braking before a bus stop: (**a**) in the direction A to B; (**b**) in the B to A direction.

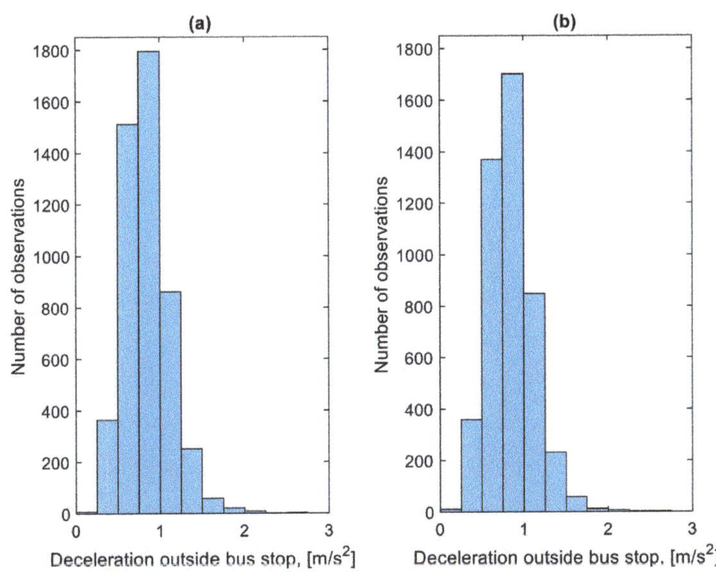

Figure 14. Distribution of decelerations during braking outside the bus stop: (**a**) in the direction A to B; (**b**) in the B to A direction.

Similarly to the braking speeds in Figures 11 and 12, an analysis of the effect of the average speed on a given course on the average deceleration on the course was also carried out. The dependence of the average deceleration during the braking process before bus stops on a course on the average speed on a course is shown in Figure 15. The dependence of the average deceleration during the braking process outside bus stops on a course on the average speed on a course is shown in Figure 16.

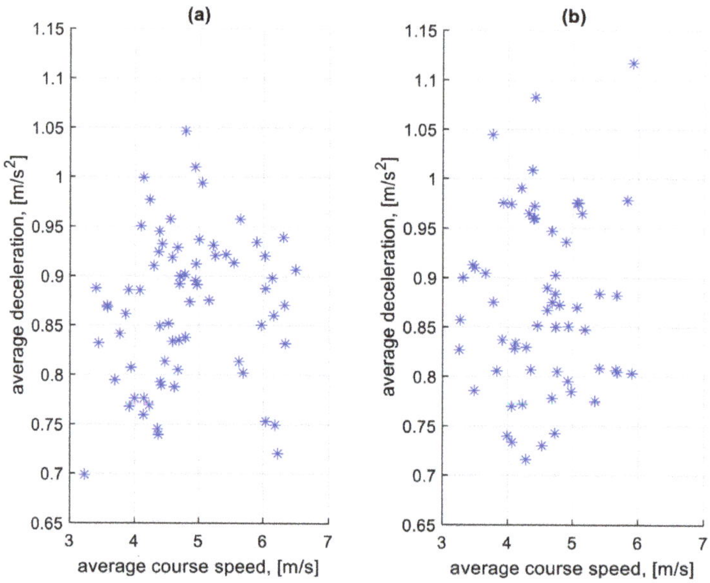

Figure 15. Average decelerations during braking before bus stops on a course depending on the average speed on a given course: (**a**) in the direction A to B; (**b**) in the B to A direction.

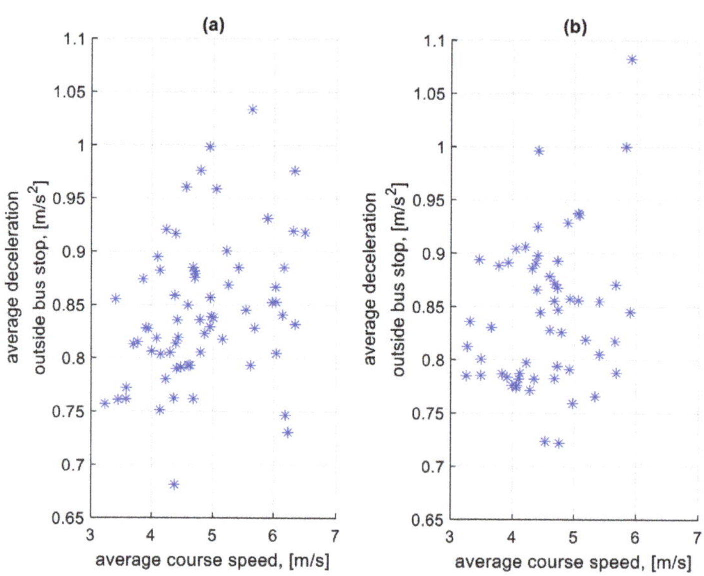

Figure 16. Average decelerations during braking outside bus stops on a course depending on the average speed on a given course: (**a**) in the direction A to B; (**b**) in the B to A direction.

Figures 15 and 16 do not indicate the effect of the average speed on a given course on the average deceleration during the braking process before a bus stop and outside a bus stop.

3.5. Analysis of Stop Times

An important element that affects the average speed on a given route is the stop times at and outside bus stops. In addition, these stop times contribute to increased fuel consumption. The total stop times at bus stops on a given course for the studied period are shown in Figure 17. The total stop times outside bus stops on a given course for the studied period are shown in Figure 18.

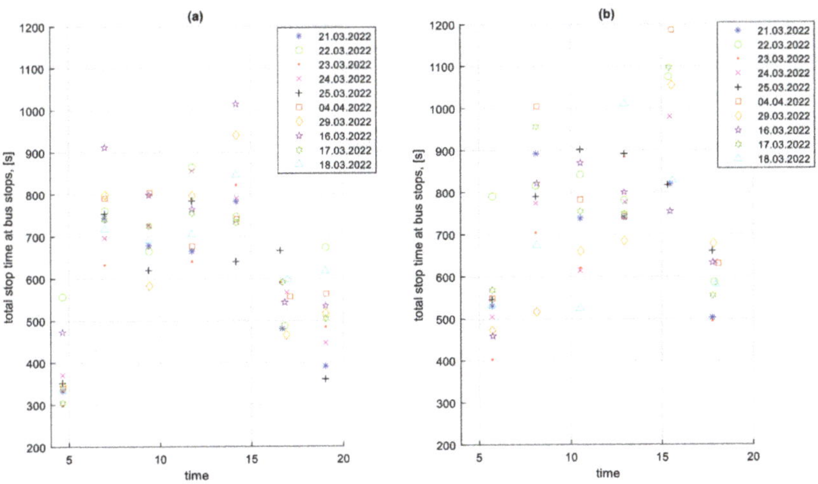

Figure 17. Total stop times at bus stops on a given course depending on the time of the start of the course: (**a**) in the direction A to B; (**b**) in the B to A direction.

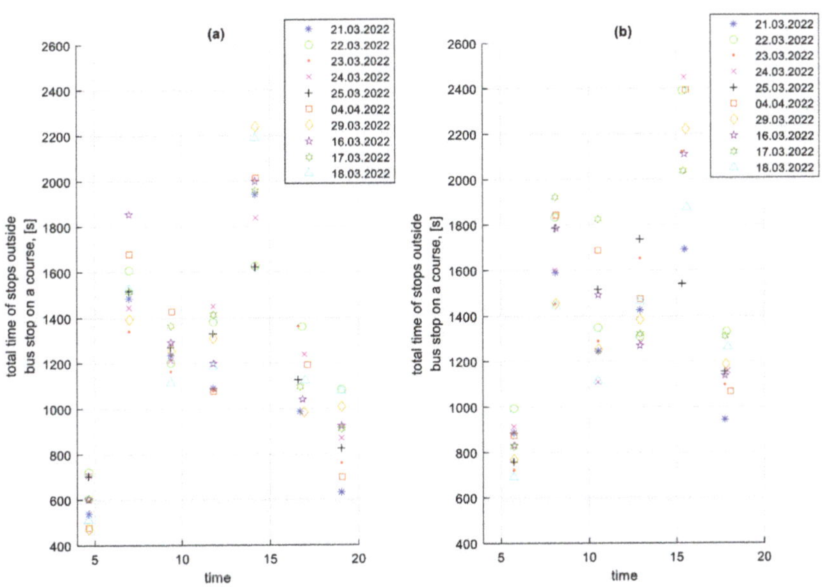

Figure 18. Total stop times outside bus stops on a given course depending on the time of the start of the course; (**a**) in the direction A to B: (**b**) in the B to A direction.

In Figure 18, the impact of rush hour on the total stop times outside bus stops for a given course is greater than on the total stop times at bus stops shown in Figure 17.

Figure 19 shows the distribution of the stop times at bus stops in the studied period. Figure 20 shows the distribution of stop times outside bus stops in the studied period.

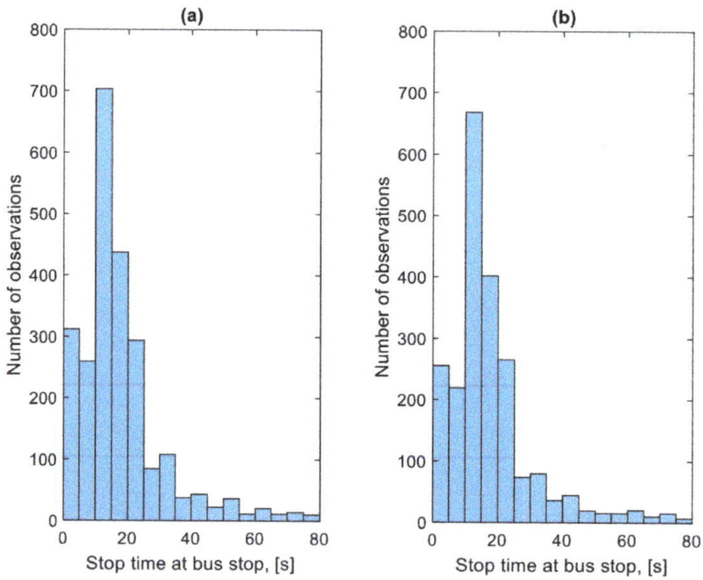

Figure 19. Distribution of stop times at bus stops: (**a**) in the direction A to B; (**b**) in the B to A direction.

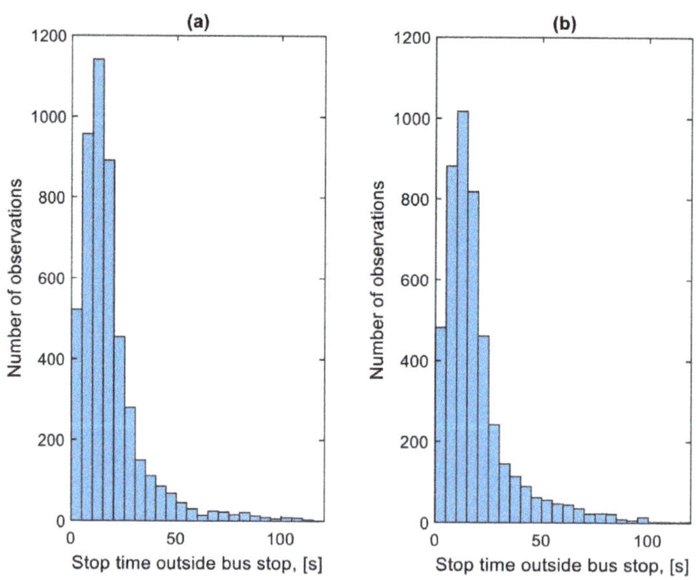

Figure 20. Distribution of stop times outside bus stops: (**a**) in the direction A to B; (**b**) in the B to A direction.

The very low stop times of approximately 5 s are visible in Figure 19, which are unrealistic from the point of view of passenger exchange. They result from the need for the bus to stop at every bus stop, regardless of the number of passengers waiting and willing to get off. This solution is imposed by the city authorities. The number of bus stops at which there was no passenger exchange on a given route during the studied period is shown in Figure 21 and the time of stops at these bus stops is shown in Figure 22.

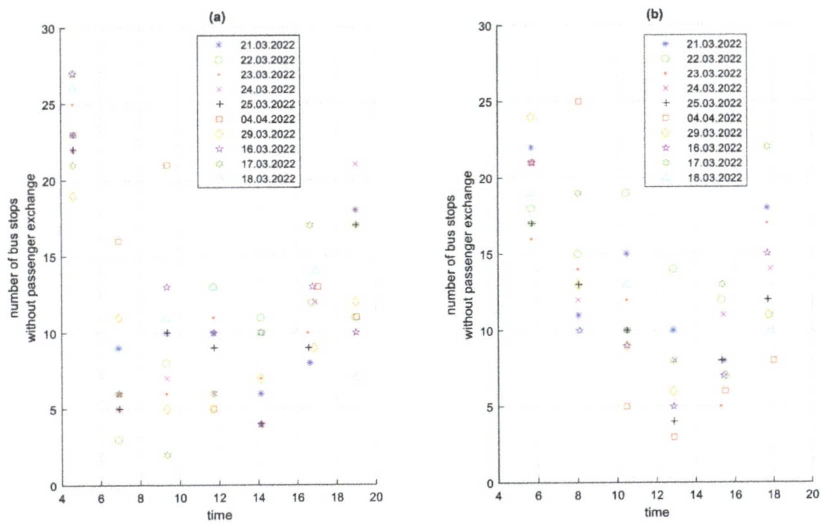

Figure 21. Number of bus stops without passenger exchange on a given course depending on the time of the start of the course: (**a**) in the direction A to B; (**b**) in the B to A direction.

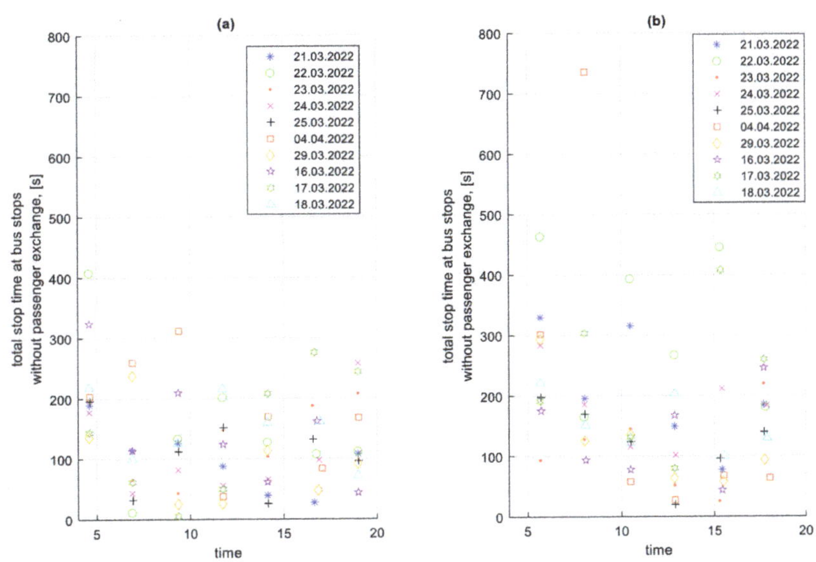

Figure 22. Total stop times at bus stops without passenger exchange on a given course; (**a**) in the direction A to B; (**b**) in the B to A direction.

Figure 23 shows the dependence of the average stop times at a bus stop on a course on the average speed on a given course. Figure 24 shows the dependence of the average stop times outside a bus stop on a course on the average speed on a given course.

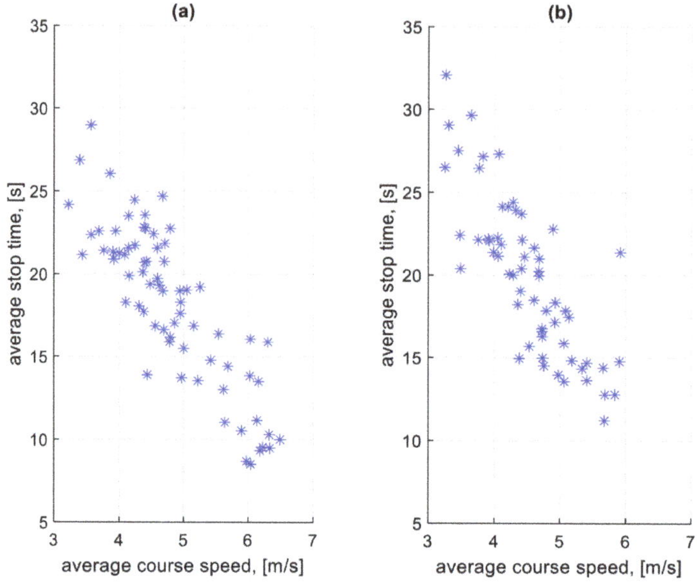

Figure 23. Average stop times at a bus stop on a course depending on the average speed on a given course: (**a**) in the direction A to B; (**b**) in the B to A direction.

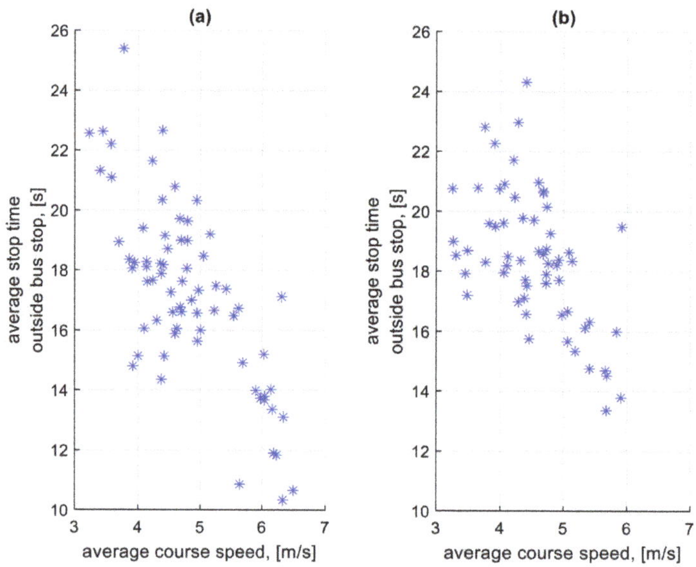

Figure 24. Average stop times outside the bus stop on a course depending on the average speed on a given course: (**a**) in the direction A to B; (**b**) in the B to A direction.

4. Determining Energy Consumption

4.1. Assumptions

According to part 2, the energy lost in breaking and at standstill contains two components. The first takes into account the braking speed. The second component depends on the stop time. In the calculations carried out, it was assumed that the mass of the vehicle with passengers m is equal to 13,000 kg. This is the mass of an empty bus that includes the mass of 15 passengers. The energy consumption at standstill can be determined from the stop times obtained from the analysis (Figures 19 and 20) and the assumed specific fuel consumption q_s. Taking into account the data obtained from the transport company and the data in the literature [5,30,31], q_s was assumed to be 0.57 g/s. This represents a consumption of approximately 2.5 L/h.

The simulation took into account the work of friction forces. However, the work of the air resistance forces was not taken into account. For an average braking speed of approximately 9 m/s and an average deceleration of approximately 1 m/s^2, this work amounts to approximately 1% of the kinetic energy, as shown in Figure 4b.

The speed at which a bus brakes, as well as the stop time, are not constant quantities. These quantities and, in particular, the speed at which the braking process begins, depend on many factors. Among the most relevant are the external traffic conditions and the associated average speeds on a given course.

The value of energy E_b lost in the fuel is

$$E_b = \frac{E_k}{\eta_n * \eta_c} + E_s \tag{13}$$

where

E_b—total energy lost;
E_k—lost kinetic energy.

$$E_k = \sum_{i=1}^{n} \frac{m * {v_b}_i^{\,2}}{2} - m * g * f_t * s_b \tag{14}$$

where

m—bus weight, m = 13,000 kg;
v_b—the speed at which the bus brakes, m/s;
g—acceleration due to gravity, g = 9.81 m/s^2;
f_t—rolling resistance coefficient, f_t = 0.01;
s_b—braking distance, m;
η_n—efficiency of the drive system, $\eta_n = 0.88$;
η_c—engine efficiency, $\eta_c = 0.36$;
E_s—energy lost during standstill.

$$E_s = \sum_{i=1}^{n} t_i * q_s * W_u \tag{15}$$

where

t—bus standstill time, s;
q_s—fuel consumption per second during standstill, g/s;
W_u—calorific value of fuel, $W_u = 42\frac{\text{kJ}}{\text{g}}$.

The average energy lost per stop $\overline{E_b}$ was calculated according to the formula

$$\overline{E_b} = \frac{\sum_{i=1}^{n} E_{bi}}{n} \tag{16}$$

where

E_{bi}—energy lost on the ith stop;
n—number of all stops.

4.2. Simulation of Energy Loss in the Braking Process

Using the above assumptions, a simulation of energy loss in the braking process was performed. The braking process and the standstill process were considered separately. For the calculations, the energy loss values due to the braking process and the stop time on a given course were determined for each braking separately. Figure 25 shows the dependence of the average braking energy before a bus stop on the course on the average speed on a given course. Figure 26 shows the dependence of the average braking energy outside the bus stop on the course on the average speed on a given course.

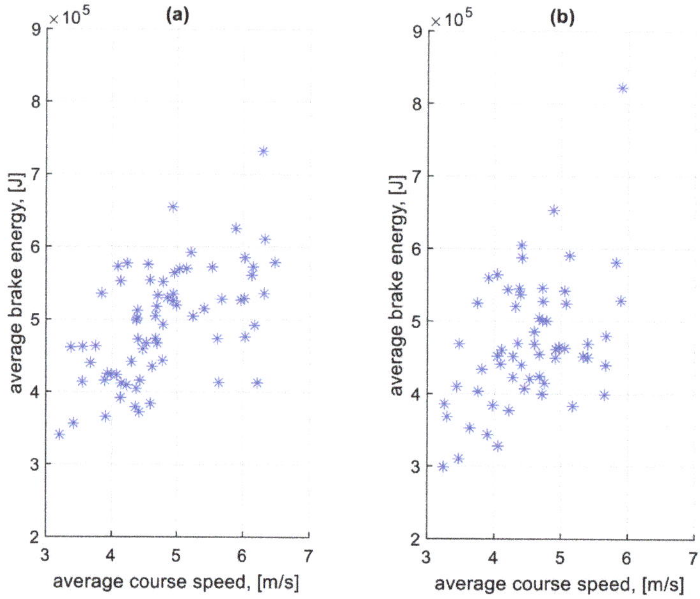

Figure 25. Average braking energy before bus stop on a course depending on the average speed on a given course: (**a**) in the direction A to B; (**b**) in the B to A direction.

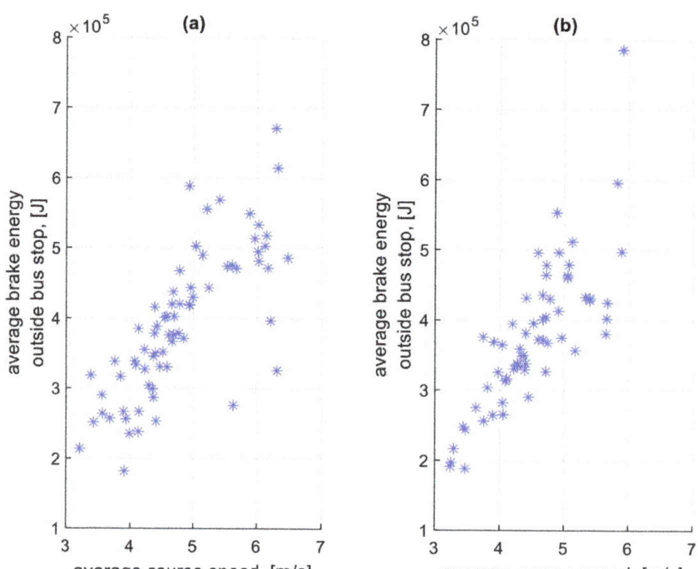

Figure 26. Average braking energy outside a bus stop on a course depending on the average speed on a given course: (**a**) in the direction A to B; (**b**) in the B to A direction.

Figure 27 shows the dependence of the average standstill energy of stopping at a bus stop on a course on the average speed of a given course. Figure 28 shows the dependence of the average standstill energy of stopping outside a bus stop on a course on the average speed of a given course.

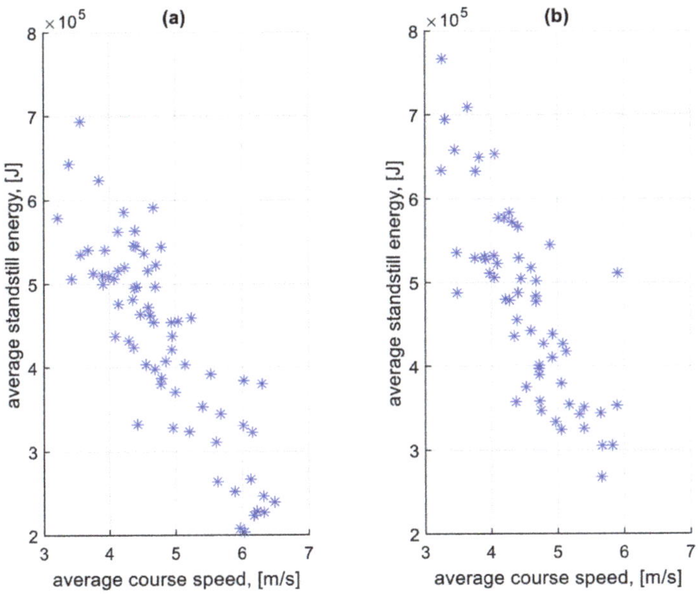

Figure 27. Average standstill energy of stopping at a bus stop on a course depending on the average speed on a given course: (**a**) in the direction A to B; (**b**) in the B to A direction.

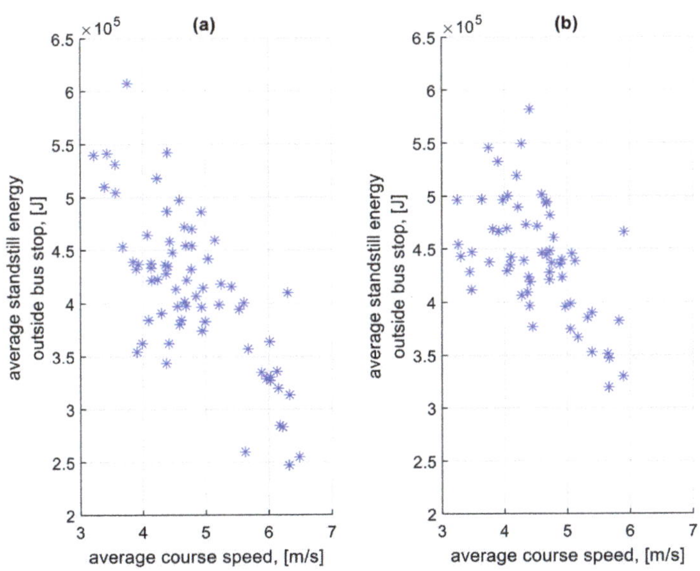

Figure 28. Average standstill energy of stopping outside a bus stop on a course depending on the average speed on a given course: (**a**) in the direction A to B; (**b**) in the B to A direction.

4.3. Total Energy Losses and Their Analysis

When summing the braking and standstill energy loss for each stop, the total braking and standstill energy loss was obtained. From this value calculated for each stop on a given course, the average total energy lost per stop on a given course was determined. Figure 29 shows the dependence of the average total energy lost per stop at a bus stop on a course on the average speed on a given course. Figure 30 shows the dependence of the average total energy lost per stop outside a bus stop on a course on the average speed on a given course.

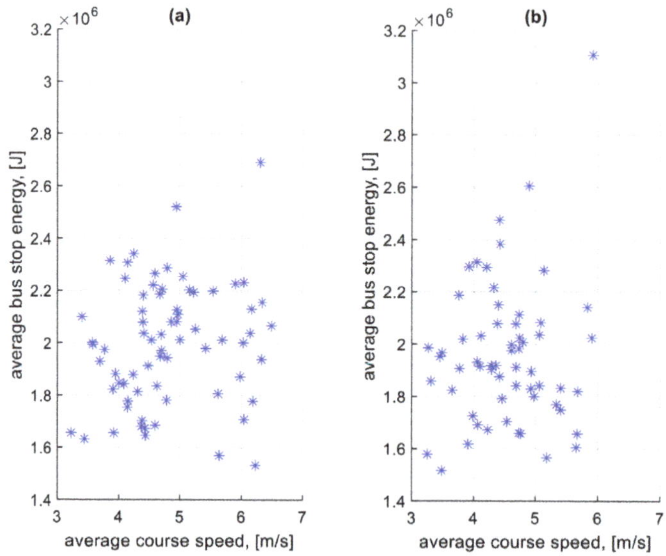

Figure 29. Average total energy lost per stop at a bus stop on a course depending on the average speed on a given course: (**a**) in the direction A to B; (**b**) in the B to A direction.

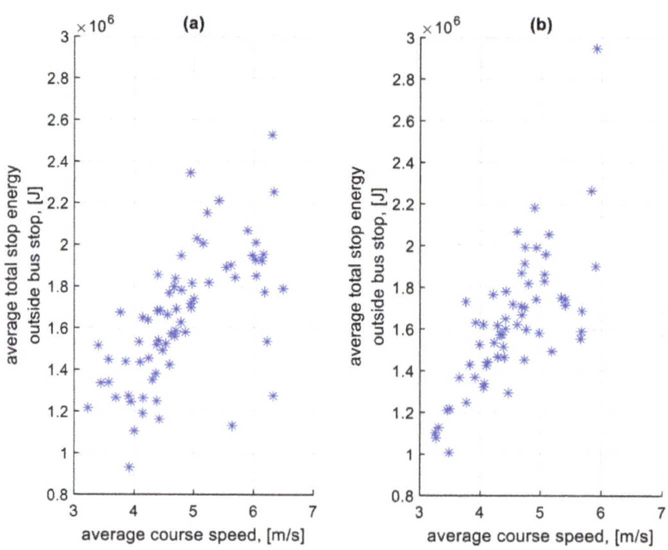

Figure 30. Average total energy lost per stop outside the bus stop on a course depending on the average speed on a given course: (**a**) in the direction A to B; (**b**) in the B to A direction.

Figure 31 shows the distribution of the total energy lost per stop at a bus stop. The red line presents the normal distribution of energy losses, which can be described by a Gaussian distribution function.

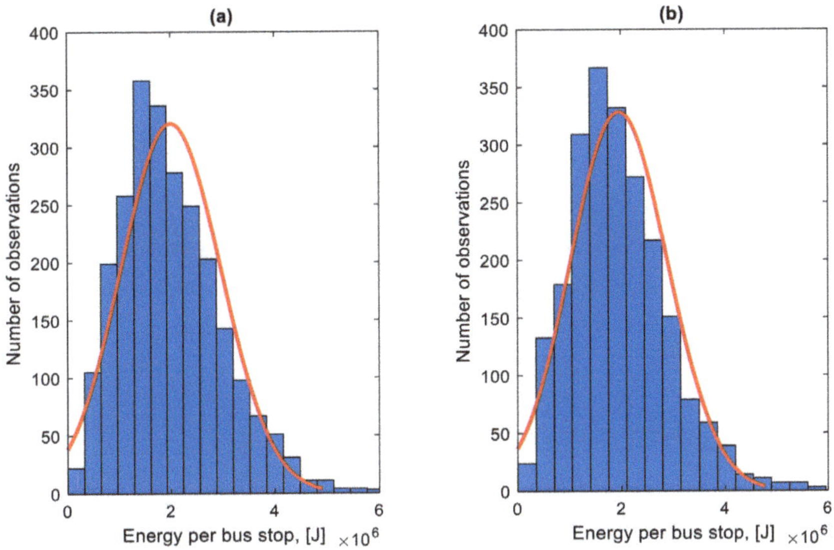

Figure 31. Distribution of the total energy lost per stop at a bus stop: (**a**) in the direction A to B; (**b**) in the B to A direction.

The average energy lost per stop at a bus stop in the direction A to B is 2.014 MJ, and in the direction B to A 1.978 MJ. During the studied period, according to data provided by the transport company, the bus was refuelled with 1020.75 L of fuel. The total energy lost during stops at bus stops in the studied period constitutes 26.15% of fuel consumption; that is, 266.97 L.

Table 1 shows the share of braking and standstill energy in the total energy lost during stops at a bus stop during the entire studied period.

Table 1. Shares of braking and standstill energy in the total energy lost during stops at a bus stop.

Direction	Share of Braking Energy [%]	Share of Standstill Energy [%]
A to B	78.35	21.65
B to A	75.89	24.11

Figure 32 shows the distribution of the total energy lost per stop outside a bus stop. The distribution presented in Figure 32 has a different character from the distribution in Figure 31. The bus brakes very often at low speed during a traffic jam, as can be seen in Figure 10.

The average energy lost per stop outside the bus stop in the direction A to B is 1.591 MJ, and in the direction B to A 1.578 MJ. The total energy lost during stops outside the bus stops constitutes 42.5% of the fuel consumption in the examined period; that is, 433.83 L.

Table 2 shows the shares of braking energy and standstill energy in the total energy lost during stops outside the bus stop during the entire studied period.

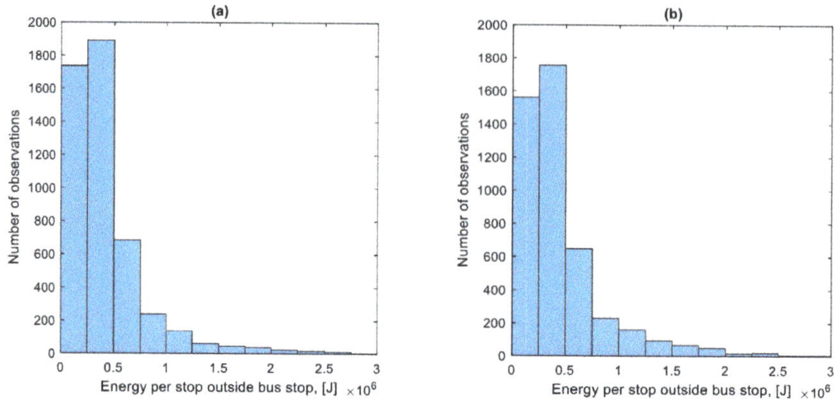

Figure 32. Distribution of the total energy lost per stop outside a bus stop: (**a**) in the direction A to B; (**b**) in the B to A direction.

Table 2. Shares of braking and standstill energy in the total energy lost during stops outside the bus stop.

Direction	Share of Braking Energy [%]	Share of Standstill Energy [%]
A to B	23.32	76.68
B to A	22.74	77.26

5. Discussion

The calculations carried out for the selected bus route showed the amount of energy lost in the process of breaking and being standstill at and outside bus stops. Within the courses analysed, the bus reached different average speeds. It can tentatively be assumed that this corresponds to the nature of operation on other routes. The amount of energy lost is an important component of the energy balance. On an average working day, the bus consumed 100 L. During the day, it made 13 courses with 473 bus stops. Fuel consumption related to stops at bus stops represents approximately 26% of total fuel consumption during the working day. The fuel consumption related to stops outside of bus stops represents approximately 42.5% of the daily fuel consumption.

The average energy lost per stop at bus stops and outside bus stops does not differ much in both directions. Traffic intensity is different in both directions, but total traffic volume is probably the same in both directions. It is indicated by almost the same average energy lost in both directions.

Most of the energy lost at stops at bus stops is braking energy. It indicates that a bus brakes from significant speed and is not standing still for long at bus stops. Most of the energy lost at stops outside of bus stops is standstill energy. It indicates that the bus spends significant time at standstill. This can be caused by traffic jams during rush hours and long waiting at intersections.

Stops at the bus stops along a given route result from the timetable. In that case, the associated energy losses do not depend on the average bus speed on a given route. The amount of these losses depends on the stop. Stop times at bus stops are determined by the number of passengers getting on and off and the number of doors. Low-floor buses with wide entrances are required on busy routes.

In order to reduce the energy consumption associated with stops at the bus stops, the number of stops could be reduced by introducing on-demand bus stops. The bus would only stop if there was a need to change passengers. Taking this into account requires further research and is part of efforts to improve existing city ITS. Those efforts should ensure the smooth running of public buses with the aim of reducing energy consumption and therefore harmful emissions. This will make public transport more environmentally friendly.

6. Conclusions

City ITS allows for smooth traffic flow and minimizes waiting time at intersections, which significantly increases the energy efficiency of vehicles in city traffic. This is especially important for public buses. Energy consumption of public buses on urban routes can be reduced by optimizing the existing network of bus stops, as well as by reducing the number of stops outside bus stops. The constant increase in the intensity of urban road traffic requires adaptive changes in the organization of road traffic and the improvement of road infrastructure through the introduction of a priority system for public transport. Such improvements of the existing ITS can be implemented only on the basis of studies of the real conditions of urban traffic, in particular, the impact of stops on a typical route of the city ITS on the energy efficiency of public buses.

The objective of this paper was to determine the energy consumption of the public bus associated with stops at bus stops and outside bus stops. Stops are an important factor of energy consumption in urban traffic. One of the main public bus routes in the city of Rzeszów was analysed. Based on the fuel consumption and GPS data, the energy model of diesel public buses was created. This model allowed the exact fuel consumption shares associated with stops at bus stops and outside bus stops to be determined. Fuel consumption shares associated with stops at bus stops and outside bus stops were determined to be 26% and 42.5%, respectively. This paper aimed to increase knowledge of how stops influence energy consumption of a public bus.

The created model was based on the movement resistance forces and energy transfer in engine and driven trains. Daily fuel consumption measurements were susceptible to errors of several litres. The method of measuring fuel consumption could be improved to reduce error to values of 1–2%.

The created model cannot be used to model losses of hybrid or battery electric buses, because the model does not take into account regenerative breaking.

As simulation studies carried out have shown, the additional energy consumption for buses related to stops outside of bus stops reaches significant values. Those values exceed the energy consumption related to stops at bus stops. These stops are the result of traffic conditions and transport infrastructure. Therefore, the reduction in the number of stops outside bus stops requires the actions of the city authorities to improve the existing ITS. Examples of such measures are changes to the traffic infrastructure consisting of separate bus lanes, adaptive changes in the organization of traffic by introducing the priority of public buses at intersections with traffic lights, and reconstruction of the road system to eliminate bottlenecks. The obtained research results are the basis for improving the ITS of the city of Rzeszow.

The second important factor is measures to reduce private vehicle traffic by taking over passenger flows by public transport. However, these measures require appropriate actions to encourage people to use public transport and the introduction of restrictions such as banning cars from entering the city centre or introducing additional entry fees.

Author Contributions: Conceptualization, M.S. and V.M.; methodology, M.S.; software, J.M.; validation, M.S. and V.M.; formal analysis, M.S.; investigation, J.M.; resources, V.M.; data curation, J.M.; writing—original draft preparation, M.S. and J.M.; writing—review and editing, V.M.; visualization, J.M.; supervision, V.M.; project administration, M.S. All authors have read and agreed to the published version of the manuscript.

Funding: This research received no external funding.

Data Availability Statement: The data presented in this study are available on request from the corresponding author due to restrictions on data made by Municipal Transport Company.

Conflicts of Interest: The authors declare no conflicts of interest.

References

1. Zheng, X.; Zhang, L. Ecodriving for reduction of bus transit emission with vehicle's Hybrid dynamic model. *Math. Probl. Eng.* **2015**, *2015*, 543429. [CrossRef]
2. Alshayeb, S.; Stevanovic, A.; Effinger, J.R. Investigating impacts of various operational conditions on fuel consumption and stop penalty at signalized intersections. *Int. J. Transp. Sci. Technol.* **2022**, *11*, 690–710. [CrossRef]
3. Abdelaty, H.; Al-Obaidi, A.; Mohamed, M.; Farag, H.E.Z. Machine learning prediction models for battery-electric bus energy consumption in transit. *Transp. Res. Part D Transp. Environ.* **2021**, *96*, 102868. [CrossRef]
4. Grabowski, L. Modeling research of city bus fuel consumption for different driving cycles. *J. Phys. Conf. Ser.* **2021**, *2130*, 012001. [CrossRef]
5. Frey, H.C.; Rouphail, N.M.; Zhai, H.; Farias, T.L.; Gonçalves, G.A. Comparing real-world fuel consumption for diesel- and hydrogen-fueled transit buses and implication for emissions. *Transp. Res. Part D Transp. Environ.* **2007**, *12*, 281–291. [CrossRef]
6. Rosero, F.; Fonseca, N.; López, J.-M.; Casanova, J. Effects of passenger load, road grade, and congestion level on real-world fuel consumption and emissions from compressed natural gas and diesel urban buses. *Appl. Energy* **2021**, *282*, 116195. [CrossRef]
7. Shahariar, G.M.H.; Bodisco, T.A.; Zare, A.; Sajjad, M.; Jahirul, M.I.; Chu Van, T.; Bartlett, H.; Ristovski, Z.; Brown, R.J. Impact of driving style and traffic condition on emissions and fuel consumption during real-world transient operation. *Fuel* **2022**, *319*, 123874. [CrossRef]
8. Rodríguez, R.A.; Virguez, E.A.; Rodríguez, P.A.; Behrentz, E. Influence of driving patterns on vehicle emissions: A case study for latin american cities. *Transp. Res. Part D Transp. Environ.* **2016**, *43*, 192–206. [CrossRef]
9. Wang, Y.; Boggio-Marzet, A. Evaluation of eco-driving training for fuel efficiency and emissions reduction according to Road Type. *Sustainability* **2018**, *10*, 3891. [CrossRef]
10. Özener, O.; Özkan, M. Fuel consumption and emission evaluation of a rapid bus transport system at different operating conditions. *Fuel* **2020**, *265*, 117016. [CrossRef]
11. Sun, R.; Chen, Y.; Dubey, A.; Pugliese, P. Hybrid electric buses fuel consumption prediction based on real-world driving data. *Transp. Res. Part D Transp. Environ.* **2021**, *91*, 102637. [CrossRef]
12. Zhang, S.; Wu, Y.; Liu, H.; Huang, R.; Yang, L.; Li, Z.; Fu, L.; Hao, J. Real-world fuel consumption and CO_2 emissions of Urban Public Buses in Beijing. *Appl. Energy* **2014**, *113*, 1645–1655. [CrossRef]
13. Rosero, F.; Fonseca, N.; López, J.-M.; Casanova, J. Real-world fuel efficiency and emissions from an urban diesel bus engine under transient operating conditions. *Appl. Energy* **2020**, *261*, 114442. [CrossRef]
14. Zhang, Z.; Ye, B.; Wang, S.; Ma, Y. Analysis and estimation of energy consumption of electric buses using real-world data. *Transp. Res. Part D Transp. Environ.* **2024**, *126*, 104017. [CrossRef]
15. Ji, J.; Bie, Y.; Zeng, Z.; Wang, L. Trip Energy Consumption Estimation for electric buses. *Commun. Transp. Res.* **2022**, *2*, 100069. [CrossRef]
16. Ghaffarpasand, O.; Talaie, M.R.; Ahmadikia, H.; Khozani, A.T.; Shalamzari, M.D.; Majidi, S. Real-world assessment of Urban Bus Transport in a medium-sized city of the Middle East: Driving behavior, emission performance, and Fuel Consumption. *Atmos. Pollut. Res.* **2021**, *12*, 113–124. [CrossRef]
17. Ma, X.; Miao, R.; Wu, X.; Liu, X. Examining influential factors on the energy consumption of electric and diesel buses: A data-driven analysis of large-scale public transit network in Beijing. *Energy* **2021**, *216*, 119196. [CrossRef]
18. Wang, A.; Ge, Y.; Tan, J.; Fu, M.; Shah, A.N.; Ding, Y.; Zhao, H.; Liang, B. On-road pollutant emission and fuel consumption characteristics of buses in Beijing. *J. Environ. Sci.* **2011**, *23*, 419–426. [CrossRef]
19. de Abreu e Silva, J.; Moura, F.; Garcia, B.; Vargas, R. Influential vectors in fuel consumption by an urban bus operator: Bus route, driver behavior or vehicle type? *Transp. Res. Part D Transp. Environ.* **2015**, *38*, 94–104. [CrossRef]
20. Macián, V.; Tormos, B.; Ruíz, S.; Ramírez, L. Potential of low viscosity oils to reduce CO_2 emissions and fuel consumption of urban buses fleets. *Transp. Res. Part D Transp. Environ.* **2015**, *39*, 76–88. [CrossRef]
21. Li, P.; Zhang, Y.; Zhang, Y.; Zhang, K.; Jiang, M. The effects of dynamic traffic conditions, route characteristics and environmental conditions on trip-based electricity consumption prediction of electric bus. *Energy* **2021**, *218*, 119437. [CrossRef]
22. Yang, X.; Liu, L. Analysis of the influence of passenger load on bus energy consumption a vehicle-engine combined model-based simulation framework. *Sci. Rep.* **2022**, *12*, 14535. [CrossRef] [PubMed]
23. Abdelaty, H.; Mohamed, M. A prediction model for battery electric bus energy consumption in transit. *Energies* **2021**, *14*, 2824. [CrossRef]
24. Stankiewicz, J.; Michalski, K. Rozwiązania Z Zakresu Inteligentnych systemów transportowych W Wybranych Miastach w Polsce. *Ekon. I Organ. Logistyki* **2018**, *3*, 79–91. [CrossRef]
25. W Rzeszowie na Wszystkich Skrzyżowaniach Z Sygnalizacją Świetlną będą Detektory Ruchu. (In Polish). Available online: https://nowiny24.pl/w-rzeszowie-na-wszystkich-skrzyzowaniach-z-sygnalizacja-swietlna-beda-detektory-ruchu/ar/c4-16359805 (accessed on 24 April 2024).
26. Kilka Słów O Rzeszowskim inteligentnym Systemie Transportowym: Wiadomości: Focus on Business—Created by Pro Progressio. (In Polish). Available online: https://focusonbusiness.eu/pl/wiadomosci/kilka-slow-o-rzeszowskim-inteligentnym-systemie-transportowym/11213 (accessed on 24 April 2024).
27. Śmieszek, M.; Mateichyk, V. Determining the fuel consumption of a public city bus in urban traffic. *IOP Conf. Ser. Mater. Sci. Eng.* **2021**, *1199*, 012080. [CrossRef]

28. Smieszek, M.; Kostian, N.; Mateichyk, V.; Mosciszewski, J.; Tarandushka, L. Estimation of the public transport operating performance: Example of a selected city bus route. *Commun.-Sci. Lett. Univ. Zilina* **2023**, *25*, B7–B21. [CrossRef]
29. Śmieszek, M.; Kostian, N.; Mateichyk, V.; Mościszewski, J.; Tarandushka, L. Determination of the model basis for assessing the vehicle energy efficiency in urban traffic. *Energies* **2021**, *14*, 8538. [CrossRef]
30. Wang, J.; Rakha, H.A. Fuel consumption model for conventional diesel buses. *Appl. Energy* **2016**, *170*, 394–402. [CrossRef]
31. Rahman, S.M.A.; Masjuki, H.H.; Kalam, M.A.; Abedin, M.J.; Sanjid, A.; Sajjad, H. Impact of idling on fuel consumption and exhaust emissions and available idle-reduction technologies for diesel vehicles—A Review. *Energy Convers. Manag.* **2013**, *74*, 171–182. [CrossRef]

energies

Article

Reliable Energy Optimization Strategy for Fuel Cell Hybrid Electric Vehicles Considering Fuel Cell and Battery Health

Cong Ji [1], Elkhatib Kamal [2,3,*] and Reza Ghorbani [4]

1 School of Energy and Environment, Southeast University, Nanjing 210096, Jiangsu Province, China; 230149312@seu.edu.cn
2 Department of Industrial Electronics and Control Engineering, Faculty of Electronic Engineering, Menoufia University, Menouf 32952, Egypt
3 Ecole Centrale de Nantes, LS2N CNRS, 44300 Nantes, France
4 Mechanical Engineering, University of Hawaii at Manoa, Honolulu, HI 96822, USA; rezag@hawaii.edu
* Correspondence: elkateb.kamal@gmail.com or elkhatib.ibrahim@ec-nantes.fr

Abstract: To enhance the fuel efficiency of fuel cell hybrid electric vehicles (FCHEVs), we propose a hierarchical energy management strategy (HEMS) to efficiently allocate power to a hybrid system comprising a fuel cell and a battery. Firstly, the upper-layer supervisor employs a fuzzy fault-tolerant control and prediction strategy for the battery and fuel cell management system, ensuring vehicle stability and maintaining a healthy state of charge for both the battery and fuel cell, even during faults. Secondly, in the lower layer, dynamic programming and Pontryagin's minimum principle are utilized to distribute the necessary power between the fuel cell system and the battery. This layer also incorporates an optimized proportional-integral controller for precise tracking of vehicle subsystem set-points. Finally, we compare the economic and dynamic performance of the vehicle using HEMS with other strategies, such as the equivalent consumption minimization strategy and fuzzy logic control strategy. Simulation results demonstrate that HEMS reduces hydrogen consumption and enhances overall vehicle energy efficiency across all operating conditions, indicating superior economic performance. Additionally, the dynamic performance of the vehicle shows significant improvement.

Keywords: hybrid electric vehicle; battery; fuel cell; energy management algorithms; optimal control; fault-tolerant control

1. Introduction

1.1. Background and Literature Survey

The high cost of oil, the limited quantity of this resource on the planet and the pollution generated by its use encourage populations, industries and governments to opt for other sources of energy for the transportation of goods and people. Therefore, the use of "hybrid electric vehicles (HEVs)" has become an urgent necessity today. On the one hand, the number of vehicles based on internal combustion engines (ICEs) is increasing; therefore, the consumption of fossil fuels is becoming more and more massive, and the concentration of greenhouse gases in the atmosphere is reaching worrying levels. Considerable research effort and considerable investment have been made in advanced battery technologies for electric vehicles. However, the major challenge remains the development of technologies and systems that ensure good performance and long range at a competitive cost. Today, one of the promising solutions is based on the design of a multi-source system, also known as a hybrid electric energy system. A hybrid energy source system is an important solution to extend the distances covered (vehicle autonomy) and to meet the power requirements of electric vehicles (EVs), especially for transient regimes (high acceleration, high deceleration and braking) [1].

Proton exchange membrane fuel cells (PEMFCs) in transportation applications are a promising solution to the energy crisis and environmental pollution because they are clean energy devices with high efficiency and low operating temperatures [2,3]. The advantages of fuel cell electric vehicles (FCEVs) are fast charging time, high driving range, noiseless operation and zero local emissions [3,4]. The disadvantages of FCEVs are a slow dynamic response and inability to recover braking power. On the other hand, lithium-ion batteries are an energy storage system with a fast dynamic response, low self-discharge, long lifetime and high energy density. Thus, the battery is usually used to compensate for the disadvantages of the pure FCEV. Thus, the fuel cell hybrid electric vehicle (FCHEV) is developed, which is powered by the PEMFC and the battery. This type of system is based on the combination of two or more energy sources. Thus, the disadvantages of one source can be compensated for by the advantages of the others and vice versa.

A lot of research is currently being done on electrical energy storage systems [5–9], including the optimization of vehicle energy management [9–12], state-of-charge (SOC) estimation techniques [13–17] and cell aging [18,19]. In addition, the scientific literature contains several works that study the value of energy storage units to improve the performance of multi-source systems dedicated to transportation applications [20–22].

More and more studies have been conducted to design energy management strategies (EMSs) for FCHEVs [23–28]. EMSs are mainly classified into rule-based and optimization-based strategies [29–31].

EMSs based on rule-based methodologies offer advantages such as low computational complexity, practicality and reliability. These primarily include state machine control (SMC) [32–35] and fuzzy logic control [36–39], among others. The formulation of control rules typically relies on "If . . . Then" based on engineering experience. For instance, the authors of [32–35] devised an EMS based on SMC comprising eight operational modes, which are selected based on the bus demand power and the auxiliary power state of charge (SOC). However, simplistic rules often fail to meet the vehicle's economic performance requirements. In another study [36–39], a two-degree-of-freedom fuzzy logic controller was implemented in an electric vehicle with an "FC + B" configuration. This approach introduces a fuzzy energy allocation method based on expert experience with respect to uncertainty and validates it using an advisor platform. The study in [40] integrated the particle swarm optimization (PSO) algorithm into an enhanced FLC strategy, optimizing the FLC membership function using PSO to enhance the fuel economy of FCHEVs. Nonetheless, the reliance on historical data and expert knowledge in rule-based strategies impedes adaptation to real-time driving conditions and does not guarantee optimal outcomes.

In recent years, EMSs based on optimization strategies have garnered significant attention and research efforts. These systems can generally be categorized into two types: global optimization and instantaneous optimization. Compared to EMSs based on rule-based methodologies, EMSs based on optimization typically exhibit superior performance in enhancing vehicle fuel economy and power durability [41,42].

Predictive and dynamic programming (DP) [43–51] is widely studied as a global optimization algorithm in hybrid power systems and is capable of obtaining optimal control outcomes for predefined operating conditions. The study in [52] proposed an enhanced DP-based control strategy aimed at minimizing system operating costs by exploring the impact of state-of-charge (SOC) penalty factors and the initial SOC. However, DP-based strategies often suffer from complexity in calculation, offline applicability only and the challenge of "dimension disaster" [53].

EMSs based on instantaneous optimization often utilize a sampling interval as the optimization window to establish an objective function for achieving instantaneous control optimization goals. One prominent strategy in this category is the equivalent consumption minimization strategy (ECMS) [54–56], which introduces an equivalent factor (EF) to convert battery power consumption into equivalent fuel consumption. Li et al. [57] implemented the ECMS strategy with a fixed EF value for FCHEVs, demonstrating its cost-effectiveness compared to rule-based EMSs. However, the fixed EF value limits the strategy's applicability to known driving cycles. Tian et al. [58] proposed an adaptive power distribution strategy based on ECMS; they regulated the EF value through a neural network velocity predictor. Li et al. [32] presented an improved ECMS strategy incorporating the lithium battery SOC that dynamically adjusts the EF value based on SOC changes to minimize total equivalent hydrogen consumption (EHC) under unknown driving conditions.

Another typical instantaneous optimization strategy is Pontryagin's minimum principle (PMP) [59], which transforms the global optimization problem into a series of minimization problems, obtaining the optimal trajectory of the costate by minimizing the Hamiltonian function at each step [60]. The costate in the PMP algorithm equates power consumption between different sources to the equivalent hydrogen consumption, similar to EF in ECMS [61]. PMP-based EMSs adjust the costate value to achieve optimal load power distribution under specific driving conditions, thereby minimizing FCHEV hydrogen consumption. Meng et al. [62] developed a hierarchical PMP-based EMS, achieving an optimal load demand power configuration under a given driving cycle and demonstrating its superiority over DP-based EMSs. However, this method is limited to specific driving conditions, and the costate value remains fixed throughout the driving cycle, restricting its applicability. Li et al. [63] proposed an adaptive PMP-based EMS for FCHEVs, adjusting the costate value online using a Markov velocity predictor. Yang et al. [64] employed a linear weighted particle swarm optimization (PSO) algorithm to update the costate value for different driving conditions, achieving a 16% improvement in system fuel economy compared to a power following control (PFC) strategy. Onori and Tribioli [65] utilized lookup table methods to select the appropriate costate, improving fuel economy by 15% compared to a state machine control (SMC) EMS.

The aforementioned EMSs typically prioritize minimal hydrogen consumption as the optimization objective to achieve satisfactory fuel economy, often overlooking system lifespan considerations.

Based on global EMSs, there are also some works that intend to ensure the continuity of the operation of the the fuel cell (FC) and the lithium-ion battery during fault actions. This is called fault-tolerant control (FTC). The fault is detected, isolated and identified thanks to a diagnostic system [66], and an adjustment to the controller's parameters is made to accommodate the fault. This is referred to as active FTC [67]. There is also passive FTC [68], which is also called robust control. The authors in [69–75] propose active FTC for a hybrid powertrain (FC/battery) applied to an urban bus. Lebreton et al. [76] developed active FTC for water management problems in the FC. However, the literature concerning the optimization of operating conditions to improve the durability of PEMFC is not yet very extensive. Furthermore, ensuring the durability of fuel cells (FCs) poses a significant challenge that must be addressed for the commercialization and mass production of FCHEVs [77]. The FC stack comprises various materials, leading to diverse reversible and irreversible degradation mechanisms within the FC. Irreversible degradation mechanisms directly impact FC performance over time, with catalyst layer microporous structure modification, carbon support corrosion, polymer membrane degradation and catalyst dissolution/redeposition being the primary contributors [78]. These complex degradation phenomena can result in performance losses in the FC stack. Notably, adverse operating conditions and dynamic power distribution dynamics are the primary causes of these degradation mechanisms [79]. Studies indicate that heavy loads, frequent load fluctuations, and prolonged start–stop times accelerate FC degradation, with frequent load fluctuations exerting the most significant impact on FC lifespan [80]. Zhang and Tao [81] devised a fuzzy logic controller with a low-pass filter for FCHEV to enhance FC durability. Similarly, Florescu et al. [82] proposed an EMS that accounts for FC degradation factors. Increasing the lifetime of PEMFCs is still one of the most important issues related to the use of this technology. There are two concomitant approaches to limit the degradation of fuel cells. The first is in the field of materials and involves finding more resistant and robust materials. The second solution is to act at the system level and, more particularly, on the operating conditions (in particular the temperature, pressure, relative humidity, sensor and actuator faults, etc.).

However, the abovementioned studies [69–82] only focus on the FC and lithium-ion battery durability and ignore the fuel economy of the FCHEV.

1.2. Objectives

The main objective of this paper is to develop an optimal management strategy to improve the energy efficiency of hybrid vehicles (fuel cell/battery). We seek to minimize as much as possible the consumption of hydrogen while maintaining the SOC of the battery. In addition, the main objective of existing battery and fuel cell management system (BFCMS) control strategies is to preserve the system and improve performance. The main aims that can be considered for developing an EMS are mentioned in Figure 1. Firstly, the energy sources used in this study, the modeling of the hybrid system and the choice of the type of power electronic converters are presented. Secondly, an improved energy management method based on dynamic programming (DP) and Pontryagin's minimum principle (PMP) are proposed. Thirdly, in order to maximize the extracted energy and the lifetimes of lithium-ion batteries and proton-exchange-membrane fuel cell (PEMFC), it is necessary to develop better battery and fuel cell management systems (BFCMSs) and thus improve SOC estimation algorithms specifically for high-capacity batteries with a large number of cells. SOC estimation is an important data point, because knowing the current capacity of the battery makes it possible to accurately estimate the number of kilometers that can be driven by the vehicle. A more accurate algorithm allows the vehicle to travel more kilometers and increases the user's sense of confidence and decrease the anxiety related to the risk of breaking down with the vehicle. Finally, a model of a hybrid vehicle is built using TruckMaker/MATLAB/Simulink software.

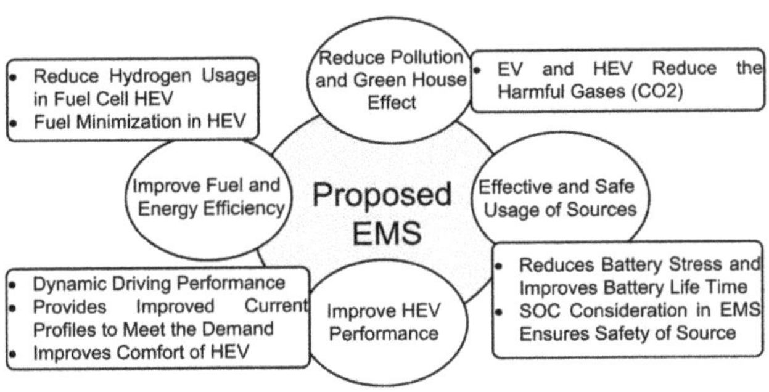

Figure 1. The main aims that can be considered for developing an EMS.

1.3. Main Contributions

Given the aforementioned considerations, this paper propose an intelligent hierarchical supervisory energy management strategy (HSEMS) (cf. Section 4) to strike a balance between fuel cell hydrogen consumption, degradation of the fuel cell lifespan and the durability of the PEMFC and lithium-ion battery of a FCHEV in the presence of sensor and/or actuator faults. The key contributions and novel aspects of this study are outlined as follows:

- The problem is the instantaneous distribution of the electrical power requested from the two energy sources while optimizing as much as possible the global consumption of hydrogen on a given mission profile. Energy management strategies based on improved energy management methods based on dynamic programming (DP) are developed. Then, we present a method of Pontryagin's minimum principle (PMP). These strategies lead the fuel cell to operate at the points of best performance while maintaining the SOC of the battery function. We then compare our method with two other strategies based on a fuzzy-logicstrategy-based EMS (FLS) and an equivalent consumption minimization strategy (ECMS).

- The estimation of the SOC of the battery is proposed. An algorithm supported by a theory from the field of advanced control is used for the first time for the estimation of the SOC. It is chosen for its accuracy and its low computational resources required, which make it well-suited for the characteristics required for light electric vehicles.
- The work presented in this manuscript concerns the implementation of a better BFCMS methodology that takes into account the occurrence of PEMFC and battery faults. The developed strategy is FTC that aims to limit the occurrence of and reduce the effects of faults. FTC based on an adaptive fuzzy observer (AFO) using the linear matrix inequality (LMI) allows for sufficiently early mitigation of the fault to reduce its consequences (performance decrease and degradations).
- Simulation results using TruckMaker/MATLAB software confirm that the proposed approach leads to optimal energy consumption of the vehicle for any unknown driving cycles and compensates for battery fault effects.

In order to implement the proposed HSEMS, firstly, a reliable model of the FCHEV corresponding to a bus that was developed using professional TruckMaker/MATLAB software (cf. Section 2.4) is proposed. A FCHEV is characterized by the same structure as a series hybrid vehicle in which the functions of the internal combustion engine (ICE)–electric generator system are performed by a PEMFC.

Secondly, an intelligent HSEMS (hierarchical supervisory energy management strategy) (cf. Section 4) is designed in order to minimize the total energy consumption; it gives better fuel cell and battery life and, therefore, increases the overall bus energy efficiency based on the merging of adaptive fuzzy logic, dynamic programming (DP) and Pontryagin's minimum principle. Combining these techniques takes advantage of their strengths while mitigating their weaknesses.

The proposed HSEMS strategy consists of two control levels. In the second level (upper level), a supervisory battery and fuel cell management system (BFCMS) is designed to generate healthy PEMFCs and the battery SOC for the first level and gives acceptable performance during fault actions.

In the first level (lower level), an advanced energy optimization strategy based on adaptive fuzzy logic, dynamic programming (DP) and Pontryagin's minimum principle is developed to minimize total energy consumption and, therefore, maximize the overall bus energy efficiency. In this level, the overall proposed strategies (improved energy management based on dynamic programming (DP) and Pontryagin's minimum principle (PMP)) are designed.

In addition, in the first level, an adaptive fuzzy-logic-controller-based proportional-integral controller is used to give optimal vehicle subsystem set-point tracking, which is generated at the second level.

The validation of the overall proposed strategies (DP and PMP) (cf. Sections 4.2 and 5.1) is performed by comparing them using the equivalent consumption minimization strategy (ECMS) given in [54] (cf. Section 5.2) and [32,51] (cf. Section 5.3).

The proposed strategy has several advantages: (i) it increases the studied bus energy efficiency and minimizes the total energy consumption; (ii) it detects and compensates for the effects of sensor and/or actuator faults of the PEMFC and battery for the studied bus; (iii) the proposed DP and PMP reduces hydrogen consumption by 9.8% and 8.86%, respectively, compared with the logic-strategy-based EMS (FLS) strategy under the same driving conditions.

1.4. Plan of the Document

After this general introduction, Section 2 presents the modeling of the hydrogen FC/battery hybrid vehicle. The formulation of the optimization problems are presented in Section 3. In Section 4, the proposed HSEMS and the details regarding its adaptation to the problem are presented. The optimized EMSs based on the fuzzy-logic-strategy-based

EMS (FLS) and the equivalent consumption minimization strategy (ECMS) are given in Section 5. Validation methods by simulation using the TruckMaker and MATLAB/Simulink software are presented in Section 6. In Section 7, a conclusion summarizes the challenges and contributions of this paper as well as suggestions for future work in order to improve the present work.

2. Modeling of the Studied Vehicle

The structure of the studied bus (cf. Figure 2) consists of a PEMFC as the main power source connected to a high-voltage DC/DC converter and battery as an auxiliary power source connected to a low-voltage bi-directional DC/DC converter, a DC bus and a traction machine. The hybrid powertrain architecture is composed of a primary source (energy source), which is a FC, and a secondary source (power source), which is a battery. The choice of these components and technologies is based on the current market and future visions of possible technologies [83].

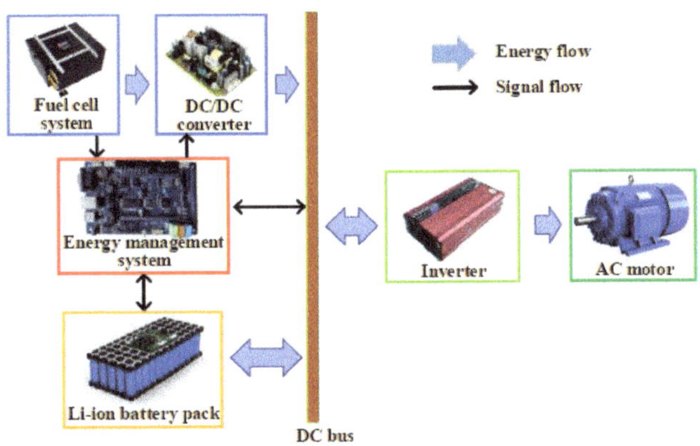

Figure 2. Structure of the bus powertrain.

2.1. Dynamics of the Vehicle

The fundamental principles of dynamics link the forces and accelerations of a solid, and according to Newton's second law, the traction force (F_t) of the vehicle can be written as [51]:

$$F_t = \delta m\frac{dv}{dt} + \frac{1}{2}\rho C_d A v^2 + mgf\cos(\alpha) + mg\sin(\alpha) \quad (1)$$

where v is the bus speed, α is the slope, ρ is the air density, m is the bus mass, A is the frontal area, and C_d is a constant called the drag coefficient that depends on the shape of the vehicle. Based on F_t, v and the wheel radius (r), the wheel torque (T_w) and its rotation speed (w_w) are given by [51]:

$$T_w = r \cdot F_t, \qquad w_w = v/r \quad (2)$$

Then, the rotational speed (w_m) and torque (T_m) are calculated by:

$$T_m = \begin{cases} T_w/(\eta_{fd}R_{fd}), & T_w \geq 0 \\ \eta_{fd} \cdot T_w/R_{fd}, & T_w < 0 \end{cases}, \quad w_m = w_w \cdot R_{fd} = \frac{v \cdot R_{fd}}{r}$$

where η_{fd} is the transmission efficiency and R_{fd} is the gear ratio. The power demand (P_t) is written as:

$$P_t = \begin{cases} T_m \cdot w_m/\eta_m, & T_m \geq 0 \\ \eta_m \cdot T_m \cdot w_m, & T_m < 0 \end{cases} \quad (3)$$

where η_m is the motor efficiency. The total power demand (P_{dem}) includes consumption by the auxiliaries (P_{aux}) and is written as:

$$P_{dem} = P_t + P_{aux} \quad (4)$$

2.2. Fuel Cell System Model

The model is composed of several cells mounted in series/parallel to obtain the characteristics necessary for the desired dimensioning. The cell is the main source of energy in the vehicle. Therefore, it must be able to provide sufficient power to the vehicle so that the vehicle can run at the maximum speed in its specifications. The instantaneous hydrogen consumption ($\dot{m}_h$) is given by [28]:

$$\dot{m}_h = N \cdot M_h / n \cdot F I_{stack} \quad (5)$$

where N is the number of cells, M_h is the molar mass of hydrogen, n is the transferred electrons, F is Faraday's constant, I_{stack} is the fuel cell stack current, and the efficiency of the PEMFC is given by

$$\eta_{FC} = P_{FC} / \dot{m}_h \cdot 120 \text{ MJ} \cdot \text{kg}^{-1} \quad (6)$$

where P_{FC} is the PEMFC power.

2.3. The Battery Model

The model used integrates unit cells connected in series/parallel. It is composed of an open-circuit voltage (E_o) of the battery, the ohmic resistance (R_{bat}) and an equivalent capacitor (C_{bat}) modeling the transient aspects of the battery's behavior (cf. Figure 3). The battery voltage (V_{bat}) is given by [74]:

$$V_{bat} = E_o - R_{bat} I_{bat} - V_{C_{bat}} \quad (7)$$

where I_{bat} is the battery current. The SOC change rate is given by [74]: *I revised all*

$$\frac{d(SOC)}{dt} = -\frac{I_{bat}}{Q_{bat}} \quad (8)$$

where Q_{bat} is the battery capacity. The battery power (P_{bat}) is given by:

$$P_{bat} = I_{bat} V_{bat} \quad (9)$$

Also, the battery SOC is given by:

$$SOC = SOC(0) - \frac{1}{C_{bat}} \int I_{bat} dt \quad (10)$$

Therefore the SOC change rate can be expressed by:

$$\frac{d(SOC)}{dt} = -\frac{E_o - \sqrt{E_o^2 - 4R_{bat} P_{bat}}}{2R_{bat} Q_{bat}} \quad (11)$$

Note that E_o is nonlinear; therefore, the battery model is nonlinear.

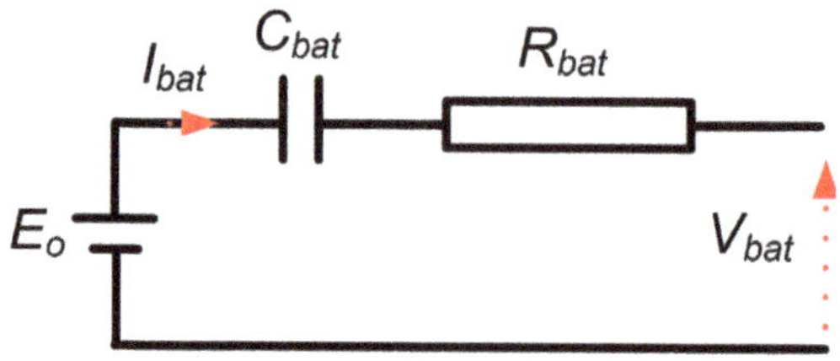

Figure 3. Electrical model of the battery.

2.4. FCHEV Design Using TruckMaker

In order to obtain precise validation of the proposed control strategies, evaluation tests of the control strategies discussed in this paper are performed on a high-fidelity simulation platform based on TruckMaker and MATLAB/Simulink software (cf. Figure 4).

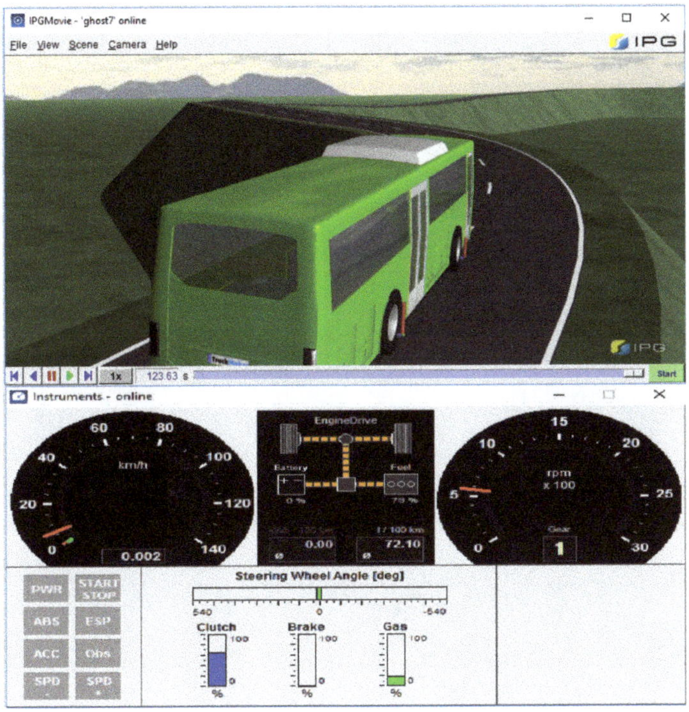

Figure 4. The studied FCHEV modeled using TruckMaker software.

The powertrain architecture of a hybrid vehicle modeled using the TruckMaker software consists of several interconnected subsystems, allowing it to provide an overall dynamic behavior of the powertrain that is comparable to the real system's operation. These subsystems can be classified into three distinct categories: (i) the powertrain's control strategy; (ii) the control units of the motors, the mechanical transmission and the electrical power supply system; and (iii) the electromechanical and hydraulic components of the powertrain.

In the proposed work, we are particularly interested in the development of simulation model of the overall high-level control strategy of the studied hybrid powertrain. In the TruckMaker simulation platform, the control layer is grouped inside a single control block containing all the high-level control subsystems. This block manages the operating states of the powertrain and generates the control set-point values for the different electromechanical components based on the inputs provided by the driver and on the current driving conditions. The structure of the control blocks in TruckMaker differs according to the powertrain's architecture. For the hybrid parallel powertrain on which our work is based, the control block must perform the following tasks: (i) braking management and monitoring the vehicle's operating state; (ii) interpretation of the accelerator pedal position; (iii) planning of the energy sharing between the motors (energy management) and generation of corresponding torque set-point values; (iv) power supply management (batteries and converters); and (v) braking management.

3. Formulation of the Optimization Problem

The decision-making problem for the vehicle is to find the best way to distribute, store and consume energy to satisfy the driver's demand while optimizing fuel consumption over the entire mission. The criterion to be minimized is the total hydrogen (H_2) consumption for a given mission. The control objective function is to minimize hydrogen consumption [1]:

$$J = min \int_{t_k+1}^{t_f+H_p} \dot{m}_h(P_{FC}(t))dt \tag{12}$$

where $\dot{m}_h$ is the instantaneous hydrogen consumption, which depends on the PEMFC power supplied (P_{FC}) and its total efficiency (η_{FC}); t_k is the current time step, and H_p is the prediction length. The total demand power P_{dem} is given by:

$$P_{dem} = P_{FC}(t).\eta_{DC-DC} + P_{bat}(t) \quad \forall t \tag{13}$$

where η_{DC-DC} is the DC/DC efficiency. Therefore (11) can be rewritten as:

$$S\dot{O}C(t) = f(SOC, P_{FC}) = \frac{-I_{bat}}{Q_{bat}} = \frac{E_o - \sqrt{E_o^2 - 4R_{bat}P_{bat}}}{2R_{bat}Q_{bat}}, \tag{14}$$

The first constraint is related to the limits of the battery SOC. We impose the following boundary conditions, as given by (15)–(17):

$$SOC_{t_f+1} = SOC_{t_f+H_p} = SOC_{ref} \tag{15}$$

where $SOCref$ are the SOC set-points, $SOC_{t_f+H_p}$ is the SOC at the end of the mission, and SOC_{t_f+1} is the SOC at the initial time.

$$\begin{aligned} P_{FC}(t) \in (&max(P_{FC,min}, (P_{dem} - P_{bat,max})/\eta_{DC-DC}), \\ &min(P_{FC,max}, (P_{dem} - P_{bat,min})/\eta_{DC-DC})) \end{aligned} \tag{16}$$

Finally, the last constraint concerns the storage element. It is necessary to restrict the evolution of the SOC in such a way that it remains within the range recommended by the equipment manufacturer. These constraint limits can be expressed as follows:

$$\begin{cases} \Delta P_{FC,min} \leq \Delta P_{FC} \leq \Delta P_{FC,max} \\ P_{bat,min} \leq P_{bat} \leq P_{bat,max} \\ SOC_{min} \leq SOC \leq SOC_{max} \end{cases} \tag{17}$$

where $P_{FC,max}$, $P_{FC,min}$, $P_{bat,max}$, $P_{bat,min}$, SOC_{max} and SOC_{min} are the maximum and minimum power of the PEMFC, battery and SOC, respectively.

4. Overall Proposed Control Architecture and Main Components

Managing the interactions between the various components of the hybrid powertrain while respecting the actuators' physical limitations and passenger comfort requirements to achieve the most efficient operation of the vehicle is the main challenge when developing a control strategy. From this perspective, the overall hierarchical supervisory energy management strategy (HSEMS) architecture proposed in this paper is designed to optimize the power distribution between energy sources and reduce torque jolts to improve powertrain reliability and passenger comfort. The proposed HSEMS strategy is given in Figure 5 and consists of two control levels.

In the upper level (second), a supervisory fuzzy fault-tolerant control (FFTC) scheme estimates a healthy SOC for the PEMFCs and the battery and gives acceptable performance during fault actions.

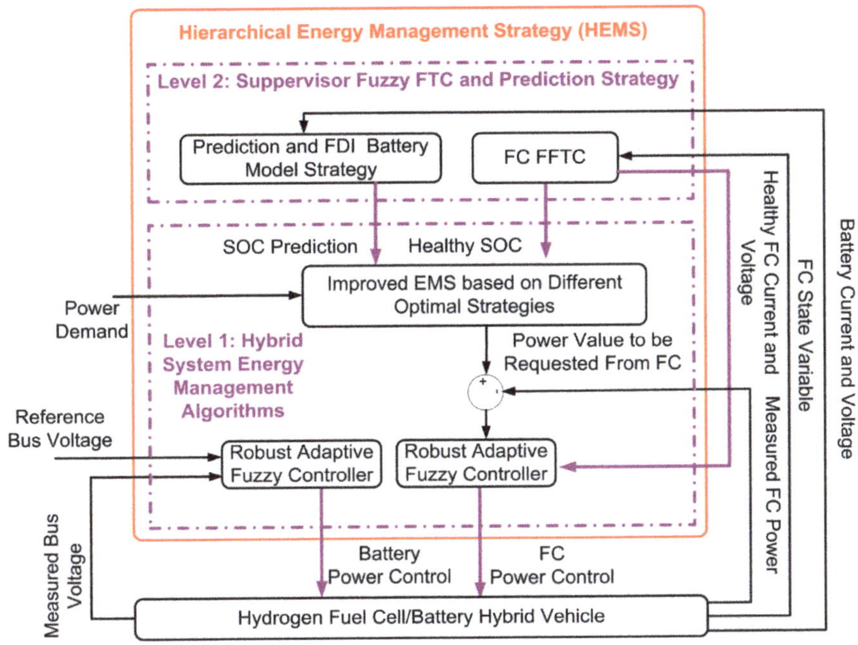

Figure 5. The proposed overall control strategy.

Energy optimization is performed at the first control level and is based on dynamic programming (DP) and Pontryagin's minimum principle (PMP); it has been developed to minimize total energy consumption and, therefore, maximize the overall bus energy efficiency. At this level, the overall proposed strategies (improved energy management based on DP and PMP) are designed (cf. Sections 4.2 and 5.1). A comparative study in terms of hydrogen consumption is made with two other strategies that are proposed for level 1: The first one is called the equivalent consumption minimization strategy (ECMS) (cf. Section 5.2) [54]. The second approach is a fuzzy logic strategy (FLS)-based EMS (cf. Section 5.3) [32], which is a very simple strategy based on the constraints imposed by the sources. The treated DP and PMP strategies present good improvement to hydrogen consumption by adopting a good management strategy for the electrical energy in the hybrid system.

4.1. Supervisory Fuzzy FTC and Prediction Strategy (Level 2: SFFTC)

4.1.1. Supervisory Fuzzy FTC

The main objective of the supervisory fuzzy fault-tolerant control (SFFTC) is to handle and regulate battery faults and establish a healthy state-of-charge (SOC) set-point, which is crucial for achieving a robust and optimal energy management system (EMS) at the first level, thereby influencing fuel cell hybrid electric vehicle (FCHEV) power optimization. The overall design of the proposed FFTC is shown in Figure 6. This section presents a comprehensive fault diagnosis and control strategy for battery cells that aims to identify current and/or voltage sensor faults and mitigate their impacts. To execute this diagnostic and control strategy, fuzzy fault-tolerant control (FFTC) utilizing a fuzzy adaptive observer is proposed to estimate and compensate for battery faults, including both current and voltage sensor faults. The parallel distributed compensation (PDC) concept [76] is applied to develop the fuzzy control and fuzzy adaptive observer using Takagi–Sugeno (TS) fuzzy models. Robust stabilization conditions are derived based on Lyapunov stability theory for voltage sensor faults, current actuator faults and state variables that are not directly measurable. These conditions are expressed as linear matrix inequalities (LMIs).

The flowchart of the proposed fuzzy logic controller (FLC)-based active fault-tolerant control system (AFTCS) is shown in Figure 6. The system assesses sensor readings and compares them to observer values to determine if they exceed a certain threshold. If no faults are detected, the fuel cell and battery operate normally. If a fault is detected in a sensor, the error signal surpasses the threshold. In this scenario, the fault detection and isolation (FDI) unit replaces the faulty sensor value with an estimated value generated by the FLC-based observer and sends this to the engine's control unit. The model receives the predicted value for the faulty sensor through analytical redundancy. It is assumed that switching and reconfiguration happen instantaneously, although in practice, there may be some delay due to controller computations. Note that the system only considers complete sensor failures and does not account for partial faults.

a. Observer Design

The process can be represented in state-space form, as shown by the observer architecture for the AFTCS system, as follows:

$$\begin{aligned} \dot{x}(t) &= \sum_{i=1}^{p} \mu_i[A_i x(t) + B_i u(t) + Ef(t)] \\ y(t) &= \sum_{i=1}^{p} \mu_i[C_i x(t)] \end{aligned} \tag{18}$$

where $x(t)$, $u(t)$ and $y(t)$ are the state, control input and the output vectors, respectively, μ are the fuzzy sets, p is the number of rules, $A_i \in \Re^{n\times n}$, $B_i \in \Re^{n\times m}$ and $C_i \in \Re^{g\times n}$ are system, input and output matrices, respectively, $f(t)$ are the actuator faults, and E is the actuator fault matrix. To estimate the state and fault of the battery and FC, the following fuzzy adaptive observer is proposed based on reference [84]:

$$\dot{\hat{x}}(t) = \sum_{i=1}^{p} \mu_i[A_i X(t) + B_i u(t) + E_i \hat{f}(t) + K_i(y(t) - \hat{y}(t))] \tag{19}$$

$$e_x(t) = x(t) - \hat{x}(t) \quad e_y(t) = y(t) - \hat{y}(t) = C_i e_x(t) \tag{20}$$

$$\dot{\hat{f}}(t) = \sum_{i=1}^{p} \mu_i L_i(\dot{e}_y(t) + e_y(t)) = \sum_{i=1}^{p} \mu_i L_{ii}(\dot{e}_x(t) + e_x(t)) \tag{21}$$

$$\hat{y}(t) = \sum_{i=1}^{p} \mu_i C_i \hat{x}(t) \tag{22}$$

In the proposed approach, the observer gains K_i and L_i are to be designed appropriately based on [84].

b. Proposed fuzzy fault-tolerant control

In this paper, an AFTCS is design based on [84].

$$u(t) = \sum_{j=1}^{p} \mu_j[G_j \hat{x}(t) - E_j \hat{f}(t)] \tag{23}$$

The controller gains G_j and E_j are designed based on [84].

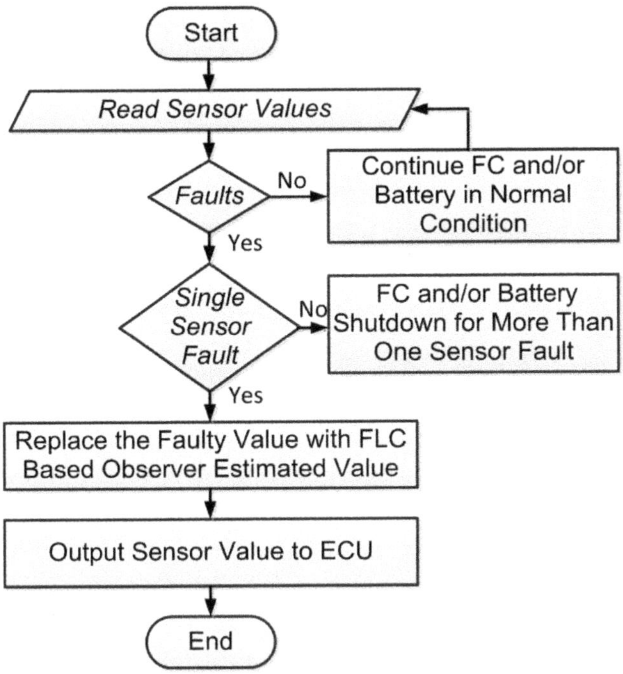

Figure 6. Schematic of the proposed FFTC.

4.1.2. Estimation of the SOC and Its Predicted Progress

Accurate state-of-charge (SOC) estimation and its anticipated evolution over a specified timeframe are critical for advanced energy management systems (EMSs). This section aims to outline the methodology proposed to address these crucial issues. The central concept involves calculating the SOC based on vehicle dynamics across various modes such as consumption and generation throughout the operational day while considering traffic conditions. For EMSs and applications like lithium-ion batteries, precise SOC estimation is paramount to prevent issues such as over-charging or over-discharging, safeguarding the battery's internal state. Given the challenges in directly measuring the SOC during battery operation, a suitable battery model becomes indispensable. Estimating the SOC using electrical models for power battery packs is particularly significant for hybrid electric vehicles as it provides essential data for efficient energy management. This subsection briefly discusses the use of an adaptive fuzzy observer strategy (AFO) based on [84] and an adaptive neuro-fuzzy inference system (ANFIS) [85] for SOC estimation. These methods are then integrated with a switching strategy to leverage their respective strengths across different operational contexts.

a. Estimation Results for the SOC Based on an AFO

The design of a battery management system is crucial for electric and hybrid vehicles and enhances performance, maintains optimal vehicle operation and detects/controls thermal runaway, which can lead to severe outcomes such as fires and explosions. Among the key parameters monitored, the state-of-charge (SOC) stands out as pivotal for these systems. However, direct measurement is impractical, necessitating estimation through analytical or machine learning techniques rather than sensors. This subsection proposes an electrical model incorporating thermal effects for lithium-ion batteries. The thermal model calculates internal heat generation, followed by the implementation of an adaptive fuzzy observer (AFO) to estimate the SOC and cell temperature within the battery system. The AFO's accuracy hinges on a precise model. The architecture leveraging the AFO for

SOC and cell temperature estimation is depicted in Figure 7. Inputs to the electro–thermal model include the current, ambient temperature and estimated SOC_k derived from the AFO, requiring a highly accurate electro–thermal model for effective implementation.

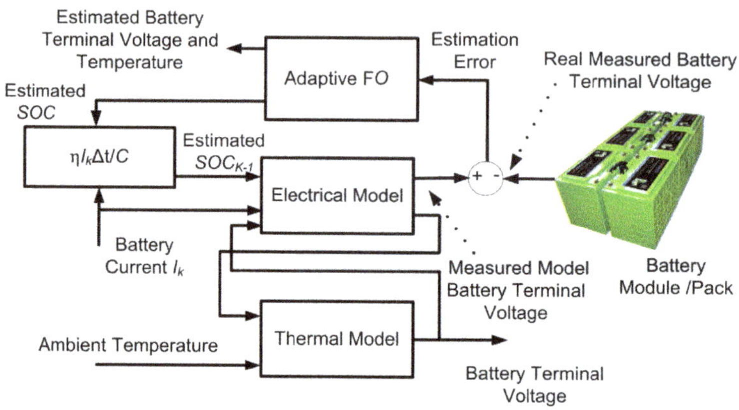

Figure 7. Block diagram for SOC prediction by adaptive fuzzy observer strategy.

b. Estimation Results for the SOC Based on the ANFIS

At the predictive level, optimal control problems are addressed based on trip forecasts. Accurate prediction of road conditions such as the slope, ambient temperature, and wind speed is crucial for energy management in hybrid systems. In [85], an adaptive neuro-fuzzy inference system (ANFIS) is proposed to predict SOC evolution under varying driving cycle conditions. The approach utilizes intelligent transportation data (such as GPS and radar) along with local sensor measurements for short-term and long-term ambient condition predictions. By analyzing locally collected data, the system can assess the battery SOC's power potential ($S\hat{O}C_{pred}$), as depicted in Figures 8 and 9. This technique primarily integrates predictions of vehicle power, real-time traffic information, vehicle state (including torque and speed), and battery conditions (such as current, voltage and temperature).

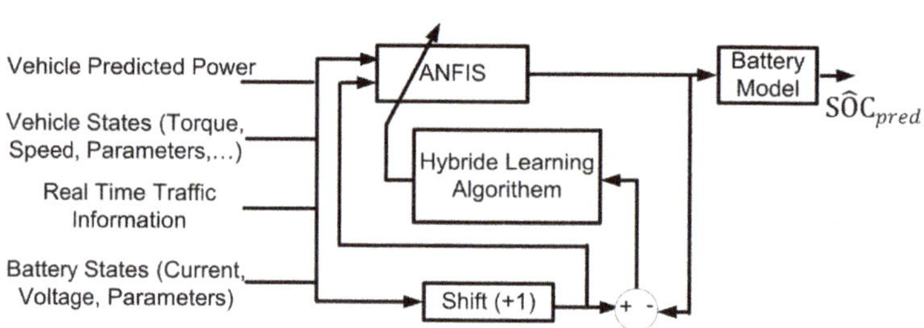

Figure 8. Block diagram for SOC prediction using ANFIS strategy.

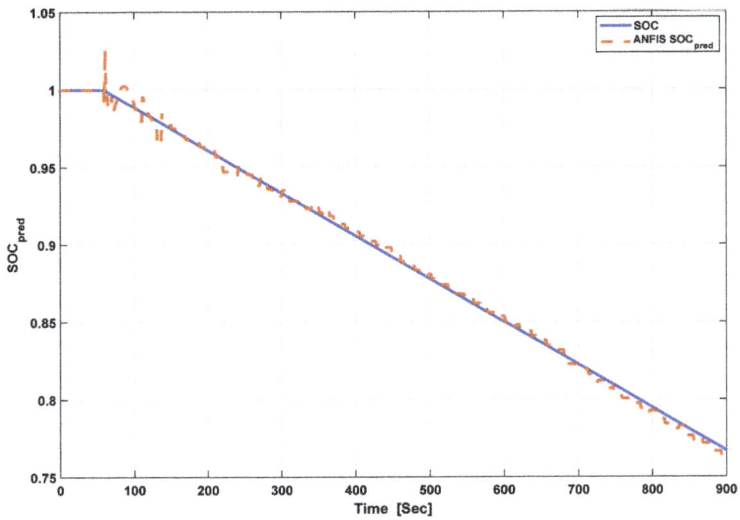

Figure 9. SOC prediction using ANFIS technique.

4.2. Hybrid System Energy Management Algorithms (Level 1: HSEMA)

The main role at this level is to achieve optimal power management between the different sources. Four different optimization strategies are implemented at this level. These optimization strategies are DP (cf. Section 4.2), PMP (cf. Section 5.1), equivalent consumption minimization strategy (ECMS) (cf. Section 5.2) and fuzzy logic strategy (FLS)-based EMS (cf. Section 5.3).

4.2.1. EMS Optimization Based on Dynamic Programming

The DP algorithm, a nonlinear programming method based on the Bellman equation, serves as a global optimization approach for multistage decision-making problems. When applied to hybrid systems with prior knowledge of the driving cycle, it yields the global optimal EMS. While real-time application is impractical, DP results serve as benchmarks for evaluating vehicle performance. Energy management poses a complex multistage decision challenge that aims to optimize fuel economy while meeting system constraints. DP is well-suited for this task and facilitates the determination of power distribution schemes that enhance FCHEV fuel economy. To derive the most economical power allocation strategy for FCHEVs, prior knowledge of the driving cycle is discretized into N stages, enabling the formulation of the system state transition equation in discrete form [86,87].

$$x(k+1) = f(x(k), u(k)), \;\; k = 0, 1, ..., N-1 \tag{24}$$

In the context of the fuel cell/battery hybrid system, the state variable at stage k, denoted as $x(k) = SOC(k)$, represents the state-of-charge (SOC) of the battery. Meanwhile, the control variable at stage k, expressed as $u(k) = P_{FC}$, corresponds to the output power of the fuel cell system (FCS) and facilitates power distribution across different stages. Consequently, the output power of the power battery is calculated as $P_{bat}(t) = P_{dem} - P_{FC}(t)$ and is based on (11). Thus, the system's state transition equation can be described as follows:

$$SOC(k+1) = SOC(k) - \frac{E_o - \sqrt{E_o^2 - 4R_{bat}(P_{dem}(k) - P_{FC}(k))}}{2R_{bat}Q_{bat}} \tag{25}$$

The primary objective of the DP algorithm is to determine the optimal control variable, $P_{FC}(k)$, at each stage to minimize the objective function, J. This objective function, as described in references [88,89], can be formulated as follows:

$$J = min \sum_{k=0}^{N} L(x(k), u(k)) \tag{26}$$

The transition cost function for each stage is denoted by $L(\dot{})$. In this study, the total hydrogen consumption of the vehicle serves as the objective function. Based on (12), the transition cost function for each stage can be expressed as follows:

$$J = \int_{k}^{k+1} \dot{m}_h(P_{FC}(t))dt \tag{27}$$

During real vehicle operation, it is imperative to ensure that constraints (17) are met. The flowchart depicting the DP process is presented in Figure 10. In the backward calculation process, one $x(k)$ is chosen from N parts at each moment k. Then, one $x(k+1)$ is selected from the reachable state set, which comprises M states. Using the state equation, $u(k)$ and the corresponding $J(k)$ are calculated for each state, and this process is repeated M times. The optimal $J^*(k)$ and $u^*(k)$ corresponding to $x(k)$ at each moment k are determined by finding the minimum $J^*(k)$. Since $u(k)$ is not discretized, it is derived from $x(k)$ and $x(k+1)$ to minimize errors and computational workload. The process involves iterating through all N parts of state x and recording the optimal solution for each state at each moment. In the forward solution process, starting from the initial value $x(1)$, the optimal control sequence $u0^{N-1} = (u(0), u(1), \ldots, u(N-1))$ and the minimum $J^*(x)$ are obtained by sequentially referencing the table.

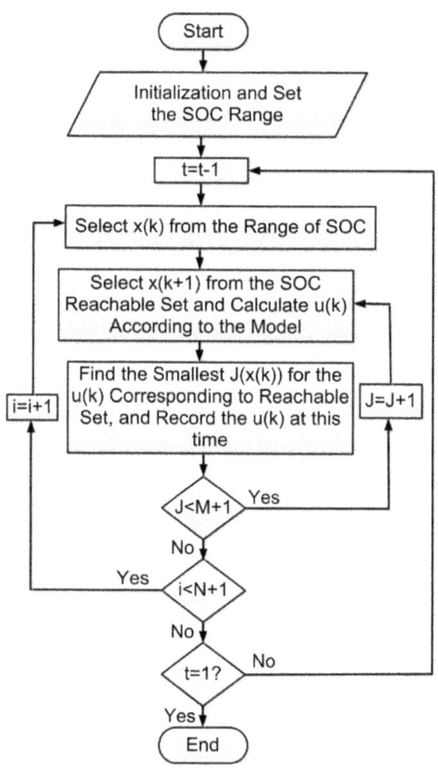

Figure 10. Flowchart of DP algorithm.

4.2.2. Robust Adaptive Fuzzy Controller

In traditional PID control methods, control performance may suffer due to constraints arising from various forms of uncertainty. To address this issue, a new PID control approach is proposed that integrates the advantages of fuzzy PID control and predictive functional control (PFC). This method is validated to give optimal vehicle subsystem set-point tracking, which is generated at the second level. Following the PFC framework, future process behavior is predicted based on the current process input signal. Subsequently, fuzzy PID control, informed by multi-step prediction, is applied to determine the optimal control strategy. The proposed strategy product with the PI controller parameters are calculated by [90].

5. EMS Optimization Based on Rule-Based Strategy and ECMS

This section presents optimal EMSs that are proposed for level 2 (cf. Figure 5). The control architectures are based on three different levels (cf. Section 4).

5.1. Improved EMS Based on PMP

Pontryagin's minimum principle (PMP), formulated by Soviet researcher Lev Pontryagin, is a fundamental algorithm in optimal control theory. According to PMP, an optimal control problem with a fixed time and a fixed endpoint can be defined as follows [91]: $P_{FC}(t)$ represents the control variable, SOC denotes the state variable, λ is the co-state, t_0 is the initial time, and t_f is the final time. The objective is to find the optimal control $P^*_{FC}(t), t \in [t_0, t_f]$ such that the system described in (28) transitions from an initial state to a terminal state while minimizing the performance index given in (29).

$$\frac{d(SOC)}{dt} = -\eta \frac{I_{bat}}{C_{bat}} \tag{28}$$

Based on the principles of PMP, the optimal output power $P^*_{FC}(t)$ of the fuel control system at each time step is given by the following equation.

$$P^*_{FC}(t) = argmin\ H \tag{29}$$

where H represents the Hamiltonian function, which is defined as shown in Equation (30) as follows:

$$H = \dot{m}_h - \lambda.S\dot{O}C(t) = \dot{m}_h - \lambda.\eta \frac{E_o - \sqrt{E_o^2 - 4R_{bat}(P_{dem} - P_{FC}\eta_{DC-DC})}}{2R_{bat}C_{bat}}, \tag{30}$$

In the equation above, λ represents the co-state, and its state function is expressed as shown in Equation (31):

$$\dot{\lambda} = \frac{dH}{dSOC} = \frac{\lambda.\eta}{C_{bat}}\left(\frac{dI_{bat}}{dE_o}\frac{dE_o}{dSOC} + \frac{dI_{bat}}{dR_{bat}}\frac{dR_{bat}}{dSOC}\right) \tag{31}$$

From the analysis above, once the initial value of λ is established, Equation (17) can be solved, and the optimal $P^*_{FC}(t)$ can be obtained. In discrete optimal control based on PMP, the optimal $P^*_{FC}(t)$ is determined by solving Equation (29) at each time step and subject to the system constraints outlined in Equation (17).

5.2. Improved EMS Based on ECMS

The studied EMS, which is based on ECMS as described in this paper, is derived from the formulation presented in [54]. Its objective is to minimize hydrogen consumption while ensuring that the SOC of the battery is maintained. ECMS operates by converting electrical energy stored in energy sources into an equivalent fuel consumption and aims to minimize the sum of instantaneous fuel consumption and equivalent fuel consumption [54].

5.3. Improved Fuzzy-Logic-Strategy-Based EMS

The fuzzy logic strategy (FLS)-based EMS discussed in this paper is built upon the approaches outlined in [32,51]. In real-world driving scenarios, vehicles operate in diverse modes, adjusting their behavior according to driving conditions and the status of the powertrain system. Following this principle, this study utilizes Stateflow within the MATLAB platform to develop a fuzzy rule-based EMS. This EMS governs and manages energy distribution within the vehicle, ensuring optimal power distribution between the fuel cell system (FCS) and the battery. Vehicle operating modes are determined based on vehicle speed (v_x) and the accelerator pedal position (p_{acc}), which is specifically categorized into idle, running and braking modes, as illustrated in Figure 11.

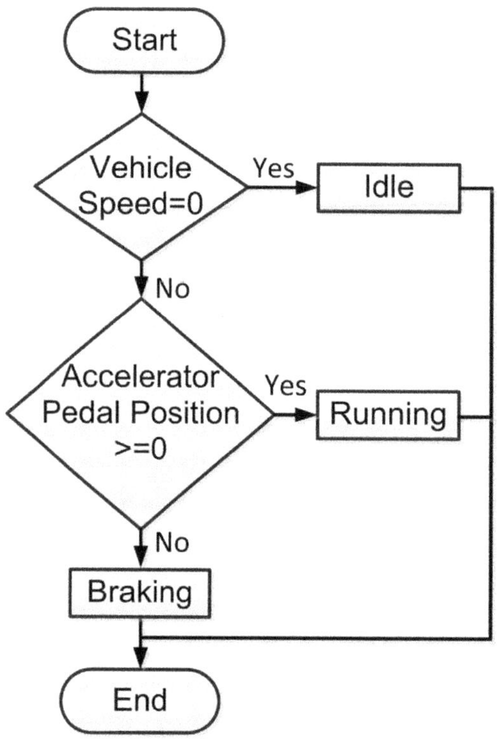

Figure 11. Algorithm of the rule-based EMS.

6. Simulation and Validation Results

This section aims to validate and to compare experimentally the main control strategies and energy management architectures described in this paper. The developed simulation platform based on TruckMaker and MATLAB/Simulink is given. In this section, two simulations are presented. The first simulation is to validate the overall HSEM during fault actions and validate the PEMFC and battery fault compensation. In the second simulation, the main proposed control strategies (DP and PMP) are compared with two other strategies based on FLS and ECMS in order to demonstrate the pros and cons of each proposed strategy.

6.1. Simulation 1: Overall HSEM Validation with and without FTC

The SOC of a hybrid vehicle's battery is critical information for the driver. The accuracy of the SOC algorithm is often limited by the computational capabilities of the electronic components. It is critical to ensure that the performance requirements obtained in the design phase are maintained during the real-time implementation. The growing demand for hybrid vehicles increases the need for low-computational-operation algorithms and low-cost battery management systems. This paper introduces an accurate SOC estimation algorithm that provides a compromise between accuracy and simplicity. The algorithm offers a simple solution that reduces the computational time while achieving performance similar to that of other well-known SOC estimation algorithms. This proposed estimator can be programmed in an embedded system to be more representative of real computational conditions. The validation of the overall control architecture with and without FTC is performed in this subsection using TruckMaker/MATLAB software. The validation is done by using a speed profile of the studied hybrid vehicle. The results indicate that the proposed observer strategy can provide a more accurate estimate and is faster than other algorithms. Figure 12 shows the driving profile (the EPA Urban Dynamometer Driving Schedule (UDDS)), and the power demand profile is shown in Figure 13. Figures 14 and 15 show the estimation of the SOC (initial condition = 68%) and the H_2 consumption, respectively, with a bias fault at the time of 600 sec, according to the proposed DP with and without FTC. Table 1 shows the evolution of the SOC and the H_2 consumption with and without FTC based on the proposed DP strategy.

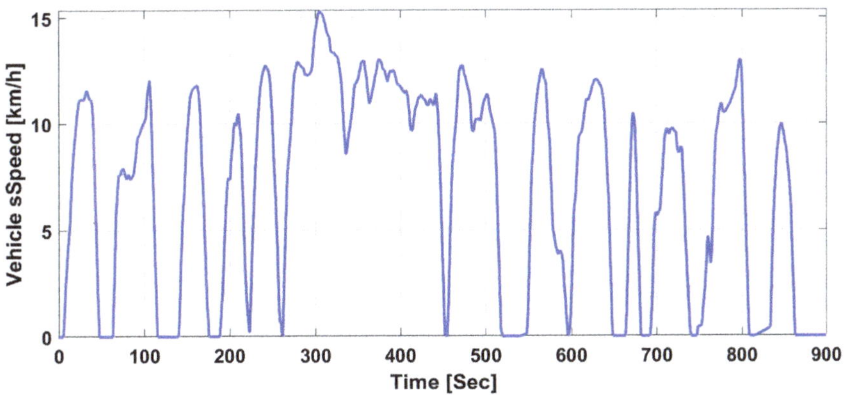

Figure 12. Speed profile of the UDDS standard velocity profile.

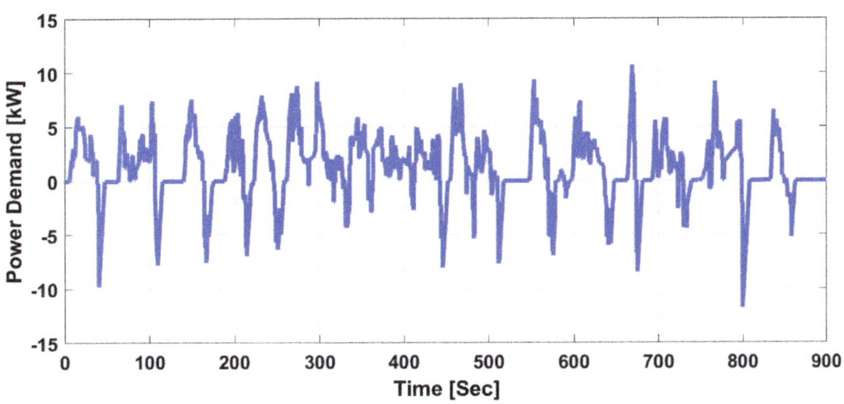

Figure 13. Power demand profile.

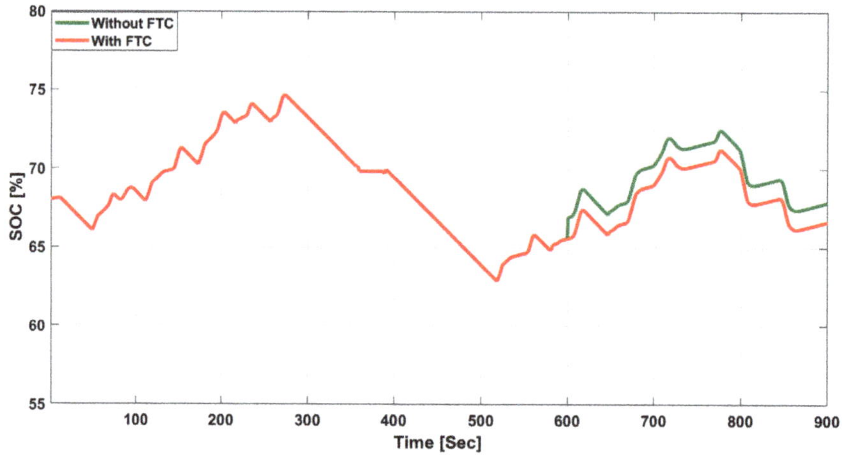

Figure 14. Estimation of the SOC according to the proposed DP strategy with and without FTC.

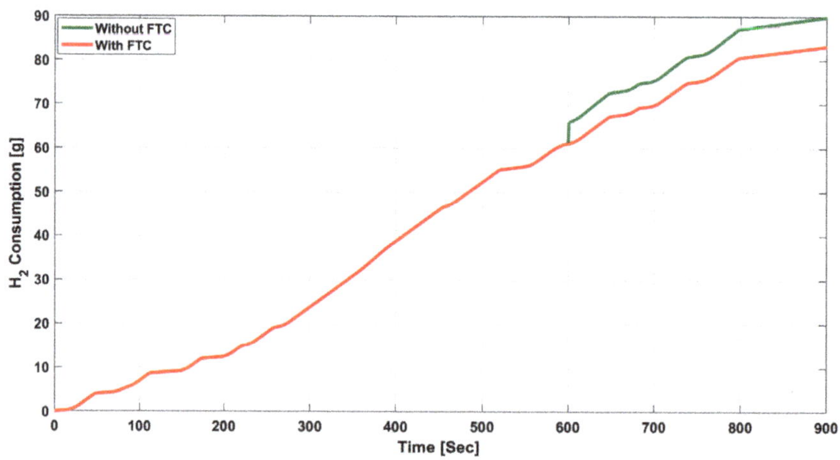

Figure 15. Evolution of H_2 consumption according to the proposed DP strategy with and without FTC.

Table 1. Overall performance obtained from simulations with the driving profile.

DP Strategy	SOC [%]	H_2 Consumption [g]
Without FTC	69.2654	90.000
With FTC	67.9912	84.3681

The state of charge (SOC) of a hybrid vehicle's battery holds crucial importance for drivers. However, the accuracy of SOC algorithms is often constrained by the computational capacities of electronic components, necessitating the preservation of performance requirements from the design to the real-time implementation. With the rising demand for hybrid vehicles, there is a growing need for low-complexity algorithms and cost-effective battery management systems.

This study presents a novel SOC estimation algorithm that strikes a balance between accuracy and simplicity, offering a streamlined solution that reduces computational overhead while maintaining comparable performance to established SOC estimation methods. This algorithm can be embedded into systems, ensuring real-world computational representation.

Validation of the control architecture, with and without fault-tolerance control (FTC), was conducted using TruckMaker/MATLAB software and a speed profile of the hybrid vehicle under study. Results indicate that the proposed observer strategy yields more accurate and faster SOC estimates compared to existing algorithms.

Figures 13–15 depict the power demand profile, SOC estimation (initial condition = 68%) and H_2 consumption, respectively with a bias fault occurring at 600 s by employing the proposed dynamic programming (DP) approach with and without FTC. Table 1 summarizes the SOC and H_2 consumption with and without FTC based on the proposed DP strategy.

From the simulation results and from Table 1, we notice that the proposed DP strategy based on FTC can save 6.26% of H_2 consumption compared to without FTC. In addition, we notice that SOC estimation is significantly improved using the proposed FTC.

6.2. Simulation 2: Comparisons of the Different Energy Management Strategies

This section aims to demonstrate the pros and cons of each proposed strategy. The main proposed control strategies (DP and PMP) are compared with two other strategies based on FLS and ECMS in this subsection. The comparison of the strategies is performed using the same power demand profile as given in Figure 13 and with the same initial conditions (initial condition = 68%) and operating limits. Figures 16 and 17 show the evolution of the SOC and H_2 consumption over time for the proposed energy management strategies (DP and PMP), which are compared to ECMS and FLS. Table 2 shows the simulation results comparison.

Figures 16 and 17 show the evolution of the SOC and hydrogen consumption under different EMSs (DP, PMP, ECMS and FLS) using the same cycle conditions with an SOC initial condition of 68% during the sensor faults. The simulation results comparison is shown in Table 2.

From Table 2, we can clearly see that DP and PMP reduce hydrogen consumption by 9.8% and 8.86%, respectively, compared with the FLS strategy under the same driving condition. From the simulation results, it can be seen that DP and PMP have the lowest H_2 consumption compared to the other EMSs (ECMS and FLS) with respect to the constraint on the final SOC and sensor faults. Therefore, an EMS based on DP can effectively improve the fuel economy.

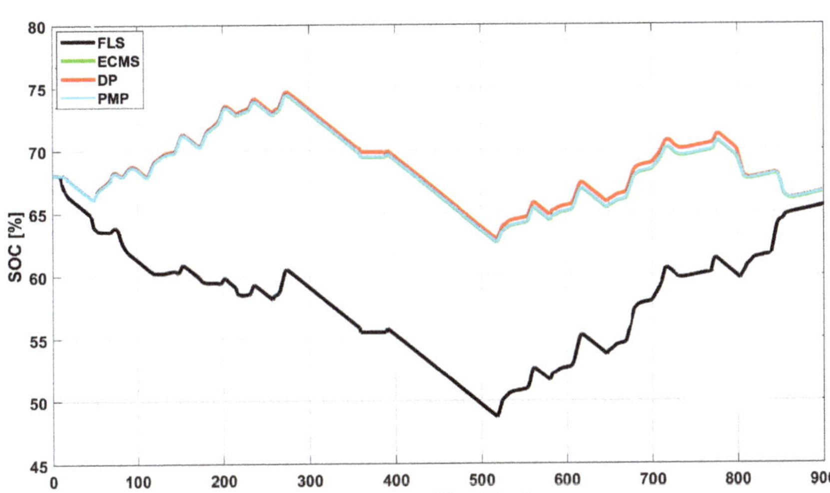

Figure 16. Evolution of SOC according to the proposed strategies.

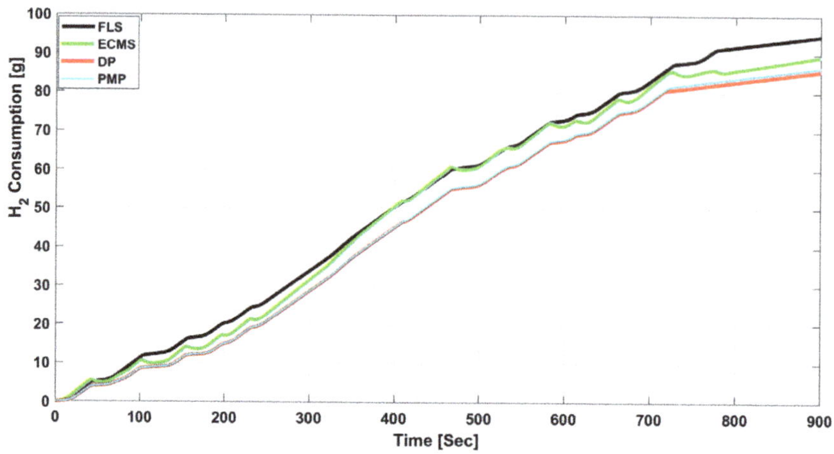

Figure 17. Evolution of H_2 consumption according to the proposed strategies.

Table 2. Comparison of different energy management strategies.

EMS Strategy	SOC [%]	H_2 Consumption [g]	Economy (Relative to FLS) %
FLS	66.8731	91.9412	-
ECMS	67.9528	85.9618	6.5035
PMP	67.9311	83.7993	8.8556
DP	67.9912	82.9303	9.8007

7. Conclusions

The control and optimization of hybrid vehicle (fuel cell/battery) power management are presented in this paper. The algorithm is validated using a dynamic bus model with TruckMaker/MATLAB software. This proposed hierarchical supervisory energy management strategy (HSEMS) consists of two control levels.

In the upper level (second level), a supervisory battery and fuel cell management system (BFCMS) is designed to generate healthy proton-exchange-membrane fuel cells (PEMFCs) and the battery state-of-charge (SOC) for the lower layer. In the lower layer, an advanced energy management system (EMS) based on adaptive fuzzy logic, dynamic programming and Pontryagin's minimum principle (PMP) is developed to minimize the total energy consumption and maximize the overall bus energy efficiency. Additionally, at this level, an adaptive proportional-integral (PI) controller is used to optimally track the bus subsystem set-points generated at the second level.

The overall validation of the proposed strategies (DP and PMP) is performed by comparing them with previous algorithms (ECMS and FLS). The proposed strategy has several advantages:

- Increasing the bus's energy efficiency and minimizing total energy consumption;
- Detecting and compensating for the effects of sensor and/or actuator faults in the PEMFC and battery;
- Reducing hydrogen consumption by 9.8% and 8.86%, respectively, compared to the FLS strategy under the same driving conditions.

In the near future, more-exhaustive studies on the robustness of the proposed strategy for various internal and external faults are planned. Additionally, the implementation of the proposed strategy on an actual platform is anticipated.

Author Contributions: Methodology, C.J., E.K. & R.G.; Software, C.J., E.K. & R.G.; Formal analysis, C.J., E.K. & R.G.; Resources, C.J., E.K. & R.G.; Writing—review & editing, C.J., E.K. & R.G. All authors have read and agreed to the published version of the manuscript.

Funding: This research received no external funding.

Data Availability Statement: The original contributions presented in the study are included in the article, further inquiries can be directed to the corresponding author.

Conflicts of Interest: The authors declare no conflict of interest.

Nomenclature

v	bus speed
F_t	traction force
α	slope
ρ	air density
m	bus mass
A	frontal area
C_d	drag coefficient (a constant)
r	wheel radius
T_w	wheel torque
w_w	rotation speed
η_{fd}	transmission efficiency
R_{fd}	gear ratio
P_t	power demand
η_m	motor efficiency
P_{dem}	total power demand
P_{aux}	consumption by the auxiliaries
N	number of cells
M_h	molar mass of hydrogen
n	transferred electrons
F	Faraday's constant
I_{stack}	fuel cell stack current
η_{FC}	efficiency of the PEMFC
$\dot{m}_h$	instantaneous hydrogen consumption
P_{FC}	PEMFC power
E_o	open-circuit voltage of the battery
R_{bat}	ohmic resistance
C_{bat}	equivalent capacitor
V_{bat}	battery voltage
I_{bat}	battery current
Q_{bat}	battery capacity
P_{bat}	battery power
H_2	total hydrogen consumption
$\dot{m}_h$	instantaneous hydrogen consumption
P_{FC}	PEMFC power supplied
H_p	prediction length
t_k	current time step
η_{FC}	total PEMFC efficiency
η_{DC-DC}	DC/DC efficiency
$SOCref$	SOC set-points
$SOC_{t_f+H_p}$	SOC at the end of the mission
SOC_{t_f+1}	SOC at the initial time
$P_{FC,max}$	PEMFC maximum power
$P_{FC,min}$	PEMFC minimum power
$P_{bat,max}$	battery maximum power
$P_{bat,min}$	battery minimum power
SOC_{max}	SOC maximum power

SOC_{min}	SOC minimum power
$x(t)$	state vector
$u(t)$	control input vector
$y(t)$	output vector
μ	fuzzy sets
p	number of rules
$A_i \in \Re^{n\times n}$	system matrix
$B_i \in \Re^{n\times m}$	input matrix
$C_i \in \Re^{g\times n}$	output matrix
$f(t)$	actuator faults
E	actuator fault matrix
K_i and L_i	observer gains
G_j and E_j	controller gains

References

1. Venkatesan , R.; Savio, A.D.; Balaji, C.; Narayanamoorthi, R.; Kotb, H.; ELrashidi, A.; Nureldeen, W. A Comprehensive Review on Efficiency Enhancement of Wireless Charging System for an Electric Vehicles Application. *IEEE Access* **2024**, *12*, 46967–46994. [CrossRef]
2. Hua, Z.; Zheng, Z.; Pahon, E.; Pera, M.-C.; Gao, F. Multi-timescale lifespan prediction for PEMFC systems under dynamic operating conditions. *IEEE Trans. Transp. Electrific.* **2022**, *8*, 345–355. [CrossRef]
3. Soumeur, M.A.; Gasbaoui, B.; Abdelkhalek, O.; Ghouili, J.; Toumi, T.; Chakar, A. Comparative study of energy management strategies for hybrid proton exchange membrane fuel cell four wheel drive electric vehicle. *J. Power Sources* **2020**, *462*, 228167. [CrossRef]
4. Chi, X.; Lin, F.; Wang, Y.-X. Disturbance and uncertainty-immune onboard charging batteries with fuel cell by using equivalent load fuzzy logic estimation-based backstepping sliding-mode control. *IEEE Trans. Transp. Electrific.* **2021**, *7*, 1249–1259. [CrossRef]
5. Tan, X.; Chen, Y.; Zeng, J.; Liao, W.; Liu, J. An Integrated Self-Modularized Battery Equalizer and Supercapacitor Charger for Hybrid Electric Vehicle Energy Storage System. *IEEE Trans. Veh. Technol.* **2024**, *73*, 9865–9877. [CrossRef]
6. Gan, J.; Li, S.; Lin, X.; Tang, X. Multi-Agent Deep Reinforcement Learning-Based Multi-Objective Cooperative Control Strategy for Hybrid Electric Vehicles. *IEEE Trans. Veh. Technol.* **2024**, 1–13. [CrossRef]
7. Zhou, J.; Zhao, J.; Wang, L. Cost-Based Research on Energy Management Strategy of Electric Vehicles Using Hybird Energy Storage System. *Tsinghua Sci. Technol.* **2024**, *29*, 684–697. [CrossRef]
8. Shen , Y.; Xie, J.; He, T.; Yao, L.; Xiao, Y. CEEMD-Fuzzy Control Energy Management of Hybrid Energy Storage Systems in Electric Vehicles. *IEEE Trans. Energy Convers.* **2024**, *39*, 555–566. [CrossRef]
9. Mehraban, A.; Ghanbari, T.; Farjah, E. AI-Based Control of Storage Capacity in High-Power-Density Energy Storage Systems, Used in Electric Vehicles. *IEEE Trans. Transp. Electrif.* **2024**, *10*, 2293–2301. [CrossRef]
10. Shuai, B.; Hua, M.; Li, Y.; Shuai, S.; Xu, H.; Zhou, Q. Optimal Energy Management of Plug-in Hybrid Electric Vehicles Through Ensemble Reinforcement Learning With Exploration-to-Exploitation Ratio Control. *IEEE Trans. Intell. Veh.* **2024**, 1–11. [CrossRef]
11. Sun, X.; Dong, Z.; Jin, Z.; Tian, X. System-Level Energy Management Optimization Based on External Information for Power-Split Hybrid Electric Buses. *IEEE Trans. Ind. Electron.* **2024**, 1–11. [CrossRef]
12. Munsi, M.S.; Chaoui, H. Energy Management Systems for Electric Vehicles: A Comprehensive Review of Technologies and Trends. *IEEE Access* **2024**, *12*, 60385–60403. [CrossRef]
13. Xu , P.; Wang, C.; Ye, J.; Ouyang, T. State-of-Charge Estimation and Health Prognosis for Lithium-Ion Batteries Based on Temperature-Compensated Bi-LSTM Network and Integrated Attention Mechanism. *IEEE Trans. Ind. Electron.* **2024**, *71*, 5586–5596. [CrossRef]
14. Xu, P.; Liu, B.; Hu, X.; Ouyang, T.; Chen, N. State-of-charge estimation for lithium-ion batteries based on fuzzy information granulation and asymmetric Gaussian membership function. *IEEE Trans. Ind. Electron.* **2022**, *69*, 6635–6644. [CrossRef]
15. Sun, Q.; Lv, H.; Wang, S.; Gao, S.; Wei, K. Optimized State of Charge Estimation of Lithium-Ion Battery in SMES/Battery Hybrid Energy Storage System for Electric Vehicles. *IEEE Trans. Appl. Supercond.* **2021**, *31*, 1–6. [CrossRef]
16. Karnehm, D.; Samanta, A.; Anekal, L.; Pohlmann, S.; Neve, A.; Williamson, S. Comprehensive Comparative Analysis of Deep Learning-based State-of-charge Estimation Algorithms for Cloud-based Lithium-ion Battery Management Systems. *IEEE J. Emerg. Sel. Top. Ind. Electron.* **2024**, *5*, 597–604. [CrossRef]
17. Bao, Z.; Nie, J.; Lin, H.; Gao, K.; He, Z.; Gao, M. TTSNet: State-of-Charge Estimation of Li-ion Battery in Electrical Vehicles with Temporal Transformer-based Sequence Network. *IEEE Trans. Veh. Technol.* **2024**, *73*, 7838–7851. [CrossRef]
18. Gao, Y.; Zhu, J.; Shi, D.; Zhang, X. A Model-Based Battery Dataset Recovery Method Considering Cell Aging in Real-World Electric Vehicles. *IEEE Trans. Ind. Inform.* **2024**, *20*, 7904–7914. [CrossRef]
19. Lucero, J.N.E.; Sujan, V.A.; Onori, S. An experimentally validated electro-thermal EV battery pack model incorporating cycle-life aging and cell-to-cell variations. *IEEE Trans. Transp. Electrif.* **2024**, 1. [CrossRef]

20. Sun, H.; Tao, F.; Fu, Z.; Gao, A.; Jiao, L. Driving-Behavior-Aware Optimal Energy Management Strategy for Multi-Source Fuel Cell Hybrid Electric Vehicles Based on Adaptive Soft Deep-Reinforcement Learning. *IEEE Trans. Intell. Transp. Syst.* **2023**, *24*, 4127–4146. [CrossRef]
21. Ali, A.M.; Moulik, B.; Söffker, D. Intelligent Real-Time Power Management of Multi-Source HEVs Based on Driving State Recognition and Offline Optimization. *IEEE Trans. Intell. Transp. Syst.* **2023**, *24*, 247–257. [CrossRef]
22. Ali, A.M.; Shivapurkar, R.; Söffker, D. Optimal Situation-Based Power Management and Application to State Predictive Models for Multi-Source Electric Vehicles. *IEEE Trans. Veh. Technol.* **2019**, *68*, 11473–11482. [CrossRef]
23. Venkatasatis, R.; Dhanamjayulu, C. Reinforcement learning based energy management systems and hydrogen refuelling stations for fuel cell electric vehicles: An overview. *Int. J. Hydrogen Energy* **2022**, *47*, 27646–27670. [CrossRef]
24. Oladosu, T.L.; Pasupuleti, J.; Kiong, T.S.; Koh, S.P.J.; Yusaf, T. Energy management strategies, control systems, and artificial intelligence-based algorithms development for hydrogen fuel cell-powered vehicles: A review. *Int. J. Hydrogen Energy* **2024**, *61*, 1380–1404. [CrossRef]
25. Wang, D.; Mei, L.; Xiao, F.; Song, C.; Qi, C.; Song, S. Energy management strategy for fuel cell electric vehicles based on scalable reinforcement learning in novel environment. *Int. J. Hydrogen Energy* **2024**, *59*, 668–678. [CrossRef]
26. Nie, Z.; Huang, J.; Lian, Y.; Yang, W. Hierarchical control-based energy management strategy of intelligent battery/supercapacitor/fuel cell hybrid vehicles. *Int. J. Hydrogen Energy* **2024**, *61*, 1092–1106. [CrossRef]
27. Machacek, D.T.; Yasar, N.O.; Huber, T.; Onder, C.H. Energy management of hydrogen hybrid electric vehicles—A potential analysis. *Int. J. Hydrogen Energy* **2024**, *58*, 1–13. [CrossRef]
28. Zhang, Y.; Ma, R.; Zhao, D.; Huangfu, Y.; Liu, W. An Online Efficiency Optimized Energy Management Strategy for Fuel Cell Hybrid Electric Vehicles. *IEEE Trans. Transp. Electrif.* **2023**, *9*, 3203–3217. [CrossRef]
29. Lü, X.; Wu, Y.; Lian, J.; Zhang, Y.; Chen, C.; Wang, P.; Meng, L. Energy management of hybrid electric vehicles: A review of energy optimization of fuel cell hybrid power system based on genetic algorithm. *Energy Convers. Manag.* **2020**, *205*, 112474.
30. Zhao, X.; Wang, L.; Zhou, Y.; Pan, B.; Wang, R.; Wang, L.; Yan, X. Energy management strategies for fuel cell hybrid electric vehicles: Classification, comparison, and outlook. *Energy Convers. Manag.* **2022**, *270*, 116179. [CrossRef]
31. Zhu, Y.; Li, X.; Liu, Q.; Li, S.; Xu, Y. Review article: A comprehensive review of energy management strategies for hybrid electric vehicles. *Mech. Sci.* **2022**, *13*, 147–188. [CrossRef]
32. Li , Q.; Su, B.; Pu, Y.; Han, Y.; Wang, T.; Yin, L.; Chen, W. A state machine control based on equivalent consumption minimization for fuel cell/supercapacitor hybrid tramway. *IEEE Trans. Transp. Electrific.* **2019**, *5*, 552–564. [CrossRef]
33. Maria, C.; Suja, S.; Sajini, M.L.M. Effective Energy Management in Grid Integrated with Renewable Resources Using Finite State Machine Model. In Proceedings of the 2022 Second International Conference on Power, Control and Computing Technologies (ICPC2T), Raipur, India, 1–3 March 2022.
34. Garcia, P.; Fernandez, L.M.; Garcia, C.A.; Jurado, F. Energy management system of fuel-cell-battery hybrid tramway. *IEEE Trans. Ind. Electron.* **2010**, *57*, 4013–4023. [CrossRef]
35. Yan, Y.; Huang, W.; Liu, J.; Li, Q.; Chen, W. The Control Strategy of Fuel Cell Hybrid Tram Based on State Machine Control. In Proceedings of the 2019 IEEE Sustainable Power and Energy Conference (iSPEC), Beijing, China, 21–23 November 2019; pp. 699–703.
36. Khanafari, A.; Alasty, A.; Kermani, M.J.; Asghari, S. Experimental study of water management of a three-cell PEM fuel cell stack using adaptive neuro-fuzzy approximation and fuzzy control. *Int J. Hydrogen Energy* **2024**, *50*, 931–944. [CrossRef]
37. Hemi, H.; Ghouili, J.; Cheriti, A. A real time fuzzy logic power management strategy for a fuel cell vehicle. *Energy Convers. Manag.* **2014**, *80*, 63–70. [CrossRef]
38. Martínez, J.S.; Mulot, J.; Harel, F.; Hissel, D.; Pera, M.C.; John, R.I.; Amiet, M. Experimental validation of a type-2 fuzzy logic controller for energymanagement in hybrid electrical vehicles. *Eng. Appl. Artif. Intell.* **2013**, *26*, 1772–1779. [CrossRef]
39. Chen, J.; Xu, C.; Wu, C.; Xu, W. Adaptive fuzzy logic control of fuel-cell-battery hybrid systems for electric vehicles. *IEEE Trans. Ind. Inform.* **2018**, *14*, 292–300. [CrossRef]
40. Shen, J.; Khaligh, A. A supervisory energy management control strategy in a battery/ultracapacitor hybrid energy storage system. *IEEE Trans. Transp. Electrific.* **2015**, *1*, 223–231. [CrossRef]
41. Duhr, P.; Balerna, C.; Onder, C.H. Time-optimal gearshift and energy management strategies for a hybrid electric race car. *Appl. Energy* **2021**, *282*, 115980. [CrossRef]
42. Biswas, A.; Emadi, A. Energy management systems for electrified powertrains: State-of-the-art review and future trends. *IEEE Trans. Veh. Technol.* **2019**, *68*, 6453–6467. [CrossRef]
43. Lin, X.; Xu, X.; Lin, H. Predictive-ECMS based degradation protective control strategy for a fuel cell hybrid electric vehicle considering uphill condition. *eTransportation* **2022**, *12*, 1–19. [CrossRef]
44. Jia, C.; Qiao, W.; Cui, J.; Qu, L. Adaptive Model-Predictive-Control-Based Real-Time Energy Management of Fuel Cell Hybrid Electric Vehicles. *IEEE Trans. Power Electron.* **2023**, *38*, 2681–2694. [CrossRef]
45. Yang, H.; Hu, Y.; Gong, X.; Cao, R.; Guo, L.; Chen, H. Energy Management Strategy for Fuel Cell Hybrid Electric Vehicles Considering the Inaccuracy of Predicted Vehicle Speed. *IEEE Trans. Transp. Electrif.* **2024**, 1. [CrossRef]
46. Hou, Z.; Chu, L.; Guo, Z.; Hu, J.; Jiang, J.; Yang, J.; Chen, Z.; Zhang, Y. A Learning-and-Tube-Based Robust Model Predictive Control Strategy for Plug-In Hybrid Electric Vehicle. *IEEE Trans. Intell. Veh.* **2024**, *9*, 579–592. [CrossRef]

47. Song, K.; Huang, X.; Xu, H.; Sun, H.; Chen, Y.; Huang, D. Model predictive control energy management strategy integrating long short-term memory and dynamic programming for fuel cell vehicles. *Int. J. Hydrogen Energy* **2024**, *56*, 1235–1248. [CrossRef]
48. Soldo, J.; Škugor, B.; Deur, J. Model Predictive Control of a Parallel Plug-In Hybrid Electric Vehicle Relying on Dynamic Programming and Extended Backward-Looking Model. *IEEE Trans. Control. Syst. Technol.* **2024**, *32*, 581–594. [CrossRef]
49. Li, Y.; Liu, M.; Wang, Y.; Xu, L.; Lei, S. Energy Management Optimization and Validation of a Hydrogen Fuel Cell-Powered Agricultural Tractor Based on Hierarchical Dynamic Programming. *IEEE Access* **2024**, *12*, 21382–21401. [CrossRef]
50. Tang, W.; Wang, Y.; Jiao, X.; Ren, L. Hierarchical energy management strategy based on adaptive dynamic programming for hybrid electric vehicles in car-following scenarios. *Energy* **2023**, *265*, 126264. [CrossRef]
51. Ma, M.; Xu, E.; Zheng, W.; Qin, J.; Huang, Q. The optimized real-time energy management strategy for fuel-cell hybrid trucks through dynamic programming. *Int. J. Hydrogen Energy* **2024**, *59*, 10–21. [CrossRef]
52. Chen, B.-C.; Wu, Y.-Y.; Tsai, H.-S. Design and analysis of power management strategy for range extended electric vehicle using dynamic programming. *Appl. Energy* **2014**, *113*, 1764–1774. [CrossRef]
53. Xu, L.; Ouyang, M.; Jianqiu, L.; Yang, F.; Lu, L.; Hua, J. Application of Pontryagin's minimal principle to the energy management strategy of plugin fuel cell electric vehicles. *Int. J. Hydrogen Energy* **2013**, *38*, 10104–10115. [CrossRef]
54. Volkan, S.; Metin, G.; Seta, B. A novel ECMS and combined cost map approach for high-efficiency series hybrid electric vehicles. *IEEE Trans. Veh. Technol.* **2011**, *60*, 3557–3570.
55. Li, S.; Chu, L.; Hu, J.; Li, J.; Hou, Z.; Sun, W. A novel A-ECMS energy management strategy based on dragonfly algorithm for plug-in FCEVs. *Sensors* **2023**, *23*, 1192. [CrossRef] [PubMed]
56. Yang, C.; Du, X.; Wang, W.; Yang, L.; Zha, M. A Rolling Convergent Equivalent Consumption Minimization Strategy for Plug-in Hybrid Electric Vehicles. *IEEE Trans. Veh. Technol.* **2024**, *73*, 3340–3353. [CrossRef]
57. Li, H.; Ravey, A.; N'Diaye, A.; Djerdir, A. A novel equivalent consumptionminimization strategy for hybrid electric vehicle powered by fuel cell, battery and supercapacitor. *J. Power Sources* **2018**, *395*, 262–270. [CrossRef]
58. Tian, X.; Cai, Y.; Sun, X.; Zhu, Z.; Xu, Y. An adaptive ECMS with driving style recognition for energy optimization of parallel hybrid electric buses. *Energy* **2019**, *189*, 116151. [CrossRef]
59. Kim, N.; Jeong, J.; Zheng, C.H. Adaptive energy management strategy for plug-in hybrid electric vehicles with pontryagin's minimum principle based on daily driving patterns. *Int. J. Precis. Eng. -Manuf.-Green Technol.* **2019**, *6*, 539–548. [CrossRef]
60. Liu, C.; Liu, L. Optimal power source sizing of fuel cell hybrid vehicles based on Pontryagin's minimum principle. *Int. J. Hydrogen Energy* **2015**, *40*, 8454–8464. [CrossRef]
61. Zheng, C.; Liang, Q. An energymanagement strategy of hybrid energy storage systems for electric vehicle applications. *IEEE Trans. Sustain. Energy* **2018**, *9*, 1880–1888. [CrossRef]
62. Meng, X.; Li, Q.; Chen, W.; Zhang, G. A layered energy management method for a fuel cell hybrid power system based on Pontryagin's principle of minimum value. *Proc. Chin. Soc. Elect. Eng.* **2019**, *39*, 782–792.
63. Li, X.; Wang, Y.; Yang, D.; Chen, Z. Adaptive energy management strategy for fuel cell/battery hybrid vehicles using Pontryagin's minimal principle. *J. Power Sources* **2019**, *440*, 227105. [CrossRef]
64. Yang, C.; Du, S.; Li, L.; You, S.; Yang, Y.; Zhao, Y. Adaptive realtime optimal energy management strategy based on equivalent factors optimization for plug-in hybrid electric vehicle. *Appl. Energy* **2017**, *203*, 883–896. [CrossRef]
65. Onori, S.; Tribioli, L. Adaptive Pontryagin's minimum principle supervisory controller design for the plug-in hybrid GM chevrolet volt. *Appl. Energy* **2015**, *147*, 224–234. [CrossRef]
66. Xu, L.; Li, J.; Ouyang, M.; Hua, J.; Li, X. Active fault tolerance control system of fuel cell hybrid city bus. *Int. J. Hydrog. Energy* **2010**, *35*, 12510. [CrossRef]
67. Li , J.; Zhou, T. Active fault-tolerant coordination energy management for a proton exchange membrane fuel cell using curriculum-based multiagent deep meta-reinforcement learning. *Renew. Sustain. Energy Rev.* **2023**, *185*, 1–16. [CrossRef]
68. Jiang, J.; Yu, X. Fault-tolerant control systems : A comparative study between active and passive approaches. *Annu. Rev. Control.* **2012**, *36*, 60–72. [CrossRef]
69. Zhong, Z.; Huo, H.; Zhu, X.; Cao, G.; Ren, Y. Adaptive maximum power point tracking control of fuel cell power plants. *J. Power Sources* **2008**, *176*, 259–269. [CrossRef]
70. Aubry, J.; Steiner, N.Y.; Morando, S.; Zerhouni, N.; Hissel, D. Fault tolerant control of a Proton Exchange Membrane Fuel Cell based on a Modified Failure Mode and Effect Analysis. In Proceedings of the 2020 IEEE Vehicle Power and Propulsion Conference (VPPC), Gijon, Spain, 18 February 2021; pp. 1–5.
71. Dijoux, E.; Benne, M.; Péra, M.-C.; Grondin-Perez, B. Experimental Validation of an Active Fault Tolerant Control Strategy Applied to a Proton Exchange Membrane Fuel Cell. *Electrochem* **2022**, *3*, 633–652. [CrossRef]
72. Wang, Z.; Guo, X.; Dong, Z.; Fan, N.; Cao, S. Fixed time adaptive fault tolerant sliding mode control of PEMFC air supply system. *Int. J. Hydrogen Energy* **2024**, *55*, 1434–1444. [CrossRef]
73. Yang, D.; Wang, Y.; Chen, Z. Robust fault diagnosis and fault tolerant control for PEMFC system based on an augmented LPV observer. *Int. J. Hydrogen Energy* **2020**, *45*, 13508–13522. [CrossRef]
74. Rout, S.; Das, S. Online State-of-Charge Estimation of Lithium-Ion Battery using a Fault Tolerant and Noise Immune Threefold Modified Adaptive Extended Kalman Filter. *IEEE Trans. Transp. Electrif.* **2024**, 1. [CrossRef]
75. Jin, H.; Zhang, Z.; Wang, Y.; Zuo, Z. Event-Triggered Sensor Fault Estimation for Lithium-Ion Battery Packs. *IEEE Trans. Circuits Syst. II Express Briefs* **2024**, *71*, 3101–3105. [CrossRef]

76. Lebreton, C.; Benne, M.; Damour, C.; Yousfi-Steiner, N.; Grondin-Perez, B.; Hissel, D.; Chabriat, J.-P. Fault Tolerant Control Strategy applied to PEMFC water management. *Int. J. Hydrogen Energy* **2015**, *40*, 10636–10646. [CrossRef]
77. Pei, P.; Chen, H. Main factors affecting the lifetime of proton exchange membrane fuel cells in vehicle applications: A review. *Appl. Energy* **2014**, *125*, 60–75. [CrossRef]
78. Robin, C.; Gérard, M.; Quinaud, M.; d'Arbigny, J.; Bultel, Y. Proton exchange membrane fuel cell model for aging predictions: Simulated equivalent active surface area loss and comparisons with durability tests. *J. Power Sources* **2016**, *326*, 417–427. [CrossRef]
79. Song , K.; Chen, H.; Wen, P.; Zhang, T.; Zhang, B.; Zhang, T. A comprehensive evaluation framework to evaluate energy management strategies of fuel cell electric vehicles. *Electrochim. Acta* **2018**, *292*, 960–973. [CrossRef]
80. Wang, T.; Li, Q.; Wang, X.; Chen, W.; Breaz, E.; Gao, F. A power allocation method for multistack PEMFC system considering fuel cell performance consistency. *IEEE Trans. Ind. Appl.* **2020**, *56*, 5340–5351. [CrossRef]
81. Zhang, R.; Tao, J. GA-based fuzzy energy management system for FC/SC-powered HEV considering H2 consumption and load variation. *IEEE Trans. Fuzzy Syst.* **2018**, *26*, 1833–1843. [CrossRef]
82. Florescu, A.; Bacha, S.; Munteanu, I.; Bratcu, A.L.; Rumeau, A. Adaptive frequency-separation-based energy management system for electric vehicles. *J. Power Sources* **2015**, *280*, 410–421. [CrossRef]
83. Behling, N.; Williams, M.C.; Managi, S. Fuel cells and the hydrogen revolution: Analysis of a strategic plan in Japan. *Econ. Anal. Policy* **2015**, *48*, 204–221. [CrossRef]
84. Zhang, K.; Jiang, B.; Cocquempot, V. Adaptive observer-based fast fault estimation. *Int. J. Control.* **2008**, *6*, 320–326.
85. Cai, C.H.; Du, D.; Liu, Z.Y. Battery state-of-charge (SOC) estimation using adaptive neuro-fuzzy inference system (ANFIS). In Proceedings of the 12th IEEE International Conference on Fuzzy Systems, FUZZ '03, St. Louis, MO, USA, 25–28 May 2003; pp. 1068–1073.
86. Lin, X.; Sun, D.; Qin, D.; Yin, Y. Development of Power-balancing Global Optimization Control Strategy for a Series-Parallel Hybrid Electric City Bus. *China Mech. Eng.* **2011**, *22*, 2259–2263.
87. Zhou, H.; Yu, Z.; Wu, X.; Fan, Z.; Yin, X.; Zhou, L. Dynamic programming improved online fuzzy power distribution in a demonstration fuel cell hybrid bus. *Energy* **2023**, *284*, 128549. [CrossRef]
88. Xie, C.; Ogden, J.; Quan, S.; Chen, Q. Optimal power management for fuel cell-battery full hybrid powertrain on a test station. *Int. J. Electr. Power Energy Syst.* **2013**, *53*, 307–320. [CrossRef]
89. Lin, W.-M.; Tu, C.-S.; Tsai, M.-T. Energy Management Strategy for Microgrids by Using Enhanced Bee Colony Optimization. *Energies* **2016**, *9*, 5. [CrossRef]
90. Ramya, A.; Imtiaz, A.; Balaji, M. Hybrid self tuned fuzzy PID controller for speed control of brushless DC motor. *Automatika* **2016**, *57*, 672–679. [CrossRef]
91. Shi , D.; Liu, S.; Cai, Y.; Wang, S.; Li, H.; Chen, L. Pontryagin's minimum principle based fuzzy adaptive energy management for hybrid electric vehicle using real-time traffic information. *Appl. Energy* **2021**, *286*, 116467. [CrossRef]

energies MDPI

Article

The Attractiveness of Regional Transport as a Direction for Improving Transport Energy Efficiency

Wojciech Miechowicz [1], Marcin Kiciński [2,*], Izabela Miechowicz [3] and Agnieszka Merkisz-Guranowska [2]

1 Doctoral School, Poznan University of Technology, Piotrowo St. 3, 61-138 Poznan, Poland; wojciech.miechowicz@doctorate.put.poznan.pl
2 Institute of Transport, Poznan University of Technology, Piotrowo St. 3, 61-138 Poznan, Poland; agnieszka.merkisz-guranowska@put.poznan.pl
3 Department of Computer Science and Statistics, Poznan University of Medical Sciences, Rokietnicka St. 7, 60-806 Poznan, Poland; iza@ump.edu.pl
* Correspondence: marcin.kicinski@put.poznan.pl; Tel.: +48-61-665-2129

Abstract: One of the ways to improve energy efficiency in transportation is through efforts aimed at increasing the usage of public transportation by residents. This, in turn, is closely related to residents' preferences. One of the most frequently cited factors influencing the attractiveness and quality of public transportation is the frequency of connections. This is important not only for urban transport but also regional transport, which has significantly lower passenger flows. This paper aims to present how the frequency of connections affects the attractiveness of regional transport. An original method for determining the attractiveness of public transport, based on the share of adult, senior, and youth passengers with single tickets and monthly passes, is introduced. The results of research on the structure of passengers and flows to/from the district center are presented. Based on the research results, attractiveness factors are calculated for each town. The statistical analysis clearly indicates there is a strong correlation between the attractiveness of regional transport and the frequency of connections ($Rs = 0.807$, $p = 0.001$). Moreover, for every connection increase, the number of adult passengers will increase by an average of 1.5. Assuming that these additional passengers switch from individual to public transportation, the resulting reduction in energy consumption due to an increase of one connection is 0.33–0.69 kWh for each kilometer traveled by these passengers.

Keywords: public bus transport; regional transport; energy consumption of public transport; modal split; transport exclusion

1. Introduction

In many urban areas, public transport is currently an attractive alternative to private car use. However, it should be noted that individual car transport has significantly higher energy consumption per passenger-kilometer (pkm) compared to that for bus or rail public transport. As indicated by Pérez-Martínez and Sorba [1], traveling by car with one person results in energy consumption ranging from 0.33 kWh/pkm to 0.49 kWh/pkm, while a standard bus consumes from 0.11 kWh/pkm to 0.13 kWh/pkm. According to data published in the Odyssee-Mure project [2], this disparity decreased in 2020–2021 due to lower occupancy rates in public bus transport during the SARS-CoV-2 pandemic. Despite this, the European Union emphasizes the importance of reducing energy consumption in, among other things, road transport. Consequently, actions have been intensified, leading to the adoption of the new Directive 2023/1791 on energy efficiency and amending Regulation (EU) 2023/955 by the European Parliament and the Council [3]. The goal of these regulations is to reduce final energy consumption at the EU level, including in the transport sector. The mentioned measures for improving the energy efficiency of this sector include implementing strategies promoting, among other things, a shift from individual to public transport. According to the authors, in addition to introducing new, more environmentally friendly means of transport, it is also necessary to simultaneously take action to increase the attractiveness of public transport. It is also significant that this topic is increasingly considered in the context of public health [4,5]. For this purpose, passenger preference surveys [6] are frequently conducted. Based on the literature, public transport users most often pay attention not only to travel time and reliability [7] but also the frequency of connections [8–11]. In urban transport, the level of service on tram, bus, or metro lines is typically determined by the frequency of services, defined as the number of trips per hour or the length of the interval between trips in minutes (frequency). However, for regional transport, which has much lower passenger flows [12] and must adapt its service offer to more specific needs, especially regarding school and work hours [13,14], using the concept of frequency as the number of trips per hour or time interval between trips is not always justified. There are, of course, short time intervals on some regional lines where the frequency can be, for example, three trips per hour (primarily between 7:00 and 8:00 a.m.). Still, there can also be periods when there is no service for an hour or longer (off-peak or evening periods). In such cases, the more appropriate measure of frequency is the number of pairs of connections per day.

In Polish regional bus transport, a decline in the level of services offered, including the number of connections served, has been observed for many years, with a gradual reduction in the number of trips [15,16]. As a result, many so-called "white spots"—places no public transport can reach—have appeared on the map of Poland. In the literature [17,18], this phenomenon is referred to as "transport exclusion" or "mobility exclusion". According to E. Beyazit [19], it is defined as "the loss of a person's ability to connect with various places (e.g., work or services) that are necessary for full participation in society". This problem has deepened further due to the SARS-CoV-2 pandemic. This is because for many years, regional bus transport in Poland was not subsidized, except for compensation given to carriers for lost revenue due to state-defined discounts. Since 2019, there have been additional opportunities to fund the restoration of regular bus services with public funds. However, the large number of organizers, the low competence of employees responsible for organizing transport, the lack of regulations forcing cooperation between organizers, tariff and schedule integration, and the allocation of state budget funds only for a year have not led to significant changes in the number of passengers who use bus transport, except in a few places. One of the fundamental decisions that organizers must make when planning the launch of a bus line is determining the number of trips that will be made on a given line. This significantly impacts operating costs on the one hand and the potential shift from car to bus transport on the other. Hence, key questions arise:

1. How does the frequency of connections, expressed as the number of pairs of trips per day, affect the attractiveness of a regional bus line?
2. Can a higher number of trips attract more people to public transport, especially those who do not need to use public transport, such as adults commuting daily to work?
3. How does the structure of passengers change with an increase in the number of trips?

Finding answers to these questions will allow the development of a method for determining energy consumption benefits in regional passenger transport in further research. This article focuses on the first stage, answering the above questions based on research conducted on 27 regional bus lines in various regions of Poland.

This article consists of five parts: This first section presents the issue of energy consumption in passenger transport and the importance of the attractiveness of regional public transport. Section 2 contains a literature review on determining the attractiveness of public transport and research in regional transport. Section 3 presents the proposed methodology, describing individual stages and steps. The course of research and the practical application of the methodology are described in Section 4. The final part presents a discussion, conclusions, and directions for further work.

2. Literature Review

The problem of the attractiveness of public transport is a broad and multifaceted concept. Most commonly, transport attractiveness is evaluated based on passengers' feelings expressed in surveys, which can encompass both satisfaction and preference (expectations). Factors influencing the attractiveness of public transport can be considered from two perspectives [20]:

- Macroscopic: The urban spatial layout, land use patterns, geographical environment, level of economic development, demographic characteristics, public policy, and service characteristics of public transport.
- Microscopic: Travelers' individual characteristics, travel demand characteristics, travel characteristics, and the service level of public transport.

Currently, many researchers focus on issues related to transport attractiveness, which is associated with concepts such as travel satisfaction or the broader quality of public transport [7–9,21–24].

According to research by Sukhov et al. [22], from a passenger's perspective, travel satisfaction is not linked to individual characteristics defining the quality of transport services but rather a whole set of these characteristics. Moreover, the same level of satisfaction can be achieved with different configurations of these characteristics. In studies, the frequency of connections is important for all groups of surveyed passengers, whereas other features, such as speed or intramodality, are only important for some.

One study [9] identified the qualitative characteristics of public transport that determine its attractiveness for passengers and the ways in which these characteristics need be changed to encourage car drivers to use public transport. The authors divide the characteristics defining transport quality into physical characteristics (e.g., frequency, reliability, speed, and cost), which can be assessed based on their impact on passengers, and perceived ones (comfort, sense of security), which require interaction with passengers for measurement. Another study [9] indicated that characteristics such as reliability and frequency are crucial in determining interest in public transport and passenger travel satisfaction. Similar conclusions were drawn earlier by Zhang et al. [21], who studied factors affecting passengers' satisfaction with public transport. However, identifying significant features that must be met becomes more complex when considering the attractiveness of public transport for car users [9]. Public transport services must ensure a basic level of accessibility, reliability, competitive cost, effective speed [23], and technical efficiency [25] to be competitive with private cars.

One of the key factors influencing the quality of public transport is frequency [26,27]. The satisfactory value of frequency for passengers depends on several factors. Some researchers [24,28] indicated that increasing the frequency of services can positively impact the number of passengers, including by attracting new passengers who previously used cars, both in cities and rural areas where the quality of public transport services is often lower. However, it should be noted that improving service frequency can have a much greater effect on increasing passenger numbers than lowering ticket prices [29]. On the other hand, it is important to consider the threshold frequency value below which a particular type of public transport becomes unattractive. For example, Hensher [30] indicated that passengers rate a decrease in frequency below one trip every half hour very poorly.

An analysis of publications on passenger preferences in regional public transport [8] shows that this type of transport is still given little attention. In the analyzed publications, the most frequently considered category is accessibility (the range of services offered in terms of geography, types of transport, operating hours, and frequency). However, the authors point out that there is a lack of studies on the operational aspects of accessibility, such as the hours and frequency of service.

Most studies on public transport frequency concern cities (highly urbanized areas), where the frequency is much higher than in regional transport, with considered values ranging from four trips per hour to six trips per hour. These frequencies are unattainable in regional bus transport, especially in countries where this transport is either not or only minimally subsidized by public funds, as is the case in Poland.

Similarly, studies on the potential of regional transport are much less frequently conducted by researchers. Some sources describe using the gravity model to determine demand for transport services in different regions, e.g., Greece [31], Slovakia [32], and the Czech Republic [33,34]. There are also studies on timetable optimization in regional transport [35], timetable evaluation [36], and the service frequency between local government units in the Czech Republic [37]. The advantages of a cyclic timetable-based transport system in rural areas, based on Swiss experiences, are also highlighted [38]. Research on the relationship between demand and additional services during off-peak periods [12], changes in demand during peak and off-peak periods [13], and factors affecting modal split in commuting to European cities [39] is of particular interest with respect to this work. In [40], the authors analyzed the impact of various characteristics of public transport, including prices, frequency, and inhabitants' incomes, on changes in demand over several decades in all types of transport in Tunisia.

Publications related to transport in Poland include analyses of rural areas' accessibility to district towns in six regions [41], case studies of rail connections and parallel bus lines in Lower Silesia [42], and regional transport planning [43,44] for transport plans.

These works mainly concern transport-planning aspects, passenger preferences, and transport exclusion. Research and analyses on regional transport mainly rely on survey studies or national censuses, which contain general information but are rather limited in accuracy to larger areas (regions). It is difficult to find studies on actual demand and passenger flows between different locations and local regional centers at different times of the day or on the ticket structure used by passengers. The topic of transport preferences, as indicated by passengers in surveys of the most desired features a transport system should offer, is widely discussed. However, we could not find publications on the impacts of various features and factors related to regional transport and its surroundings on its actual use by different passenger groups and thus on the attractiveness of regional bus transport. This is significant because, ultimately, the attractiveness of a transport system is indicated by the number of passengers using it, especially those with alternative travel options.

3. Methodological Framework

Achieving the set goal required a step-by-step approach. A diagram of the procedure employed is shown in Figure 1a. The research work assumed two main tasks:

- Task I—study the impact of service frequency on the attractiveness of regional bus transport;
- Task II—assess the energy efficiency benefits of regional public transport while considering its attractiveness.

This article details Task I (Figure 1b), which consists of eight stages, two of which (Stages 6 and 7) include four steps each: preparing input data, conducting the Shapiro–Wilk test, determining the Pearson linear correlation coefficients, and determining the Spearman rank correlation.

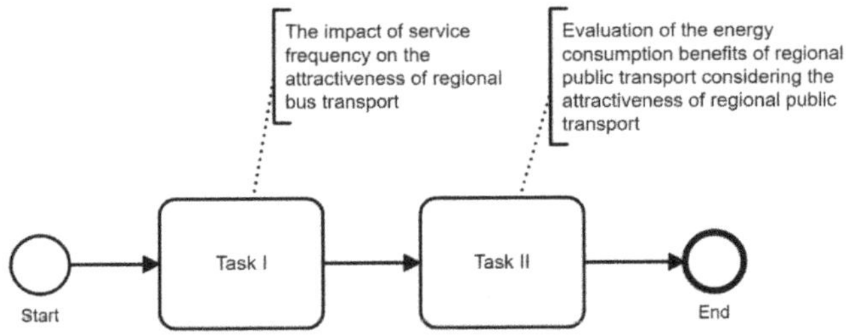

(a)

Figure 1. *Cont.*

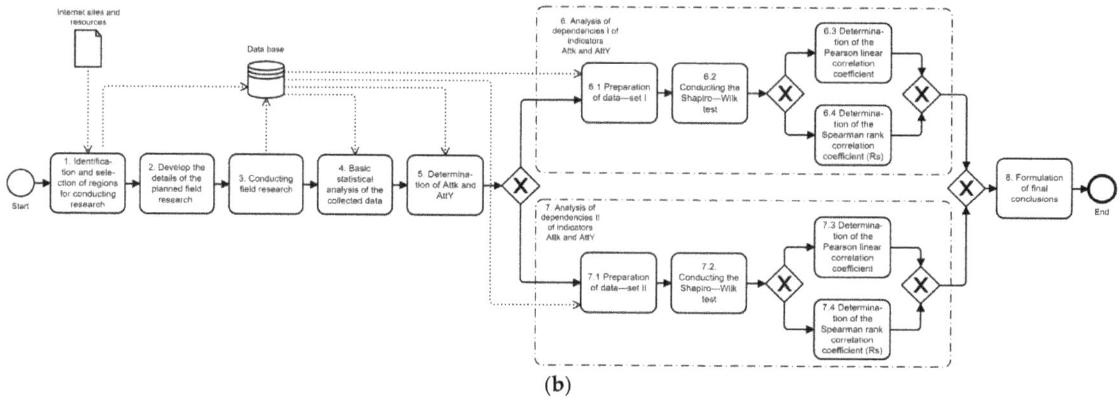

Figure 1. Concept of the methodology: (**a**) main stages; (**b**) steps in Stage I that are implemented and presented in the article.

In the first planned stage—the identification and selection of regions for research—it was assumed that information would be collected on regional bus transport services in a specific area (in this case, Poland). This is related to the specificity of this type of transport, as according to our previous studies, the main group of passengers is youth commuting to high schools [14] located in regional centers (district capitals). Moreover, these buses also cater to other, less numerous but significant groups, such as adults and seniors who use public transport occasionally and adults who regularly commute to work. The basic goal at this stage was to collect information from open sources (websites, reports, and travel planners). The result was the development of a database on regional transport services in the country.

The next task was to develop the details for the planned field research. Due to the lack of publicly available data on the number and structure of passengers, it was decided to conduct direct observations in regional buses. Additionally, a schedule and locations for these observations were planned. To ensure the highest possible comparability of the results, it was assumed that the observations would be conducted on weekdays, excluding Mondays and Fridays, and days before or after holidays.

Next, in Stage 3 of the proposed concept, field research was conducted, dividing passengers as follows:

- Three age groups: youth, adult, and senior;
- Ticket type: regular single ticket, discounted single ticket, monthly pass without discount, and discounted monthly pass (49%).

This division was based on the applicable Polish law, the Act of 20 June 1992, on entitlements to discounted travel on public collective transport [45].

It was assumed that field research would take the form of direct observations, where at each stop, the number of boarding and alighting passengers would be counted, and their structure and ticket type would be recorded (according to the previously presented division). The data would be recorded in an observation sheet.

Additionally, the following assumptions were made:

- The studies would be conducted on bus lines connecting the district seat with the commune seat and other localities within the administrative boundaries of the same district.
- Only bus lines that are the sole service for a given route would be included.

Based on the conducted field research, basic statistical analyses were planned (Stage 4), including the following:

- Determining the shares of individual passenger groups using public regional transport in 1 h time intervals;

- Determining the average number of passengers boarding per service in 1 h time intervals;
- Comparing the number of passengers boarding in the direction of the commune and the district in 1 h time intervals.

In the next step (5), the authors planned to calculate two indicators of transport attractiveness for a given locality:

- Adults and seniors—Att_K;
- Youth—$AttY_K$.

Both indicators allow for assessing the impact of service frequency in regional transport, expressed in the number of daily pairs of trips in locality *K*, on the attractiveness of this type of transport for different passenger groups. While the first indicator determines the attractiveness for adult passengers, who can often use other means of transport, especially individual transport, the second indicator refers to young people who cannot commute to high school by car.

For the Att_K indicator, the following formula was proposed (1):

$$Att_K = \frac{\sum_{i=1}^{n} \left[WSS_K \cdot \left(SS_{Ki}^B + SS_{Ki}^A\right) + WSA_K \cdot \left(SA_{Ki}^B + SA_{Ki}^A\right) + WMA_K \cdot \left(MA_{Ki}^B + MA_{Ki}^A\right)\right]}{P_K}, \quad (1)$$

The notations used above are defined below:

- *i*—the index of the service passing through locality *K* (round trip), where *i* = 1, 2, 3, . . ., *n*, and *n* denotes all services passing through locality *K*;
- *K*—the locality index;
- *A*—the index of alighting passengers;
- *B*—the index of boarding passengers;
- SS_{Ki}^A—the number of seniors with single tickets alighting in locality *K* coming from the direction of the district [–];
- SS_{Ki}^B—the number of seniors with single tickets boarding in locality *K* coming from the direction toward the district [–];
- SA_{Ki}^A—the number of adults (excluding seniors) with single tickets alighting in locality *K* coming from the direction of the district [–];
- SA_{Ki}^B—the number of adults (excluding seniors) with single tickets boarding in locality *K* in the direction toward the district [–];
- MA_{Ki}^A—the number of adults (excluding seniors) with monthly passes alighting in locality *K* in the direction of the district [–];
- MA_{Ki}^B—the number of adults (excluding seniors) with monthly passes boarding in locality *K* in the direction facing the district [–];
- WSS_K—the weighting factor for the number of seniors (not included in the group of adults) with single tickets boarding and alighting in locality *K* [–], $WSS_K = 1$;
- WSA_K—the weighting factor for the number of adults (not included in the group excluding seniors) with single tickets boarding and alighting in locality *K* [–], $WSA_K = 2$;
- WMA_K—the weighting factor for the number of seniors (not included in the group of adults) with monthly tickets boarding and alighting in locality *K* [–], $WMA_K = 4$;
- P_K—the number of residents in locality *K*.

For the $AttY_K$ indicator, the following formula was proposed (2):

$$AttY_K = \frac{\sum_{i=1}^{n} \left[WSY_K \cdot \left(SY_{Ki}^B + SY_{Ki}^A\right) + WMY_K \cdot \left(MY_{Ki}^B + MY_{Ki}^A\right)\right]}{P_K}, \quad (2)$$

The notation used in the formula above is explained below:

- SY_{Ki}^A—the number of children and youth with single tickets alighting in locality *K* in the direction of the district [–];

- SY^{B}_{Ki}—children and youth with single tickets boarding in locality K in the direction toward the district [–];
- MY^{A}_{i}—the number of children and youth with monthly passes alighting in locality K in the direction of the district [–];
- MY^{B}_{Ki}—the number of children and youth with monthly passes boarding in locality K in the direction toward the district [–];
- WSY_K—the weighting factor for the number of children and youth with single tickets boarding and alighting in locality K [–], $WSY_K = 1$;
- WMY_K—the weighting factor for the number of children and youth with monthly tickets boarding and alighting in a given locality K [–], $WMY_K = 2$.

The remaining notation is the same as that in Formula (1).

The indicators Att_K and $AttY_K$ are calculated for localities K with access to a town whose functions and size qualify it as a local center of a given area. The basic amenities that such a town should provide are as follows:

- Secondary education (one or more high schools);
- Commerce (marketplace, large stores, and shopping malls);
- Employment concentration;
- Healthcare (health center, hospital, and laboratory);
- Entertainment (cinema or theater).

In Poland, such centers are usually district towns with populations ranging from 12,000 to 25,000.

Weighting factors WSS_K, WSA_K, and WMA_K in Formula (1) and WSY_K and WMY_K are based on the assumption that the higher the weight, the greater the possibility of using an alternative means of transport, such as a passenger car. The highest weight is for adults with monthly passes, but the lowest is for seniors with single tickets.

The next two stages, 6 (Dependency Analysis I) and 7 (Dependency Analysis II), involve determining the relationship between the transport attractiveness indicators Att_K and $AttY_K$ and the number of connections. Selected statistical methods were used for this purpose, with the first step being the preparation of two sets of input data.

- Set I for stage 6 (Step 6.1): For each locality K, the attractiveness indicators Att_K and $AttY_K$ are determined by the number of pairs of trips toward the town, the average number of adult passengers boarding and alighting in locality K, and the average number of children and youth boarding and alighting in locality K.
- Set II for stage 7 (step 7.1): The data from set I are aggregated into classes based on the number of round trips per day, and the attractiveness indicators Att_K and $AttY_K$ are recalculated for each class as weighted averages, using the locality population as the weight. This means that for each class, the attractiveness indicators are the sum for all localities of boarding and alighting passengers with respective weights divided by the total number of residents in all localities of that class.

Aggregating the data in Step 7.1 means that further analyses consider weighted transport attractiveness indicators for the following:

- Adults and seniors—Att^{wgh};
- Youth—$AttY^{wgh}$.

For Att^{wgh}, the following formula is proposed:

$$Att^{wgh} = \frac{\sum_{K} P_K \cdot Att_K}{\sum_{K} P_K}, \tag{3}$$

where the notation is the same as that in Formulas (1) and (2). The weighted travel attractiveness indicator for youth $AttY^{wgh}$ is calculated using the following formula:

$$AttY^{wgh} = \frac{\sum\limits_{K} P_K \cdot AttY_K}{\sum\limits_{K} P_K}, \tag{4}$$

where the notation is the same as that in Formulas (1) and (2).

Next, in Steps 6.2 and 7.2, the Shapiro–Wilk test is conducted, which, according to Guzik and Więckowska [46], allows one to arrive at the most accurate answer to whether the distribution of each variable is consistent with the normal distribution. The workflow is based on the status of the answer:

- If positive, proceed to Step 6.3 for Stage 6 or Step 7.3 for Stage 7—determining the Pearson linear correlation coefficients between variables with a normal distribution.
- If negative, proceed to Step 6.4 for Stage 6 or Step 7.4 for Stage 7—determining the Spearman rank correlation coefficients to examine the relationships between variables without a normal distribution.

The last stage (8) involves formulating conclusions based on Stages 1–7 (previously completed).

4. Implementation of the Research

4.1. Characteristics of the Research Areas

According to the methodology developed in Section 3 of this article, in Stage 1, regions where direct observations were possible due to public bus transport were identified. Publicly available information sources were used for this purpose: websites of carriers, local government units, and travel planners. As a result of the work in Stages 1 and 2, the research was selected and planned in 19 districts:

- Kolbuszowa District (powiat kolbuszowski);
- Zwoleń District (powiat zwoleński);
- Szczytno District (powiat szczycieński);
- Myszków District (powiat myszkowski);
- Wąbrzeźno District (powiat wąbrzeski);
- Przysucha District (powiat przysuski);
- Chełmno District (powiat chełmiński);
- Iława District (powiat iławski);
- Goleniów District (powiat goleniowski);
- Stargard District (powiat stargardzki);
- Włoszczowa District (powiat włoszczowski);
- Strzelin District (powiat strzeliński);
- Krotoszyn District (powiat krotoszyński);
- Kutno District (powiat kutnowski);
- Mielec District (powiat mielecki);
- Oborniki District (powiat obornicki);
- Lipno District (powiat lipnowski);
- Szamotuły District (powiat szamotulski);
- Chodzież District (powiat chodzieski).

The mentioned districts cover a total area of 17.6 thousand square kilometers, representing 5.6% of Poland's area. Considering the population, the area of the districts where the observations were made is inhabited by a total of 3.4% of Poland's population [47]. However, in the selected districts, over half of the people (54%) live in rural areas, where access to public transport is limited. For comparison, in Poland, this value is 40.5% [47]. The locations of these districts are shown in Figure 2.

The bus lines where the research was planned covered a total of 328 localities. Their size, expressed by the number of inhabitants, aggregated into seven classes, is shown in Table 1. For each class, the average number of inhabitants was calculated (inhs).

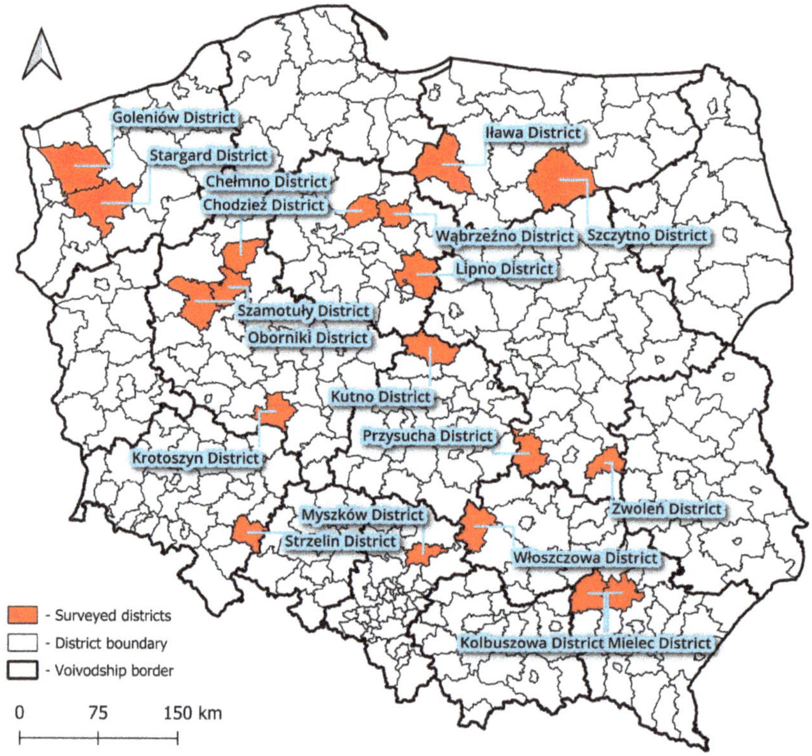

Figure 2. The districts where the research was conducted.

Table 1. The size structure of localities covered by the surveyed lines. Source: commune offices.

Class of Town Size [inhs]	Count	Share [%]	Average Town Size [inhs]
More than 10,000	16	4.9	26,259
Between 5000 and 9999	5	1.5	7745
Between 2000 and 4999	11	3.4	2932
Between 1000 and 1999	15	4.6	1449
Between 500 and 999	38	11.6	663
Between 100 and 499	187	57.0	260
Less than 100	56	17.1	61
Total	328	100.0	

By analyzing the size structures of the localities covered by the surveyed lines, the following conclusions can be drawn:

- Most localities (187) had between 100 and 499 inhabitants (57%);
- The fewest localities (5) had between 5000 and 9999 (1.5%);
- In the category of the largest localities (more than 10,000), there are only cities that are the seats of districts;
- In the second category (5000–9999), only one locality is not the seat of a district.

4.2. Limitations

Considering, on the one hand, the variable ridership from day to day and, on the other hand, the need to maintain comparability of research results on different lines, observations were carefully planned throughout a 3-month period. Due to the extensiveness of the research, only one observation of each connection was possible. Most observations took place from Tuesday to Thursday because these days have the lowest passenger fluctuations. However, in a few cases, for logistical reasons, it was necessary to conduct research on Mondays and Fridays. The research was conducted only during working days, excluding days right before and after holidays. Bus lines with no competing connections were selected for observation; however, in sporadic cases, there were other single parallel connections on part of the route.

In this study, only one factor influencing the attractiveness of public transport is considered (frequency). Still, there are other factors, like ticket prices, travel times, the number of cars per 1000 inhabitants, unemployment rate, etc. Multifactor analysis will be the subject of future research.

4.3. Observation Results

The travel direction and passenger structure research in regional bus transport (Stage 3) was conducted between 12 September and 1 December 2023, covering 27 bus lines connecting the district seats with the commune seats and other localities within the administrative boundaries of the same district.

A total of 6248 passengers were recorded during the study, for whom the travel structure is shown in Table 2.

Table 2. The number of passengers in individual groups on all surveyed lines.

Description	Single Tickets				Monthly Passes		Total
	Discount	Youth	Adult	Senior	Youth	Adult	
Number [-]	36	407	1160	320	3955	370	6248
Ratio [%]	0.6	6.5	18.6	5.1	63.3	5.9	100.0

The main group of passengers consisted of children and youth with monthly tickets—amounting to about 63%. The second largest group of passengers (about 19%) was adults with single tickets. Youth with single tickets accounted for about 6.5% of passengers, and adults with monthly tickets accounted for about 6%. A slightly smaller group of about 5% was seniors with single tickets. To summarize, all young people, both with single and monthly tickets, constitute about 70% of all passengers. The remaining 30% are adults. About 70% of all passengers are regular users of regional bus lines holding monthly passes. It should be mentioned that none of the observed lines served the function of transporting people to primary schools, as this task is conducted by communes as separate transport for children. Hence, there is no separate category for children in this research; they are included in the youth group of passengers.

While Table 1 shows the passenger structure as an average for the entire day, Figure 3a,b show the changes in this structure in 1 h intervals throughout the day for the direction toward the commune and the district, respectively. Detailed data are provided in Tables A1 and A2.

Figure 3a shows that the largest share of students with monthly tickets to the commune, reaching 80%, is found between 14:00 and 15:59. This share is also high in the intervals of 13:00–13:59 and 16:00–17:59. In the evening hours, the share of passengers with normal monthly tickets commuting to work dominates. Adults with single tickets comprise the largest share in the morning and afternoon. In the direction of the district, the situation is reversed (Figure 3b): students with monthly tickets travel in the morning, with the highest (80%) share recorded in the interval of 7:00–7:59 and a slightly lower share in the intervals of 6:00–6:59 and 8:00–8:59, while adults with monthly tickets comprise the largest share early in the morning in the interval of 4:00–5:59. Adult passengers with single tickets comprise the largest share in this direction in the noon and evening hours. Seniors travel most frequently between 8:00 and 14:59 in both directions. The missing data for intervals 21:00–21:59 in Figure 3a and 19:00–19:59 in Figure 3b are related to the lack of bus services on all lines on which the research was conducted.

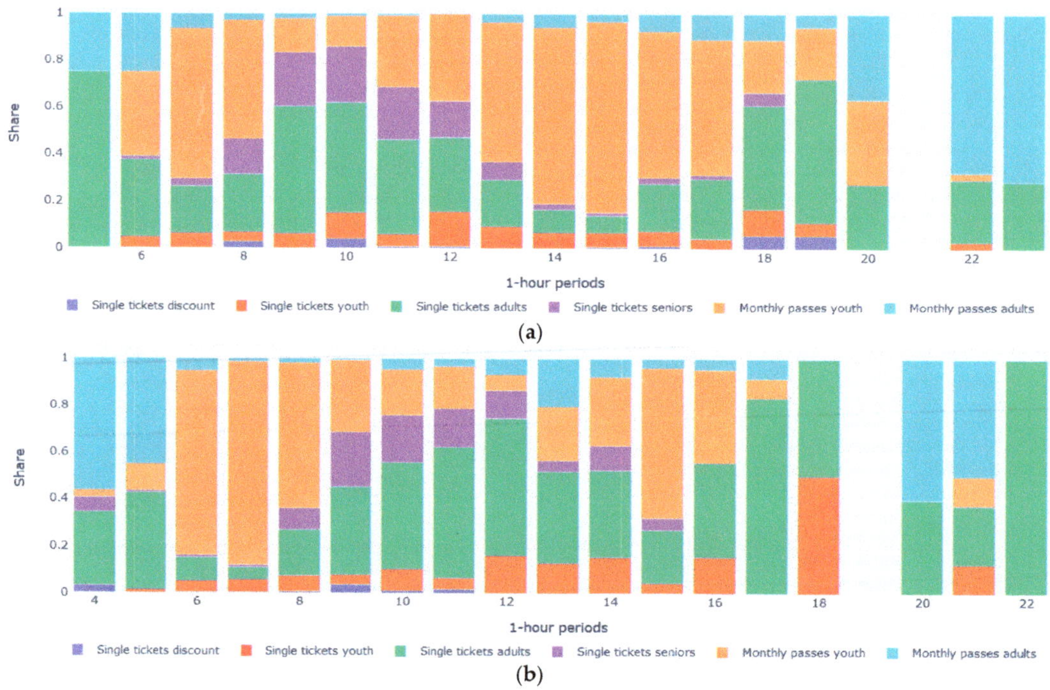

Figure 3. The share of individual passenger groups in time intervals: (**a**) toward the commune; (**b**) toward the district.

Figure 4a,b show the number of passengers boarding toward the commune and the district, as well as the average number of passengers boarding per trip in 1 h intervals.

The charts (Figure 4a,b) show that the most passengers (1283 people) boarded buses between 7:00 and 7:59 for travel toward the commune and between 15:00 and 15:59 for travel toward the district (832 people). However, slightly fewer were also in the previous time interval, i.e., 14:00–14:59 (702 people). Interestingly, the 14:00–14:59 interval had a higher value for the average number of passengers per trip than the 15:00–15:59 interval. This is because more bus trips took place in the latter interval. In the morning peak, travel toward the district prevails roughly until 9:59. Between 10:00 and 12:59, the numbers of passengers boarding in both directions are similar. After 13:00, travel toward the commune begins to dominate, except for the 21:00–21:59 interval, when trips to the district for the night shift take place. After 18:00, the number of passengers in both directions is the lowest for the entire day, which also results from the small number of connections at that time, as most carriers end service around 16:00–17:00. This is visible in the average number of passengers per trip in the evening hours. However, after 22:00, there is a fairly large increase in the number of boarding passengers and the average number of passengers per trip toward the commune due to returns from work from the afternoon shift, which most often ends at 22:00.

The smaller number of connections with a relatively high number of passengers boarding in the district's direction during 6:00–6:59 and 8:00–8:59 resulted in a fairly high average number of passengers in these intervals, which is not as noticeable in Figure 4. A similar situation occurred in the direction from the commune in the interval of 22:00–22:59; the small number of connections resulted in high average numbers of passengers per trip, although the number of passengers was not as high as that at other times of the day.

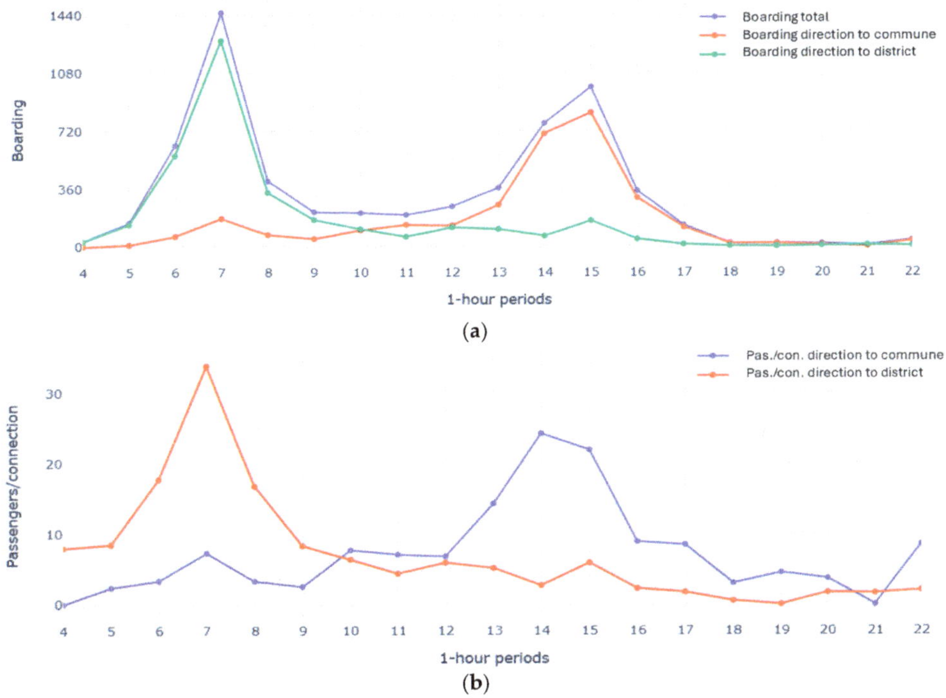

Figure 4. The number of passengers boarding in the direction of the commune and the district in time intervals: (**a**) total in all surveys; (**b**) average from all measurements.

The data on the number of passengers boarding during the afternoon peak, departing from the district town, are concentrated in the time interval corresponding to the scheduled departure time. In contrast, in the case of the morning commute to the district town, passengers gradually boarded in successive localities, and if the bus departed from them before 7:00, the passengers boarding there were assigned to the time interval 6:00–6:59. If arrival times in the district town were considered for the morning hours, then for the time interval 7:00–7:59, even higher values would be obtained at the expense of the 6:00–6:59 interval. The difference between the peak values in the morning and afternoon peak would also be greater.

4.4. Analysis of the Attractiveness Index

In accordance with the proposed methodology, in Step 5, the attractiveness indices Att_K (Formula (1)) and $AttY_K$ (Formula (2)) were determined. Subsequently, analyses of their dependence on the number of routes connecting this locality with the district seat were conducted (Stages 6 and 7). First, in Step 6.1, Data Set I was prepared. This set included a total of 296 localities, for which the following were also calculated:

- The average number of adults and seniors per trip—A_{trip}^{avg} [pax];
- The average number of children and youth per trip—Y_{trip}^{avg} [pax].

The determined index values for the selected localities, arranged in alphabetical order, are presented in Table 3. As can be observed, there is significant variability among the individual localities; e.g., the values for $AttY_K$ range from 0.000 to 0.389. This variability is due to the differing numbers of passengers boarding and alighting in each locality, as well as their age structure and the types of tickets the passengers held.

Table 3. A fragment of the data summary prepared in Step 6.1. Source: commune offices.

No.	Locality—K	Trips [-]	P_{trip}^{avg} [inhs]	Att_K [-]	$AttY_K$ [-]	A_{trip}^{avg} [pax]	Y_{trip}^{avg} [pax]
1	Bartlewo	9	250	0.144	0.114	1.056	1.611
2	Bartodzieje	8	510	0.000	0.034	0.000	1.313
3	Bebelno-Kolonia	7	408	0.077	0.042	1.692	1.308
4	Bebelno-Wieś	7	375	0.024	0.008	0.615	0.231
5	Bedlno	14	377	0.130	0.130	1.393	1.857
6	Bętlewo	8	251	0.000	0.068	0.000	1.125
7	Bielejewo	2	74	0.000	0.122	0.000	2.250
8	Binino	5	363	0.014	0.056	0.500	2.100
9	Bliżyce	2	542	0.004	0.041	0.667	7.333
10	Błachta	9	94	0.011	0.021	0.056	0.111
11	Bobolice	6	79	0.013	0.051	0.091	0.364
...	...	...	...	...	...	...	...
291	Zielonka	4	639	0.006	0.027	0.750	2.250
292	Żarki	30	4431	0.037	0.091	2.864	7.186
293	Żarnowo	1	355	0.000	0.014	0.000	3.000
294	Żarnówko	1	81	0.000	0.000	0.000	0.000
295	Żdżary	7	277	0.000	0.000	0.000	0.000
296	Żychlin	23	7484	0.033	0.016	4.913	2.783
	Min	1	10	0.000	0.000	0.000	0.000
	Max	30	7484	0.201	0.389	5.000	25.000

Table 4, in turn, presents the results for Data Set II after its aggregation into classes of the number of pairs of trips. The individual calculated quantities are as follows:

- P_{trip}^{avg}—the average size of the locality in a given class of pairs of trips [inhs];
- occur—the number of occurrences [–];
- Att^{wgh}—the weighted attractiveness index of transportation for adults and seniors [–];
- $AttY^{wgh}$—the weighted attractiveness index of transportation for youth [–];
- PAS_{trip}^{avg}—the average number of passengers per trip overall [pax];
- PAS_{town}^{avg}—the average number of passengers boarding and alighting per locality overall [pax];
- A_{town}^{avg}—the average number of adults and seniors boarding and alighting per locality overall [pax];
- Y_{town}^{avg}—the average number of children and youth boarding and alighting per locality overall [pax].

The remaining notation is the same as that in Table 3.

In Table 4, classes not considered in the statistical analysis are marked in red. This pertains to the classes of the number of pairs of trips, 18, 23, and 30, because of the following:

- All these classes include single cases (one pair of trips—occur = 1);
- For 23 and 30, the number of inhabitants significantly deviates from the average number of inhabitants for other classes (23 trips—7484.0 [inhs]; 30 trips—4431.0 [inhs]).

In accordance with Step 6.2 of Stage 6, the Shapiro–Wilk statistical test was conducted while considering the following:

- The independent variable: the number of pairs of trips per day in a given locality (trips);
- The dependent variables: transportation attractiveness indices for adults and seniors (Att_K) and children and youth ($AttY_K$); the average number of adults and seniors per trip; and the average number of children and youth per trip (Y_{trip}^{avg}).

Table 4. Data set after its aggregation into classes of the number of pairs of trips.

No.	Trips [-]	P_{trip}^{avg} [inhs]	Occur [-]	Att^{wgh} [-]	$AttY^{wgh}$ [-]	PAS_{trip}^{avg} [pax]	A_{trip}^{avg} [pax]	Y_{trip}^{avg} [pax]	PAS_{town}^{avg} [pax]	A_{town}^{avg} [pax]	Y_{town}^{avg} [pax]
1	1	201.6	45	0.001	0.024	2.733	0.200	2.600	5	0	5
2	2	331.8	27	0.004	0.017	1.870	0.444	1.537	7	1	6
3	3	266.3	3	0.002	0.020	1.000	0.111	0.889	6	0	5
4	4	424.3	66	0.009	0.020	1.716	0.519	1.220	13	4	9
5	5	302.9	9	0.011	0.034	1.422	0.378	1.089	14	3	11
6	6	327.3	45	0.024	0.033	1.685	0.815	0.881	20	9	10
7	7	498.8	24	0.024	0.021	1.613	0.768	0.863	22	10	12
8	8	497.9	33	0.014	0.041	1.633	0.451	1.208	26	7	19
9	9	566.6	30	0.017	0.046	1.970	0.481	1.519	35	8	27
10	10	390.3	7	0.024	0.053	2.686	1.429	1.257	53	28	25
11	11	770.9	10	0.019	0.054	2.555	0.600	1.964	56	13	43
12	12	846.0	9	0.022	0.030	1.685	0.620	1.083	40	15	25
13	14	288.9	8	0.060	0.064	1.223	0.536	0.696	34	14	19
14	18	774.0	1	0.036	0.085	2.556	0.722	1.833	91	25	66
15	23	7484.0	1	0.033	0.016	7.696	4.913	2.783	354	226	128
16	30	4431.0	1	0.037	0.091	9.900	2.833	7.067	593	169	424
Min [1]	1	201.6	1	0.001	0.016	1	0.111	0.696	5	0	5
Max [1]	30	7484	66	0.06	0.091	9.9	4.913	7.067	593	226	424

[1] Values for all data from 1 to 16.

The results of the statistical analysis for this step for Data Set I are presented in Table 5 and Figure 5a–c. In Step 6.2, none of the variables were found to conform to a normal distribution. Therefore, to examine the relationships between the variables, Spearman's rank correlation coefficient, Rs, was used (until Step 6.4 in Stage 6).

Table 5. The summary of the relationships between the number of trips and the variables from Data Set I.

N = 296	Variable	R Spearman	t (N-2)	p
Trips &	Att_K	0.510	10.18	<0.001 [1]
	A_{trip}^{avg}	0.315	5.69	<0.001 [1]
	$AttY_K$	0.298	5.36	<0.001 [1]
	Y_{trip}^{avg}	−0.047	−0.81	0.418

[1] a statistically significant result.

Upon analyzing the results obtained in Steps 6.2 and 6.4, regarding the dependencies of the indices Att_K, $AttY_K$, and A_{trip}^{avg} on the number of trips, it becomes clear that the correlation coefficients marked ([1]) in Table 5 are significant. The strongest dependency on the number of trips is shown by index Att_K; the Spearman correlation coefficient, in this case, is +0.510, which means that the higher the number of pairs of trips per day, the higher the attractiveness index. Nonetheless, this dependency is moderate. Positive correlations are also shown by the indices $AttY_K$ and A_{trip}^{avg}, although the dependencies are weaker than those for Att_K. However, there is no correlation between the number of trips and the average number of children and youth per trip.

In the analysis of Data Set II (Step 7.2 of Stage 7), the following were considered:

- The independent variable: the number of pairs of trips per day in a given locality (trips);
- The dependent variables: weighted transportation attractiveness indices for adults and seniors (Att^{wgh}) and children and youth ($AttY^{wgh}$), the average number of adults per trip (A_{trip}^{avg}), the average number of children and youth per trip (Y_{trip}^{avg}), the average

number of adults boarding and alighting per locality (A^{avg}_{town}), and the average number of children and youth boarding and alighting per locality (Y^{avg}_{town}).

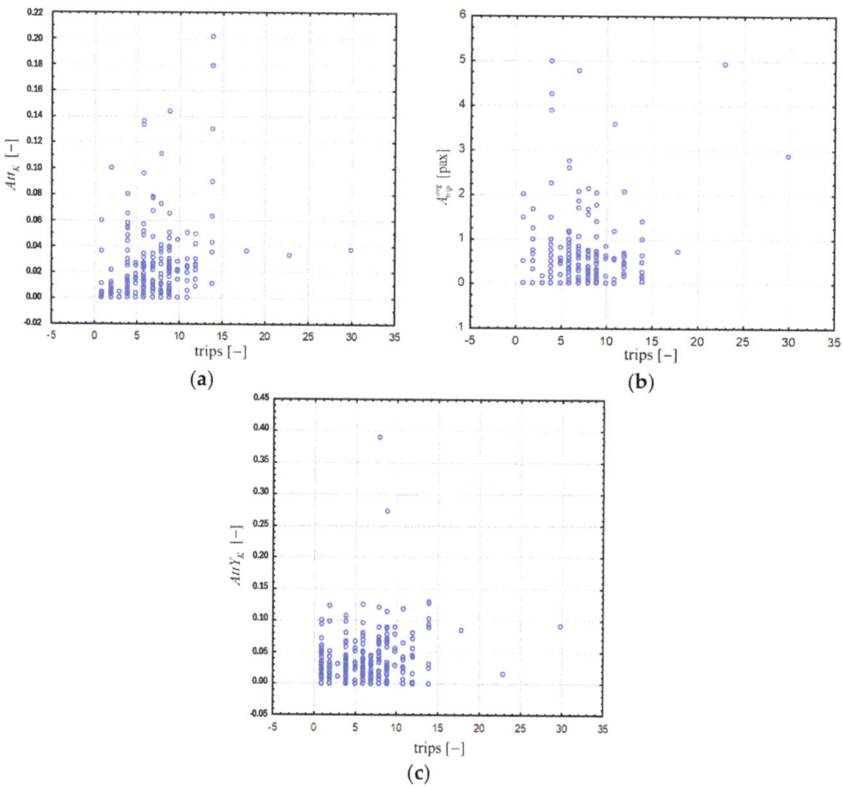

Figure 5. Dependency plots comparing (**a**) the transportation attractiveness index for adults and seniors (Att_K) and the number of trips; (**b**) the average number of adults and seniors per trip (A^{avg}_{trip}) and the number of trips; and (**c**) the attractiveness index for children and youth ($AttY_K$) and the number of trips.

Non-conformity with the normal distribution only occurred for the index Att^{wgh}. Therefore, to examine the dependencies on the number of trips, Spearman's rank correlation coefficient Rs was used (Step 7.4); the results are presented in Table 6 and Figure 6.

Table 6. The results of the statistical analysis for Data Set II—index Att^{wgh} (Step 7.4).

Variable	R Spearman	t (N-2)	p
Trips and Att^{wgh}	0.807	4.53	0.001 [1]

[1] a statistically significant result.

There is a strong positive correlation between the attractiveness index Att^{wgh} and the number of pairs of trips (Spearman's R coefficient of 0.807). This means the higher the number of trips, the higher the attractiveness index. The graph in Figure 6 also indicates that when the number of pairs of trips is below four per day, the attractiveness index is close to 0. Given this, one can conclude that public transportation for adults starts to become attractive when the number of pairs of trips per day is at least four.

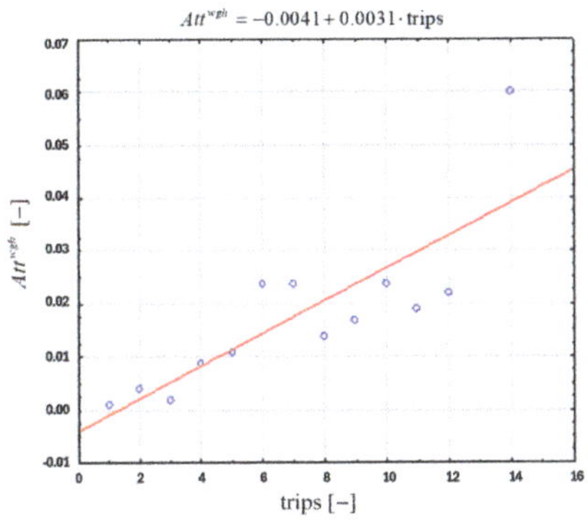

Figure 6. The dependency between the attractiveness index (Att^{wgh}) and the number of trips.

For the other dependent variables ($AttY^{wgh}$, A_{trip}^{avg}, Y_{trip}^{avg}, A_{town}^{avg}, and Y_{town}^{avg}), conformity with a normal distribution was observed. Therefore, Pearson's linear correlation coefficient r was used to examine the dependencies between these variables and the number of trips (Step 7.3); the results are shown in Table 7 and Figure 7a–c.

Table 7. The results of the statistical analysis for Data Set II—Step 7.3.

N = 13	Variable	r Pearson	t	*p*
Trips	$AttY^{wgh}$	0.813	4.63	0.001 [1]
	A_{trip}^{avg}	0.471	1.77	0.105
	Y_{trip}^{avg}	−0.346	−1.22	0.247
	A_{town}^{avg}	0.773	4.04	0.002 [1]
	Y_{town}^{avg}	0.783	4.17	0.002 [1]

[1] a statistically significant result.

Regarding the dependencies of the average number of adults per trip (A_{trip}^{avg}) and the average number of children and youth per trip (Y_{town}^{avg}), the obtained probability *p* value is above the accepted significance level of 0.05. Therefore, the correlation coefficients are insignificant. For the remaining variables, the correlation coefficients are significant; their dependency on the number of trips is shown in Figure 7a–c.

For the dependent variable $AttY^{wgh}$, the correlation coefficient r indicates a very strong positive dependency on the number of trips (r = 0.812). Strong dependencies on the number of trips were also observed for the average number of adults and seniors per locality A_{town}^{avg} (r = 0.773) and the average number of children and youth per locality Y_{town}^{avg} (r = 0.783).

The weighted transportation attractiveness index for adults and seniors Att^{wgh}, as well as the average number of adults and seniors per locality (A_{town}^{avg}), shows that for fewer than four trips, there is minimal interest in using public transportation. In contrast, for children and youth, the variables $AttY_K$ and Y_{town}^{avg} indicate some interest in public transportation even with just one pair of trips per day. This is because, on the studied bus lines, when there was a small number of pairs of trips (1–3 daily), the schedules were adjusted to the start and end times of high school classes.

For the dependencies that conformed to a normal distribution, linear regression equations could be determined, as shown in Figure 7a–c. Based on the equation for the dependency of the average number of inhabitants per locality (A_{trip}^{avg}) on the number of trips, for every increase of one trip, the number of adult passengers will increase by an average of 1.5. Assuming that the additional passengers switched from individual transportation to public transportation, the resulting reduction in energy consumption due to an increase of one trip is 0.33–0.69 kWh (the difference between the energy consumption of 1 km of travel by one passenger by car and by bus multiplied by 1.5) for each kilometer traveled by these passengers.

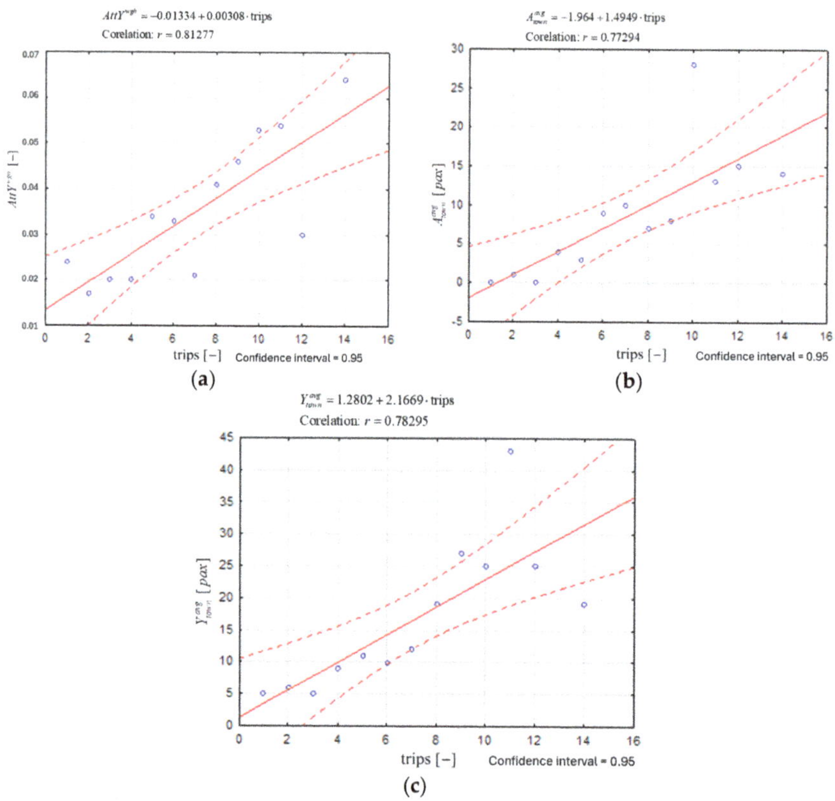

Figure 7. Dependency plots comparing (**a**) the weighted attractiveness index for children and youth ($AttY^{wgh}$) and the number of trips; (**b**) the average number of adults and seniors per locality (A^{avg}_{town}) and the number of trips; and (**c**) the average number of children and youth per locality (Y^{avg}_{town}) and the number of trips.

5. Discussion and Conclusions

This article presents Task I of the proposed approach for determining the potential energy consumption benefits based on the attractiveness of public transport by bus. Due to a lack of similar studies and analyses, it is difficult to compare the obtained results with those of other authors. Only the variability in the number of passengers throughout the day in the direction of a larger urban center and returns from it (shown in Figure 4a) is similar to what was described in [13]. The results presented in this article clearly indicate a strong correlation between the frequency of bus trips and the attractiveness of public transport for both adults and the group of children and youth. Such correlations, although obtained in a somewhat different and more general approach, are also indicated by experience and studies from countries where the policy of ensuring a high-quality public transport offer in rural areas, i.e., less urbanized areas, plays a significant role [12,38].

Although statistical analysis indicates there is a linear correlation for some variables that is dependent on the number of trips, it seems that the correlations are mostly nonlinear. After a certain number of connections is reached, further increases will cause a progressively smaller increase in the number of passengers, both adults and children and youth. In the studies, the maximum number of pairs of trips per day was 16 (excluding the discarded single cases for 18, 23, and 30 pairs of trips in Data Set II) due to the limited possibility of finding lines with a larger number of trips operating under similar conditions in Poland (outside of large agglomerations).

Due to the poor transport offer of regional transport in Poland, in most cases, it was possible to identify bus lines that were the only ones serving a given transport corridor and the analyzed localities. In other countries with better-developed regional transport, there are many localities with alternative attractive travel options to other significant urban centers, attracting residents of smaller localities.

Due to the conducted studies and the proposed approach, the authors believe that it is possible to determine the benefits related to the energy consumption of travel provided by regional public transport compared to individual car travel. In this case, the focus was on factors related to the attractiveness of public transport, which affects the number of passengers, especially adults, who often have an alternative option for individual transport.

The studies also found that the size of the district town influences the number of passengers. However, further analytical work is required to verify this. Similarly, dependencies on other factors, such as the distance of a given locality from the district town, travel time, ticket price, unemployment rate, income level of residents, or the motorization rate in the region, also need to be considered.

Author Contributions: Conceptualization, W.M. and M.K.; methodology, W.M. and M.K.; software, W.M., I.M. and M.K.; validation, W.M. and I.M.; formal analysis, W.M. and I.M.; investigation, W.M. and I.M.; resources, W.M.; data curation, W.M.; writing—original draft preparation, W.M. and I.M.; writing—review and editing, M.K.; visualization, M.K., W.M. and I.M.; supervision, A.M.-G.; project administration, M.K.; funding acquisition, M.K. and A.M.-G. All authors have read and agreed to the published version of the manuscript.

Funding: This research was funded by the Ministry of Education and Science at the Poznan University of Technology (Poland). Grant title: management of land transport systems in the aspect of their efficiency and safety, grant number 0416/SBAD/0005.

Data Availability Statement: The datasets presented in this article are not readily available because the data are part of an ongoing study. Requests to access the datasets should be directed to wojciech.miechowicz@doctorate.put.poznan.pl.

Conflicts of Interest: The authors declare no conflicts of interest.

Appendix A

Table A1. Passenger share and number in individual groups in time intervals in the commune's direction.

Hours	Single Tickets [%] and [–]			Monthly Passes [%] and [–]			Total [%] and [–]
	Discount	Youth	Adults	Seniors	Youth	Adults	
4:00–4:59	– (0)	– (0)	– (0)	– (0)	– (0)	– (0)	– (0)
5:00–5:59	0.0 (0)	0.0 (0)	75 (9)	0.0 (0)	0.0 (0)	25 (3)	100.0 (12)
6:00–6:59	0.0 (0)	4.7 (3)	32.8 (21)	1.6 (1)	35.9 (23)	25 (16)	100.0 (64)
7:00–7:59	0.0 (0)	6.3 (11)	20 (35)	3.4 (6)	64 (112)	6.3 (11)	100.0 (175)
8:00–8:59	2.7 (2)	4.1 (3)	24.7 (18)	15.1 (11)	50.7 (37)	2.7 (2)	100.0 (73)
9:00–9:59	0.0 (0)	6.3 (3)	54.2 (26)	22.9 (11)	14.6 (7)	2.1 (1)	100.0 (48)
10:00–10:59	4.0 (4)	11.0 (11)	47.0 (47)	24.0 (24)	13.0 (13)	1.0 (1)	100.0 (100)
11:00–11:59	0.8 (1)	5.2 (7)	40.3 (54)	22.4 (30)	30.6 (41)	0.8 (1)	100.0 (134)
12:00–12:59	0.8 (1)	14.7 (19)	31.8 (41)	15.5 (20)	37.2 (48)	0.0 (0)	100.0 (129)
13:00–13:59	0.0 (0)	9.3 (24)	19.8 (51)	7.8 (20)	59.5 (153)	3.5 (9)	100.0 (257)
14:00–14:59	0.1 (1)	6.7 (47)	9.5 (67)	2.6 (18)	75.4 (529)	5.7 (40)	100.0 (702)
15:00–15:59	0.6 (5)	6.1 (51)	7.2 (60)	1.3 (11)	81.5 (678)	3.3 (27)	100.0 (832)
16:00–16:59	1.0 (3)	6.3 (19)	20.2 (61)	2.7 (8)	62.6 (189)	7.3 (22)	100.0 (302)
17:00–17:59	0.0 (0)	4.2 (5)	25.4 (30)	1.7 (2)	57.6 (68)	11.0 (13)	100.0 (118)
18:00–18:59	5.6 (1)	11.1 (2)	44.4 (8)	5.6 (1)	22.2 (4)	11.1 (2)	100.0 (18)

Table A1. *Cont.*

Hours	Single Tickets [%] and [–]				Monthly Passes [%] and [–]		Total [%] and [–]
	Discount	Youth	Adults	Seniors	Youth	Adults	
19:00–19:59	5.6 (1)	5.6 (1)	61.1 (11)	0.0 (0)	22.2 (4)	5.6 (1)	100.0 (18)
20:00–20:59	0.0 (0)	0.0 (0)	27.3 (3)	0.0 (0)	36.4 (4)	36.4 (4)	100.0 (11)
21:00–21:59	– (0)	– (0)	– (0)	– (0)	– (0)	– (0)	– (0)
22:00–22:59	0.0 (0)	2.9 (1)	26.5 (9)	0.0 (0)	2.9 (1)	67.7 (23)	100.0 (34)
23:00–23:59	0.0 (0)	0.0 (0)	28.6 (2)	0.0 (0)	0.0 (0)	71.4 (5)	100.0 (7)

Table A2. Passenger share and number in individual groups in time intervals in the district's direction.

Hours	Single Tickets [%] and [–]				Monthly Passes [%] and [–]		Total [%] and [–]
	Discount	Youth	Adults	Seniors	Youth	Adults	
4:00–4:59	3.1 (1)	0.0 (0)	31.3 (10)	6.3 (2)	3.1 (1)	56.3 (18)	100.0 (32)
5:00–5:59	0.0 (0)	1.5 (2)	41.2 (56)	0.7 (1)	11.8 (16)	44.9 (61)	100.0 (136)
6:00–6:59	0.4 (2)	4.8 (27)	10.1 (57)	0.9 (5)	78.8 (447)	5.1 (29)	100.0 (567)
7:00–7:59	0.3 (4)	5.4 (69)	5.1 (66)	1.0 (13)	86.9 (1115)	1.3 (16)	100.0 (1283)
8:00–8:59	0.6 (2)	6.9 (23)	19.4 (65)	9.3 (31)	62.1 (208)	1.8 (6)	100.0 (335)
9:00–9:59	3.6 (6)	4.2 (7)	37.4 (62)	23.5 (39)	30.7 (51)	0.6 (1)	100.0 (166)
10:00–10:59	0.9 (1)	9.3 (10)	45.4 (49)	20.4 (22)	19.4 (21)	4.6 (5)	100.0 (108)
11:00–11:59	1.6 (1)	4.9 (3)	55.7 (34)	16.4 (10)	18.0 (11)	3.3 (2)	100.0 (61)
12:00–12:59	0.0 (0)	16.1 (19)	58.5 (69)	11.9 (14)	6.8 (8)	6.8 (8)	100.0 (118)
13:00–13:59	0.0 (0)	13.0 (14)	38.9 (42)	4.6 (5)	23.2 (25)	20.4 (22)	100.0 (108)
14:00–14:59	0.0 (0)	15.4 (10)	36.9 (24)	10.8 (7)	29.2 (19)	7.7 (5)	100.0 (65)
15:00–15:59	0.0 (0)	4.4 (7)	22.6 (36)	5.0 (8)	64.2 (102)	3.8 (6)	100.0 (159)
16:00–16:59	0.0 (0)	15.6 (7)	40 (18)	0.0 (0)	40 (18)	4.4 (2)	100.0 (45)
17:00–17:59	0.0 (0)	0.0 (0)	83.3 (10)	0.0 (0)	8.3 (1)	8.3 (1)	100.0 (12)
18:00–18:59	0.0 (0)	50 (1)	50 (1)	0.0 (0)	0.0 (0)	0.0 (0)	100.0 (2)
19:00–19:59	– (0)	– (0)	– (0)	– (0)	– (0)	– (0)	– (0)
20:00–20:59	0.0 (0)	0.0 (0)	40.0 (2)	0.0 (0)	0.0 (0)	60.0 (3)	100 (5)
21:00–21:59	0.0 (0)	12.5 (1)	25 (2)	0.0 (0)	12.5 (1)	50.0 (4)	100.0 (8)
22:00–22:59	0.0 (0)	0.0 (0)	100 (4)	0.0 (0)	0.0 (0)	0.0 (0)	100.0 (4)
23:00–23:59	– (0)	– (0)	– (0)	– (0)	– (0)	– (0)	– (0)

Table A3. The number of passengers boarding and the average number of passengers per trip in the direction of the commune and district in individual time intervals.

Hours	Direction to Commune (Gmina)		Direction to District (Powiat)	
	Boarding Passengers	Passenger/Connection	Boarding Passengers	Passenger/Connection
4:00–4:59	0	0.00	32	8.00
5:00–5:59	12	2.40	136	8.50
6:00–6:59	64	3.37	567	17.72
7:00–7:59	175	7.29	1283	33.76
8:00–8:59	73	3.32	335	16.75
9:00–9:59	48	2.53	166	8.30
10:00–10:59	100	7.69	108	6.35
11:00–11:59	134	7.05	61	4.36
12:00–12:59	129	6.79	118	5.90
13:00–13:59	257	14.28	108	5.14
14:00–14:59	702	24.21	65	2.71
15:00–15:59	832	21.89	159	5.89
16:00–16:59	302	8.88	45	2.25
17:00–17:59	118	8.43	12	1.71
18:00–18:59	18	3.00	2	0.50
19:00–19:59	18	4.50	0	0.00
20:00–20:59	11	3.67	5	1.67

Table A3. *Cont.*

Hours	Direction to Commune (Gmina)		Direction to District (Powiat)	
	Boarding Passengers	**Passenger/Connection**	**Boarding Passengers**	**Passenger/Connection**
21:00–21:59	0	0.00	8	1.60
22:00–22:59	34	8.50	4	2.00
23:00–23:59	7	3.50	0	0.00

References

1. Pérez-Martínez, P.J.; Sorba, I.A. Energy Consumption of Passenger Land Transport Modes. *Energy Environ.* **2010**, *21*, 577–600. [CrossRef]
2. Sectoral Profile—Transport. Specific Consumption. Available online: https://www.odyssee-mure.eu/publications/efficiency-by-sector/transport/specific-consumption.html (accessed on 10 March 2024).
3. Parlamento Europeo. Directive (EU) 2023/1791 of the European Parliament and of the Council of 13 September 2023 on Energy Efficiency and Amending Regulation (EU) 2023/955. *Off. J. Eur. Union* **2023**, *50*, 1–111.
4. Masoumi, H.; Rooijen, M.V.; Sierpiński, G. Children's Independent Mobility to School in Seven European Countries: A Multinomial Logit Model. *Int. J. Environ. Res. Public Health* **2020**, *17*, 9149. [CrossRef] [PubMed]
5. Del Rosario, L.; Laffan, S.W.; Pettit, C.J. Physical Activity Associated with Public Transport Commutes and Potential Feeder Mode Shifts. *J. Transp. Health* **2024**, *35*, 101751. [CrossRef]
6. Deng, T.; Nelson, J.D. The Perception of Bus Rapid Transit: A Passenger Survey from Beijing Southern Axis BRT Line 1. *Transp. Plan. Technol.* **2012**, *35*, 201–219. [CrossRef]
7. Hoo, W.; Waheeda, A.; Reesha, A. Factors Influencing the Passenger Satisfaction at Public Transport in Kuala Lumpur, Malaysia. *J. Proj. Manag. Pract.* **2023**, *3*, 77–94. [CrossRef]
8. Hansson, J.; Pettersson, F.; Svensson, H.; Wretstrand, A. Preferences in Regional Public Transport: A Literature Review. *Eur. Transp. Res. Rev.* **2019**, *11*, 38. [CrossRef]
9. Redman, L.; Friman, M.; Gärling, T.; Hartig, T. Quality Attributes of Public Transport That Attract Car Users: A Research Review. *Transp. Policy* **2013**, *25*, 119–127. [CrossRef]
10. Göransson, J.; Andersson, H. Factors That Make Public Transport Systems Attractive: A Review of Travel Preferences and Travel Mode Choices. *Eur. Transp. Res. Rev.* **2023**, *15*, 32. [CrossRef]
11. De Oña, J.; De Oña, R. Quality of Service in Public Transport Based on Customer Satisfaction Surveys: A Review and Assessment of Methodological Approaches. *Transp. Sci.* **2015**, *49*, 605–622. [CrossRef]
12. Hansson, J.; Pettersson-Löfstedt, F.; Svensson, H.; Wretstrand, A. Patronage Effects of Off-Peak Service Improvements in Regional Public Transport. *Eur. Transp. Res. Rev.* **2022**, *14*, 19. [CrossRef]
13. Metelka, S.; Janoš, V. Demand Variation in Regional Transport. In Proceedings of the 2021 Smart City Symposium Prague (SCSP), Prague, Czech Republic, 27–28 May 2021; pp. 1–5.
14. Miechowicz, W.; Kiciński, M. Passenger flows analysis in regional bus transport—A case study of the Chodzież Region. In *Annały Inżynierii Ruchu i Planowania Transportu*; Stowarzyszenie Inzynierów i Techników Komunikacji Rzeczypospolitej Polskiej Oddział w Poznaniu: Poznan, Poland, 2023; Volume 5, ISBN 978-83-960628-5-7.
15. Taylor, Z.; Ciechański, A. What Happened to the National Road Carrier in a Post-Communist Country? The Case of Poland's State Road Transport. *Transp. Rev.* **2008**, *28*, 619–640. [CrossRef]
16. Gitkiewicz, O. *Nie zdążę*; Wydawnictwo Dowody na Istnienie: Warszawa, Poland, 2019; ISBN 978-83-65970-39-8.
17. Kiciński, M.; Bieńczak, M.; Zmuda-Trzebiatowski, P. *Ocena Zagrożenia Wykluczeniem Społecznym Związanym z Transportem w Powiatach Wielkopolski*; Statistical Office in Poznan: Poznań, Poland, 2015; pp. 13–14.
18. Zmuda-Trzebiatowski, P. Linkage between transport accessibility and participation in activities, poverty and transport related social exclusion. *Autobusy Tech. Eksploat. Syst. Transp.* **2016**, *17*, 754–759.
19. Beyazit, E. Evaluating Social Justice in Transport: Lessons to Be Learned from the Capability Approach. *Transp. Rev.* **2011**, *31*, 117–134. [CrossRef]
20. Zhang, C.; Wang, M.; Dong, J.; Lu, W.; Liu, Y.; Ni, A.; Yu, X. Factors and Mechanism Affecting the Attractiveness of Public Transport: Macroscopic and Microscopic Perspectives. *J. Adv. Transp.* **2022**, *2022*, 5048678. [CrossRef]
21. Zhang, C.; Liu, Y.; Lu, W.; Xiao, G. Evaluating Passenger Satisfaction Index Based on PLS-SEM Model: Evidence from Chinese Public Transport Service. *Transp. Res. Part A Policy Pract.* **2019**, *120*, 149–164. [CrossRef]
22. Sukhov, A.; Lättman, K.; Olsson, L.E.; Friman, M.; Fujii, S. Assessing Travel Satisfaction in Public Transport: A Configurational Approach. *Transp. Res. Part D Transp. Environ.* **2021**, *93*, 102732. [CrossRef]
23. Schnieder, M. Effective Speed: Factors That Influence the Attractiveness of Cost Effective and Sustainable Modes of Transport in Cities. *Sustainability* **2023**, *15*, 8338. [CrossRef]
24. Levinson, H.S. Transit Cooperative Research Program. In *Bus Rapid Transit*; National Research Council (U.S.), Ed.; TCRP Report; Transportation Research Board: Washington, DC, USA, 2003; ISBN 978-0-309-08751-3.

25. Yao, D.; Xu, L.; Li, J. Does Technical Efficiency Play a Mediating Role between Bus Facility Scale and Ridership Attraction? Evidence from Bus Practices in China. *Transp. Res. Part A Policy Pract.* **2020**, *132*, 77–96. [CrossRef]
26. Daganzo, C.; Ouyang, Y. *Public Transportation Systems: Principles of System Design, Operations Planning and Real-Time Control*; World Scientific: Hackensack, NJ, USA, 2019; ISBN 978-981-322-408-7.
27. Ceder, A. *Public Transit Planning and Operation: Modeling, Practice and Behavior*, 2nd ed.; CRC Press/Taylor & Francis Group: Boca Raton, FL, USA; London, UK, 2019; ISBN 978-0-367-86760-7.
28. De Vos, J.; Waygood, E.O.D.; Letarte, L. Modeling the Desire for Using Public Transport. *Travel Behav. Soc.* **2020**, *19*, 90–98. [CrossRef]
29. Brechan, I. Effect of Price Reduction and Increased Service Frequency on Public Transport Travel. *J. Public Transp.* **2017**, *20*, 139–156. [CrossRef]
30. Hensher, D.A. A Service Quality Index for Area-Wide Contract Performance assessment. *Res. Transp. Econ.* **2007**, *18*, 259–276. [CrossRef]
31. Stefanouli, M.; Polyzos, S. Gravity vs Radiation Model: Two Approaches on Commuting in Greece. *Transp. Res. Procedia* **2017**, *24*, 65–72. [CrossRef]
32. Gasparik, J.; Dedik, M.; Cechovic, L.; Blaho, P. Estimation of Transport Potential in Regional Rail Passenger Transport by Using the Innovative Mathematical-Statistical Gravity Approach. *Sustainability* **2020**, *12*, 3821. [CrossRef]
33. Juřík, D.; Janoš, V. Using Census Data for the Gravity Model in Smart Transport Planning. In Proceedings of the 2022 Smart City Symposium Prague (SCSP), Prague, Czech Republic, 26–27 May 2022; pp. 1–6.
34. Janoš, V.; Kříž, M. Pragmatic Approach in Regional Rail Transport Planning. *Sci. J. Silesian Univ. Technol. Ser. Transp.* **2018**, *100*, 35–43. [CrossRef]
35. Yang, X.; Qi, Y. Research on Optimization of Multi-Objective Regional Public Transportation Scheduling. *Algorithms* **2021**, *14*, 108. [CrossRef]
36. L'upták, V.; Droździel, P.; Stopka, O.; Stopková, M.; Rybicka, I. Approach Methodology for Comprehensive Assessing the Public Passenger Transport Timetable Performances at a Regional Scale. *Sustainability* **2019**, *11*, 3532. [CrossRef]
37. Tesla, J.; Horak, J.; Ivan, I. Frequency Analysis of Public Transport between Czech Municipalities. In Proceedings of the 15th International Multidisciplinary Scientific GeoConference SGEM, Albena, Bulgaria, 18–24 June 2015.
38. Petersen, T. Watching the Swiss: A Network Approach to Rural and Exurban Public Transport. *Transp. Policy* **2016**, *52*, 175–185. [CrossRef]
39. Santos, G.; Maoh, H.; Potoglou, D.; von Brunn, T. Factors Influencing Modal Split of Commuting Journeys in Medium-Size European Cities. *J. Transp. Geogr.* **2013**, *30*, 127–137. [CrossRef]
40. Daldoul, M.; Jarboui, S.; Dakhlaoui, A. Public Transport Demand: Dynamic Panel Model Analysis. *Transportation* **2016**, *43*, 491–505. [CrossRef]
41. Guzik, R.; Kołoś, A. The 2019 level of accessibility of Poland's county cities by public transport from rural areas. *Prz. Geogr.* **2021**, *93*, 181–206. [CrossRef]
42. Smolarski, M.; Jurkowski, W.; Raczyk, A. Bus and Train Connections between Towns in Lower Silesia under Different Operational Models: Competition or Complementarity? *Morav. Geogr. Rep.* **2019**, *27*, 31–40. [CrossRef]
43. Bieńczak, M.; Fierek, S.; Kiciński, M.; Kwaśnikowski, J.; Sawicki, P. Regional transportation model for the plan of sustained public transport development. *Logistyka* **2014**, 1673–1682.
44. Bieńczak, M.; Fierek, S.; Kiciński, M.; Kwaśnikowski, J.; Sawicki, P. Regional transportation model for the plan of sustained public transport development. *Pr. Nauk. Politech. Warszawskiej. Transp.* **2015**, *105*, 5–20.
45. Sejm, RP. The Act of 20 June 1992 on Entitlements to Discounted Travel on Public Collective Transport. Journal of Laws 1992.54.254; 1992. Available online: https://isap.sejm.gov.pl/isap.nsf/download.xsp/WDU19920540254/U/D19920254Lj.pdf (accessed on 10 May 2024).
46. Guzik, P.; Więckowska, B. Data Distribution Analysis—A Preliminary Approach to Quantitative Data in Biomedical Research. *J. Med. Sci.* **2023**, *92*, e869. [CrossRef]
47. Statistics Poland—Local Data Bank. Available online: https://bdl.stat.gov.pl (accessed on 15 May 2024).

energies

MDPI

Article

High-Capacity Energy Storage Devices Designed for Use in Railway Applications

Krystian Woźniak [1,2], Beata Kurc [3], Łukasz Rymaniak [4,*], Natalia Szymlet [4], Piotr Pielecha [2] and Jakub Sobczak [4]

[1] Łukasiewicz Research Network, Electrotechnical Testing Laboratory, Institute of Technology, Warszawska 181, 61055 Poznan, Poland; krystian.wozniak@doctorate.put.poznan.pl
[2] Faculty of Civil and Transport Engineering, Doctoral School, Poznan University of Technology, Piotrowo 3, 60965 Poznan, Poland; piotr.pielecha@doctorate.put.poznan.pl
[3] Faculty of Chemical Technology, Institute of Chemistry and Electrochemistry, Poznan University of Technology, Berdychowo 4, 60965 Poznan, Poland; beata.kurc@put.poznan.pl
[4] Faculty of Civil Engineering and Transport, Institute of Powertrains and Aviation, Poznan University of Technology, Piotrowo 3, 60965 Poznan, Poland; natalia.szymlet@put.poznan.pl (N.S.); jakub.sobczak@student.put.poznan.pl (J.S.)
* Correspondence: lukasz.rymaniak@put.poznan.pl

Abstract: This paper investigates the application of high-capacity supercapacitors in railway systems, with a particular focus on their role in energy recovery during braking processes. The study highlights the potential for significant energy savings by capturing and storing energy generated through electrodynamic braking. Experimental measurements conducted on a diesel–electric multiple unit revealed that approximately 28.3% to 30.5% of the energy could be recovered from the traction network, regardless of the type of drive used—whether electric or diesel. This research also explores the integration of starch-based carbon as an electrode material in supercapacitors, offering an innovative, sustainable alternative to traditional graphite or graphene electrodes. The carbon material was obtained through a simple carbonization process, with experimental results demonstrating a material capacity of approximately 130 F/g. To quantify the energy recovery, calculations were made regarding the mass and power requirements of the supercapacitors. For the tested vehicle, it was estimated that around 28.7% of the energy could be recovered during the braking process. To store 15 kWh of energy, the total mass of the capacitors required is approximately 245.1 kg. The study emphasizes the importance of increasing voltage levels in railway systems, which can enhance energy transmission and utilization efficiency. Additionally, the paper discusses the necessity of controlled energy discharge, allowing for the flexible management of energy release to meet the varying power demands of trains. By integrating high-voltage supercapacitors and advanced materials like starch-based carbon, this research paves the way for more sustainable and efficient railway systems, contributing to the industry's goals of reducing emissions and improving operational performance. The findings underscore the crucial role of these capacitors in modernizing railway infrastructure and promoting environmentally responsible transportation solutions.

Keywords: supercapacitor; energy; rail vehicle; recovery braking; starch

1. Introduction

1.1. Possibility of Energy Saving in Transport

The environment when it comes to vehicles is a major challenge in the development of transport. Manufacturers of machines and vehicles are constantly taking steps to minimize adverse effects on the environment. A helpful tool in this type of work are tests involving real operation conditions, which concern various groups of vehicles [1–3]. Determining energy consumption in real test conditions provides reliable results regarding the energy flow and operation of individual elements of the drive system [4,5]. One of the most promising solutions is regenerative braking. The use of braking energy can relieve the load on the electric motors in the drive or the combustion engine. The combustion engine produces emissions that affect and pollute the environment [6–8]. These capacitors also support power quality by stabilizing voltage levels, reducing voltage sags and surges, and minimizing harmonic distortion. This ensures a stable power supply for train systems, improving both operational reliability and safety [9–12].

As railway electrification expands, the integration of high-voltage supercapacitors is crucial for optimizing energy use, promoting environmental sustainability, and driving innovation in the rail industry. Their role in energy storage and power management will continue to support the development of efficient and eco-friendly rail transportation systems [13–17].

1.2. Materials Used in Supercapacitors

Supercapacitors, also known as ultracapacitors or electric double-layer capacitors (EDLCs), store energy through electrostatic charge accumulation rather than electrochemical reactions. The efficiency and performance of a supercapacitor are largely determined by the materials used for the electrodes, as well as the electrolyte that separates them. The basic characteristics of an ideal electrode material include high surface area, good electrical conductivity, and stability under cyclic conditions. Activated carbon is one of the most commonly used electrode materials for supercapacitors due to its high surface area and porosity. Its porous structure allows for maximum charge accumulation on the electrode surface. However, its relatively low conductivity and moderate energy density limit its performance compared to other materials.

Carbon nanotubes (CNTs) are another common electrode material due to their exceptional electrical conductivity, high surface area, and mechanical strength. CNTs enable faster charging and discharging, which is crucial for some high-power applications. However, their high production costs and challenges in scaling up the synthesis are significant drawbacks [18–20].

Graphene is a promising material for supercapacitors due to its excellent electrical conductivity, large surface area, and mechanical strength. It allows for both high energy and power densities. Despite these advantages, graphene synthesis can be expensive, and large-scale production remains a challenge [21–25].

Recently, starch-based carbon materials have gained attention as a potential alternative to conventional carbon materials. These materials are synthesized via starch carbonization, offering an environmentally friendly, low-cost option for electrode production. They can offer good electrochemical performance, although their energy and power densities are still being optimized for practical applications [26].

The pressing issue of increases in worldwide use of fossil fuels, in spite of environmental pollution has led to a growing need for solutions. For this reason, rapid growth concerning renewable energy sources and efficient energy storages has started. These include highly effective energy storage devices such as supercapacitors. These are used in everyday electronics as well as in military devices. Evaluating the performance of supercapacitors involves examining their electrochemical properties, which depend on the materials used for the electrodes and electrolytes [27]. The recent advancements in supercapacitor technology focus on charge storage mechanisms and materials for electrodes and electrolytes. The main takeaway is the inclusion of biomaterials in energy storing technologies due to their sustainability, eco-friendliness, degradability, cost-effectiveness, and promising potential (Figure 1).

Biomaterials have emerged as a promising and environmentally friendly option when it comes to developing supercapacitors, providing a contribution to the energy storage technology. These natural or synthetic materials, often derived from living organisms or inspired by biological systems, offer unique structural and chemical properties that make them attractive for use in supercapacitors [28].

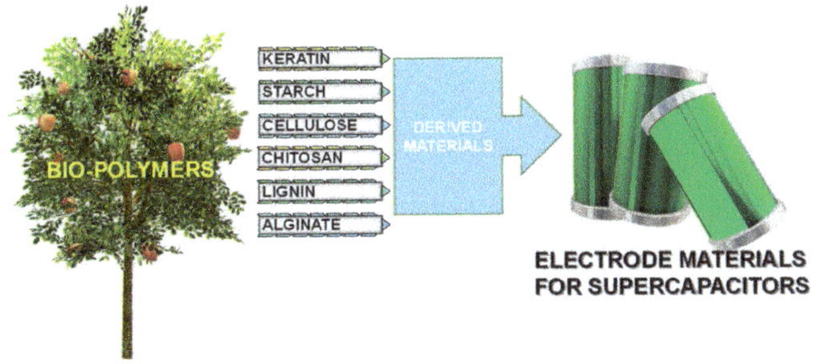

Figure 1. Application directions for biomass in a broad sense.

One key advantage of biomaterials in supercapacitors is their sustainability. As renewable resources, they can be sourced in an eco-friendly manner, reducing ecological impact in comparison with traditional materials. Additionally, the implementation of biomaterials fits the global trends highlighting green and sustainable energy solutions.

Biomaterials also exhibit remarkable properties that can enhance the performance of supercapacitors. Their high surface area and porosity allow for efficient ion adsorption and charge storage, leading to improved energy storage capabilities. Moreover, some biomaterials possess excellent electrical conductivity and stability, further enhancing the overall efficiency and lifespan of supercapacitor devices. Their ability to break down naturally after their useful life reduces the burden of electronic waste.

The motivation for using starch as a material for supercapacitor electrodes lies not primarily in energy density, but rather in its sustainability, cost-effectiveness, and availability. Although starch does have a lower energy density compared to some conventional electrode materials, it offers several advantages:

- ➢ Eco-friendliness and renewability—starch is a biodegradable, renewable material, making it an environmentally friendly choice. This is particularly valuable in reducing the ecological impact of energy storage devices, which is a growing concern in material science and engineering.
- ➢ High power density—starch-derived materials can deliver high power density, which is advantageous in applications where rapid charging and discharging are more critical than overall energy storage capacity.
- ➢ Cost-effectiveness—starch is an abundant and low-cost material. This can reduce the overall production cost of supercapacitors, making them more accessible for broader applications.

In summary, while starch may have limitations in terms of energy density, it brings unique benefits that align with the goals of sustainability, affordability, and high power output. These qualities make starch a promising candidate for specific supercapacitor applications where these factors outweigh the need for higher energy density.

Metal oxides (e.g., manganese dioxide, nickel oxide) and conductive polymers (e.g., polyaniline, polypyrrole) are also used in hybrid supercapacitors to enhance capacitance and energy density. While these materials provide improved energy storage, they often suffer from limited cycle life and lower power densities than carbon-based electrodes.

1.3. Challenges of Using Supercapacitors in Transport, Particularly Rail

While supercapacitors offer great potential for energy storage in transportation systems, especially in applications such as regenerative braking, there are several challenges that arise when implementing them in rail transportation.

- ➢ Energy density limitations

Supercapacitors typically have a lower energy density compared to batteries. This means that while they can charge and discharge quickly, they cannot store as much energy in a given volume or weight. For rail applications where large amounts of energy need to be stored and delivered (especially in long trains or heavy operations), this limitation may require the use of additional energy storage systems, such as batteries, in addition to supercapacitors [29].

➢ Cost

While supercapacitors are efficient and have a long cycle life, their cost can be prohibitive, especially for large-scale transportation systems. The use of advanced materials such as graphene or carbon nanotubes can further increase the cost of production. In addition, the need for a large number of capacitors to store sufficient energy for transportation applications increases the initial investment, which can be a barrier to widespread adoption [30–32].

➢ Voltage balancing

Supercapacitors often need to be connected in series to achieve higher voltage levels. However, when connected in series, individual capacitors can become unbalanced, reducing the overall system efficiency and lifespan. To address this problem, effective voltage equalization systems are required, which increases the complexity and cost of the system [33].

➢ Temperature sensitivity

Supercapacitors, like all electronic components, are sensitive to extreme temperatures. High temperatures can increase the rate of degradation and shorten the life cycle of supercapacitors, while very low temperatures can affect their performance. Rail transportation systems often operate under variable environmental conditions, making temperature stability a critical issue for the reliability and longevity of supercapacitors [34].

➢ Energy recovery efficiency

While supercapacitors are very efficient at recovering energy, the overall system efficiency is affected by the regenerative braking technology and the type of power conversion systems used. In rail systems, this can be complicated by the need to adapt supercapacitors to both electric and diesel–electric drives with different energy characteristics [35].

➢ Space and weight considerations

In rail transport, space and weight are often at a premium. The use of supercapacitors requires significant installation space, and the weight of the capacitors can also become a factor, especially in heavy trains where large amounts of energy need to be stored. In addition, large capacitors can affect the design of trains, especially in urban transit systems where space is limited [36].

➢ Integration with existing infrastructure

Rail networks are often built on older infrastructure, which may not be optimized for the integration of modern energy storage systems. Integrating supercapacitors into existing rail systems can require significant upgrades to both the electrical systems and control software, which can increase costs and require additional planning and testing.

➢ Controlled energy discharge

Supercapacitors discharge energy very quickly, which makes them ideal for short bursts of power, such as during braking. However, controlled and efficient energy discharge is crucial, particularly in rail systems, where trains have varying energy demands depending on acceleration, speed, and braking conditions. Without sophisticated energy management systems, ensuring that stored energy is effectively used or released back into the grid can be challenging.

While supercapacitors have significant potential for energy recovery and storage in rail transport, they pose a number of challenges that need to be addressed to optimize their performance. Selecting the right electrode materials, such as activated carbon, carbon nanotubes, or starch-based carbon, plays a key role in determining their performance. However, issues such as limited energy density, cost, temperature sensitivity, and integration with existing infrastructure need to be carefully managed. With ongoing advances in materials science and systems engineering, supercapacitors could play an increasingly important role in the future of sustainable rail transport.

2. Methods and Materials

2.1. Synthesis of Carbon Material from Starch

Starch-derived carbonaceous materials are obtained by the carbonisation process. The synthesis takes place in an environment of inert gas such as nitrogen by using a Nabertherm pipe furnace. The obtaining procedure starts with heating the furnace and then carbonisation takes 4 h. Every hour the temperature is increased by 300 °C. The whole process takes just over 6 h. During carbonization, usage of nitrogen gas reaches 50 L/h. In addition, the air in the chamber is replaced by an inert gas before the process begins. A measuring system indicated that technical gas stood for 99.998% of atmosphere during the process. The illustration (Figure 2) depicts the actual structure of the carbon material produced post-starch carbonization, as well as the appearance of the tube after the completion of the process.

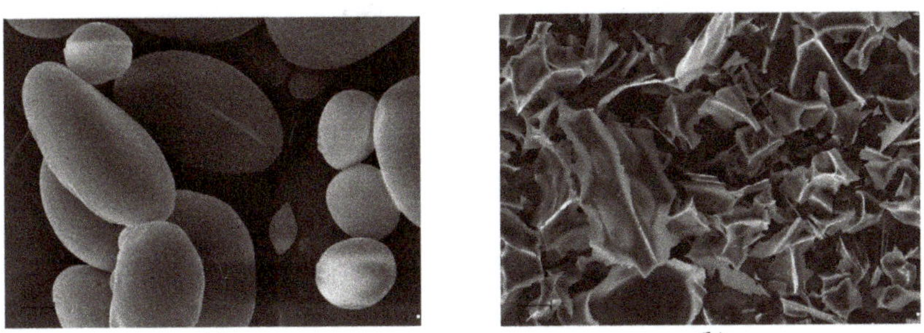

(a) (b)

Figure 2. SEM of starch: (**a**) before, (**b**) after carbonization.

2.2. Research Object

The estimation of the volume of electric energy recoverable by electrodynamic braking was taken on actual runs on the test track. A diesel–electric multiple unit, shown in Figure 3, was chosen for testing. The general characteristics and selected parameters of the test vehicle are shown in Table 1.

Figure 3. Three car body diesel–electric multiple unit.

Table 1. Technical characteristics selected for testing.

Technical Data	Value
Track width [mm]	1435
Overall length with bumpers [mm]	78,600
Top speed—electric drive [km/h]	160
Top speed—diesel drive [km/h]	120
Axle arrangement	2′Bo′ + Bo′2′+ Bo′2′
Traction motor power [kW]	6 × 300
Total power of combustion engine power [kW]	900
Braking resistors maximum power (RH1, RH2, RH3) [kW]	3 × 1000
Number of seats for passengers	156
Mass of vehicle [kg]	171,332

The test was performed in real operating conditions on the test track. Based on the tests performed, the necessary data for the selection of supercapacitors were defined. The tests were carried out on the experimental track of the Railway Institute in Zmigrod. Tests were carried out for electric traction up to 160 km/h and diesel traction up to 120 km/h. During each test, the vehicle accelerated at maximum power and then decelerated with maximum braking force to a stop. This provided information on the volume of electrical energy drawn from the overhead line and the energy that could be recovered during electrodynamic braking. The energy recovered in the braking process was transferred to the braking resistors and converted to heat. The recuperation of electrical energy into the overhead line was blocked. The measuring system consisted of current and voltage transducers connected to digital recorders prod. DEWESoft DewesoftX 2024.4. The measuring system included the following:

- Current and voltage at the pantograph before the main switch;
- Current and voltage at braking resistors;
- Current and voltage at the output of traction rectifier;
- Speed from the GPS device and from sensors installed on the axles.

2.3. Electrochemical Properties

The following components were used to form carbon electrodes: carbon material, binder: polyvinylidene fluoride (PVdF—Sigma Aldrich—St. Louis, MO, USA), acetylene black (Sigma Aldrich—USA). To obtain equal electrode masses, the mixture is formed in an agate mortar. The next step is to transfer the mass to copper interceptors with a unit weight of 4 mg. The electrodes are subjected to a temperature of 105 °C and paired together according to the criterion of the nearest mass value.

To prepare the supercapacitor, the samples were placed in a Swagelok® (Solon, OH, USA) measuring cell. In order to ease electrolyte (2 M Na_2SO_4—Sigma Aldrich) distribution by gravity flow, a working electrode with lower mass was placed in the position of the lower collector. To avoid unwanted contact between the electrodes a glass fibre isolator is used. After topping up with electrolyte, the system is sealed by adding and using the current collector of the counter electrode [37]. The components of the cell were connected using a dedicated Swagelok® connector.

Using a multi-channel Potentiostat/Galvanostat ZRA 6750 from Gamry Industries (Warminster, PA, USA), the electrochemical properties were assessed. In addition to the conventional method of assessing electrochemical properties by cyclic voltammetry, the equipment also offers other methods such as: linear voltammetry, stripping voltammetry, pulsed voltammetry, chronoamperometry, and more. A two-electrode symmetric system using processed starch was used to determine the results [37].

Cyclic voltammetry serves as a valuable tool to assess the reversibility of a redox reaction, categorizing it as either fully reversible, partially reversible, or irreversible. A fully reversible system is typically indicated when each successive measurement's current-versus-voltage graph aligns with the previous one. Conversely, when this alignment is absent, it suggests a partially reversible or irreversible reaction. Cyclic voltammetry finds extensive utility in various applications, including the following:

1. Identifying surface contaminants.
2. Estimating the relative specific surface area and surface roughness.
3. Determining the potential at which oxidation–reduction reactions take place.
4. Qualitatively analyzing specific substances.
5. Assessing the capacitance of the electrode.
6. Evaluating the kinetics of electron transfer.

Scanning electron microscopy is used to determine the morphological microstructure of grains. The studies were performed using an EVO40 (scanning electron microscope) (Zeiss, Jena, Germany). The microscope accelerates electrons in the range of 0.2 to 30 kV, allowing for a satisfactory distribution of the substance being scanned.

2.4. Supercapacitor Energy and Power

Supercapacitor energy refers to the stored energy in an electrical supercapacitor, a fundamental component in electronic circuits capable of storing and releasing electrical charge. The energy deposited in a condenser is a function of the electric charge it can hold and the voltage applied across its terminals. This energy storage capability makes capacitors essential in various electronic applications, including power supply filtering, energy storage, and signal coupling.

The energy required to charge a supercapacitor can be quantified as the work done in transferring the charge. Since the charge is transferred incrementally (from zero to the final charge, q), the differential component of the supercapacitor charging energy, dG, can be expressed as follows:

$$dG = \frac{\alpha q}{C} \cdot dq \tag{1}$$

where αq is the portion of charge accumulated in a given stage of charging the supercapacitor cell ($0 < \alpha < 1$), and C—the capacitance of the supercapacitor.

Taking into account that

$$dq = q \cdot d\alpha \tag{2}$$

the charging energy G (Gibbs energy) is equal to the following:

$$\Delta G = \int_0^q dG = \int_C^{\alpha q} q d\alpha \tag{3}$$

$$\Delta G = \frac{q^2}{C} \int_0^1 \alpha \cdot d\alpha \tag{4}$$

where the energy is equal to ($\Delta G = E_{MAX}$).

$$E = \frac{1}{2} \frac{q^2}{C} \tag{5}$$

The capacitance of the supercapacitor (C) is as follows:

$$C = \frac{dQ}{dU} \text{ or } \frac{q}{U} \tag{6}$$

where U—voltage between the electrodes [V].

Transforming Equation (6) and substituting it into Equation (5), we obtain the following:

$$E = \frac{1}{2} CU \tag{7}$$

The power (P) in an electrical circuit containing a supercapacitor can be described as the rate of change of energy stored in the supercapacitor with respect to time.

$$P = U \cdot I \quad (8)$$

The studied system was composed of two electrodes with an electrolyte between them; therefore, the supercapacitor cell can be considered a system of two series-connected supercapacitors, the capacity of which is expressed by the following formula:

$$\frac{1}{C} = \frac{1}{C_1} + \frac{1}{C_2} \quad (9)$$

where C_1 and C_2 are the capacities of individual electrodes.

Assuming that $C_1 = C_2 = C_{electrode}$ and after substituting into Formula (9), the following relationship is obtained:

$$C_{electrode} = 2C \quad (10)$$

The specific capacity per unit mass of activated carbon can be calculated from the following formula:

$$C_{sp} = \frac{C_{electrode}}{m} \quad (11)$$

where m—mass of a single electrode (mass of activated carbon) [g].

The capacity of a condenser cell is given by the following equation:

$$C = \frac{I}{dU/_{dt}} \quad (12)$$

where C—capacitance of the supercapacitor [F], I—current [mA], and dU/dt—voltage change over time [mV/s].

This formula indicates that the power in a capacitor circuit is directly proportional to the rate of change of charge with respect to time. Capacitors store energy in an electric field and this is associated with the flow of energy into or out of the capacitor over time. However, it is important to note that while capacitors store energy in an electric field, they cannot retain electrical energy for extended periods. Supercapacitors, on the other hand, operate on a different principle, generating charge through the formation of a double-electric layer. This distinction highlights the different mechanisms of charge accumulation and storage between capacitors and supercapacitors.

The data obtained from voltametric measurements were analyzed using Equations (10)–(12). In the capacitance calculations, consideration was given to the current value in the circuit being tested once the potential reached the midpoint of the predetermined level.

In a parallel supercapacitor setup, all supercapacitors share the same voltage across them, denoted as follows:

$$V_1 = V_2 = \ldots = V_n \quad (13)$$

where V_1 to V_n—the voltage at each respective condenser.

This voltage equals the input voltage that is applied to the parallel supercapacitor arrangement via the input wiring. Nevertheless, the charge stored in every supercapacitor varies and is contingent upon the capacitance of each supercapacitor, following the following formula:

$$Q_n = C_n + V_n \quad (14)$$

where Q_n—accumulated load in a condenser, C_n—the capacity of the condenser, and V_n stands for the voltage applied to the supercapacitor, equivalent to the voltage applied to the entire parallel supercapacitor block.

The total stored charge in the supercapacitor block, represented by Q, is distributed among all supercapacitors within the circuit:

$$Q = Q_1 + Q_2 + \cdots + Q_n \tag{15}$$

The equation used to ascertain the equivalent capacity for a parallel arrangement of several supercapacitors is as follows:

$$C_{eq} = \frac{Q}{V} = C_1 + C_2 + \cdots + C_n \tag{16}$$

where C_{eq}—equivalent capacitance of the parallel supercapacitor arrangement, V—voltage applied to the supercapacitors via the input wires, and Q_1 to Qn—the charges kept at each respective supercapacitor.

This leads to a pivotal point, which is the following:

$$C_{eq} = C_1 + C_2 + \cdots + C_n \tag{17}$$

This implies that the equivalent capacitance of a parallel supercapacitor connection is equal the total of the respective capacitances. This outcome makes intuitive sense, as supercapacitors in parallel can be conceptualized as a singular supercapacitor with a plate area equivalent to the sum of individual plate areas.

3. Results and Discussion

3.1. Measurement Results of Railway Vehicle Electricity Consumption Estimation

The test results are presented in Table 2 and Figure 4. The purpose of the measurements is to estimate the energy consumption that can be recovered during electrodynamic braking for a single cycle, consisting in accelerating a stopped vehicle to the maximum speed and stopping it using electrodynamic braking.

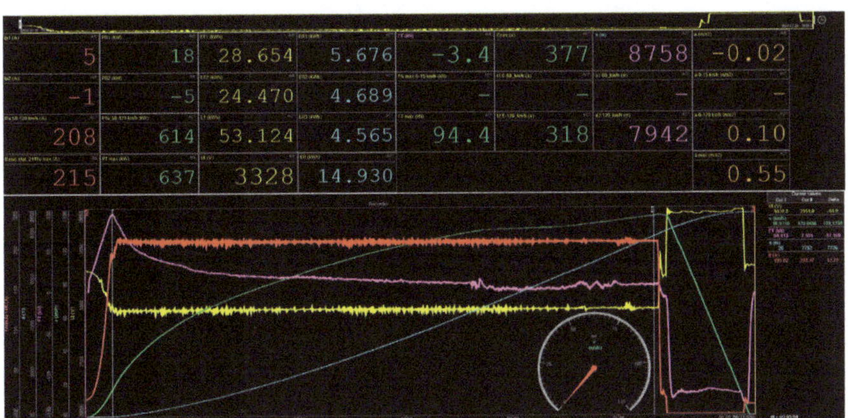

Figure 4. Measurement results for sample.

The total electrical energy ET was the energy consumed by the vehicle from the traction network ET and ER released during the recuperation process. Based on experimental measurements, the volume of energy that could be recovered in the electrodynamic braking process was estimated. In the case of the tested vehicle, regardless of the type of electric or diesel drive used, it was possible to recover approximately 15 kWh of the energy taken from the traction network in the electrodynamic braking process. Measurements carried out for a two-unit diesel–electric multiple unit according to the described methodology produced similar results. The energy recoverable through electrodynamic braking was 28.5%.

Table 2. Traction parameters for load condition.

No.	Specific Parameters							Time (t) and Distance (s)				Total Distance	Energy Consumed by the Vehicle E_T and Returned to the Braking Resistors $E_{R1} + E_{R3} + E_{R3} = E_R$ Brake Setting "P"—Passanger						
								60 km/h		120 km/h									
	I_{Tu} (1)	P_{Tmax} (2)	P_{Tu} (3)	F_{Nmax} (4)	a_{max} (5)	a_v (6)	a_v (7)	t_1	s_1	t_2	s_2	s	E_T	E_{R1}	E_{R2}	E_{R3}	E_R	$E_T - E_R$	E_R/E_T
	A	kW	kW	kN	m/s^2	m/s^2	m/s^2	s	m	s	m	m	kWh	kWh	kWh	kWh	kWh	kWh	%
1	208	636	612	96	0.56	0.30	0.11	54	513	306	7 518	8 522	50.72	5.83	4.89	4.63	15.35	35.37	30.27
2	207	636	611	95	0.55	0.28	0.10	55	513	318	7 839	8 776	52.29	5.74	4.83	4.64	15.21	37.07	29.09
3	208	646	616	96	0.56	0.30	0.10	56	528	324	8 120	8 938	53.95	5.70	4.80	4.53	15.02	38.93	27.84
4	208	640	615	94	0.55	0.29	0.11	56	521	313	7 869	8 669	51.91	5.81	4.87	4.59	15.27	36.64	29.41
5	208	642	613	92	0.53	0.30	0.10	57	563	319	7 790	8 731	53.66	5.66	4.90	4.64	15.20	38.47	28.32
6	208	637	614	94	0.55	0.30	0.10	57	550	318	7 742	8 758	53.12	4.69	4.57	4.57	13.82	39.31	26.01
7	208	640	613	95	0.55	0.30	0.11	56	533	317	7 743	8 747	52.79	5.65	4.82	4.57	15.03	37.75	28.48
Av. values	208	640	613	95	0.55	0.30	0.10	56	532	316	7 803	8 734	52.63	5.58	4.81	4.60	14.98	37.65	28.49

(1) Average current during start-up (for speed 50 ÷ 120 km/h); (2) the maximum value of the power consumed by the vehicle during starting (for speed 0 ÷ 120 km/h); (3) average power value during start-up (for speed 50 ÷ 120 km/h); (4) maximum value of the driving force during the start-up (for speed 0 ÷ 120 km/h); (5) maximum acceleration value during start-up (for speed 0 ÷ 120 km/h); (6) average acceleration value during start-up (for speed 0 ÷ 15 km/h); (7) average acceleration value during start-up (for speed 0 ÷ 120 km/h).

3.2. Development of Electrode Material from Starch

Devices featuring starch-based electrodes, much like their carbon–graphite counterparts, exhibit operational capabilities up to a maximum voltage of 2 V [38–41]. Notably, a distinct alteration in the cyclic voltammetry (CV) curves of supercapacitors employing starch-based electrodes becomes evident, clearly signifying an enhancement in charge propagation within each device following the introduction of acetonitrile into the electrolyte (Figure 5).

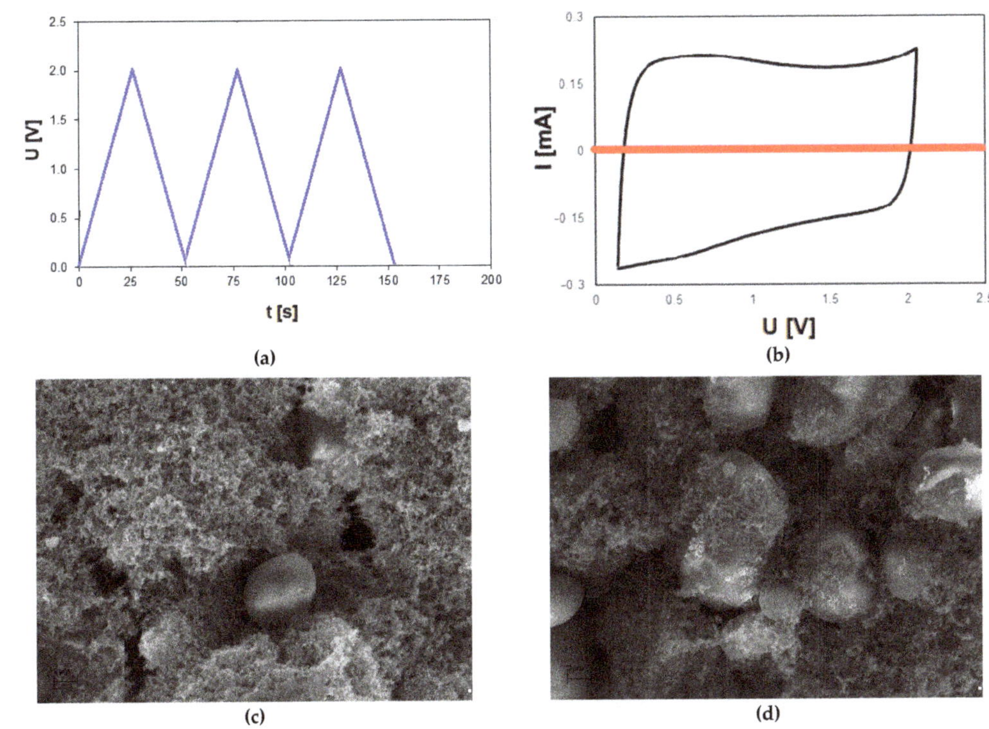

Figure 5. Electrochemical characterization of the supercapacitors. (**a**) Galvanostatic charging and discharging curves of the supercapacitor measured at current density from 100 mA/g. (**b**) Cyclic voltammogram curves measured at a scan rate of 100 mV/s (**c**). (**d**) SEM images of carbon materials.

This observable change in the CV curves is indicative of an improved charge transfer process within these devices. Furthermore, the incorporation of ionic liquid leads to an augmentation in the capacity of these devices and a noticeable enhancement in their charge retention capabilities at higher potential cycling rates.

The obtained material has a capacity of up to 130 F/g. Activated carbons, especially commercial ones, are attractive electrode materials for EDLC supercapacitors due to economic reasons and the possibility of obtaining a highly developed internal surface with a controlled pore size distribution. Depending on the porous texture and chemical nature of the surface—factors influenced by the type of raw material and preparation conditions—the relative capacity of the electrical double layer of activated carbons varies widely, within the range of 15–50 $\mu F/cm^2$. Assuming a theoretical mean level of 25 $\mu F/cm^2$ and an internal surface area of 1000 m^2/g, it should be possible to achieve a capacity of 250 F/g. In practice, however, much lower values are obtained because of the limited availability of the largely microporous carbon surface for the electrolyte.

The internal area determined by the gas sorption method is very different compared to the electrochemically reactive surface accessible to electrolyte ions. For aqueous electrolytes, it can be assumed that pores are greater than 0.5 nm, determined by the nitrogen sorption method, and should be electrochemically accessible. This is due to the similar average (thermodynamic) size of the nitrogen molecule at 77 K and the hydrated ions (e.g., H^+, OH^-, and K^+). For aprotic media, such as propylene carbonate, the dimensions of the solvated ions are much larger, e.g., ~2 nm for BF_4^- or ~5 nm for $(C_2H_5)_4N^+$, and the micropores do not contribute to the capacity of the bi-layer. Therefore, the type of electrolyte must be matched to the pore characteristics of the material (or vice versa).

In practice, the situation is more complicated. The porous texture of the material, in addition to the "molecular sieve" effect described above, affects the overall ionic conductivity of the system. The movement capacity of ions in the pores (electrochemically accessible) varies from their mobility in the electrolyte volume. The mobility of ions in narrow pores is problematic. The most significant improvements are observed in devices utilizing electrolytes derived from ionic liquids, which inherently possess the higher specific conductivity in their pure form.

Supercapacitors can be connected with other components in series or in parallel. At times, it proves beneficial to join multiple supercapacitors in parallel to create a functional unit, as illustrated in the diagram. In such scenarios, it is crucial to understand the equivalent capacity of the parallel arrangement. This discussion is going to delve into the analysis of parallel supercapacitor connections and explore potential applications for such configurations.

Electrical energy is stored in supercapacitors in the form of electrical charge. By paralleling multiple supercapacitors, the resultant circuit can store greater amount of energy [42–45], given that the equivalent capacitance is the sum of the individual capacitances of all involved capacitors. This principle finds application in various scenarios:

- DC supplies may incorporate multiple parallel supercapacitors to enhance signal filtration and eliminate AC ripple. This method enables the utilization of smaller supercapacitors with superior ripple characteristics while achieving higher capacitance values.
- Certain applications demand capacity rates surpassing those offered by commercially available supercapacitors. Supercapacitor banks address this need, such as those employed for power factor correction with inductive loads or energy storage in automotive applications like KERS (Kinetic Energy Recovery System) for regenerative braking.

When connecting supercapacitors in parallel, it must be taken into account that the maximum rated voltage of the parallel connection of supercapacitors is equal to the lowest rated voltage of all supercapacitors involved. Therefore, care must be taken to ensure system compatibility and security. To determine how many supercapacitors are needed to store a total energy of 15 kWh, we can follow these steps:

- 1 kWh = 3600 kJ
- Therefore, 15 kWh = 15 × 360,015 × 3600 kJ = 54,000 kJ = 54,000,000 J.
- The energy stored in a supercapacitor is given by Equation (7), where E is the energy, C is the capacitance, and U is the voltage.
- Given that the capacitance C is 130 F/g, we need to know the voltage U at which the supercapacitors operate.
- The energy stored in one supercapacitor $E = 1/2 \times 130\,F \times U^2$; $U = 1.8$ V
- Simplifying, $E = 65 \times 1.82 = 220.32$ J/g.
- To determine the number of supercapacitors needed, we carry out the following:
    1. Total energy required = 54,000,000 J
    2. Energy per supercapacitor = $65 \times U^2$ J (assuming 1 g per capacitor)
    3. Number of supercapacitors N:

$$N = \frac{54,000,000}{65\,U^2} = \frac{54,000,000}{65\,1.8^2} = 256,041 \tag{18}$$

Thus, approximately 256,410 supercapacitors, each with a capacity of 130 F/g at 1.8 V, are needed to store 15 kWh of energy.

- To determine the total mass of supercapacitors, we consider the following:
  1. Total energy = 54,000,000 J
  2. Energy stored per gram = 220.32 J/g
  3. Total mass M:

$$M = \frac{54,000,000\ J}{220.32\frac{J}{g}}$$
$$M = 245,098.04\ \text{g} = 245.1\ \text{kg} \quad (19)$$

The total mass of the supercapacitors required to store 15 kWh of energy is approximately 245.1 kg.

4. Conclusions

This study presents a comprehensive exploration of energy storage using starch-derived carbon materials for supercapacitors, along with an analysis of energy recovery systems in railway vehicles and their potential applications. Starch carbonization provides an economical and efficient route to the production of carbon materials with favourable properties for energy storage applications.

The resulting materials exhibit capacitances of up to 130 F/g, making them competitive for use in supercapacitors. Starch-derived carbon materials show significant improvements in charge transfer and retention when combined with ionic liquids and optimized electrolytes. This indicates the importance of tailoring the electrolytes to the porous structure of the material. The internal surface area and pore size distribution play a key role in determining the specific capacitance, with changes in material processing and electrolyte selection required to maximize performance.

Measurements from electrodynamic braking tests in diesel–electric trains reveal that about 28.8% of the consumed energy can be recovered during braking. This highlights the potential of regenerative systems to improve energy efficiency. Supercapacitors are a viable solution for storing this recovered energy due to their rapid charge and discharge capabilities and long lifetime. This study found that approximately 256,041 supercapacitors (each with a capacitance of 130 F/g at 1.8 V) would be required to store 15 kWh of energy. This would correspond to a total mass of 245.1 kg. Parallel connections of supercapacitors enable capacitance aggregation, enabling practical applications such as DC signal filtering, power factor correction, and kinetic energy recovery systems.

While starch-derived materials are promising, the discrepancy between theoretical and practical capacitance values highlights the need for further research into optimizing pore accessibility and electrolyte compatibility. Further research should focus on increasing the energy density of supercapacitors to reduce the total mass and volume required for large-scale energy storage applications.

This study highlights the feasibility of integrating supercapacitor technologies into renewable energy systems and regenerative braking applications. Future work should investigate the scaling of these technologies and the development of hybrid systems combining supercapacitors with other energy storage technologies to address diverse application needs.

Author Contributions: Conceptualization, K.W., B.K., Ł.R. and N.S.; methodology, K.W. and B.K.; software, K.W. and B.K.; validation, formal analysis, K.W., B.K., Ł.R. and N.S.; investigation, B.K., J.S. and P.P.; data curation, K.W., B.K. and Ł.R.; writing—original draft preparation, K.W., B.K. and Ł.R.; writing—review and editing, N.S., J.S. and P.P.; visualization, K.W. and B.K.; supervision, Ł.R. and N.S.; project administration, K.W., B.K. and Ł.R.; funding acquisition, Ł.R. and N.S. All authors have read and agreed to the published version of the manuscript.

Funding: The study presented in this article was performed within statutory research (No. 0415/SBAD/0351).

Data Availability Statement: The original contributions presented in the study are included in the article. Further inquiries can be directed to the corresponding authors.

Conflicts of Interest: The authors declare no conflict of interest.

References

1. Andrych-Zalewska, M.; Chłopek, Z.; Merkisz, J.; Pielecha, J. Research on the Results of the WLTP Procedure for a Passenger Vehicle. *Maint. Reliab./Eksploat. I Niezawodn.* **2024**, *26*, 1–15. [CrossRef]
2. Rymaniak, Ł.; Merkisz, J.; Szymlet, N.; Kamińska, M.; Weymann, S. Use of Emission Indicators Related to CO2 Emissions in the Ecological Assessment of an Agricultural Tractor. *Eksploat. I Niezawodn.* **2021**, *23*, 605–611. [CrossRef]
3. Ziółkowski, A.; Fuć, P.; Jagielski, A.; Bednarek, M.; Konieczka, S. Comparison of the Energy Consumption and Exhaust Emissions between Hybrid and Conventional Vehicles, as Well as Electric Vehicles Fitted with a Range Extender. *Energies* **2023**, *16*, 4669. [CrossRef]
4. Pielecha, I.; PIelecha, J. Simulation Analysis of Electric Vehicles Energy Consumption in Driving Tests. *Eksploat. I Niezawodn.* **2020**, *22*, 130–137. [CrossRef]
5. Pielecha, I.; Szwajca, F. Cooperation of a PEM Fuel Cell and a NiMH Battery at Various States of Its Charge in a FCHEV Drive. *Eksploat. I Niezawodn.* **2021**, *23*, 468–475. [CrossRef]
6. Kęska, A. The Actual Toxicity of Engine Exhaust Gases Emitted from Vehicles: The Development and Perspectives of Biological and Chemical Measurement Methods. *ACS Omega* **2023**, *8*, 24718–24726. [CrossRef]
7. Merkisz, J.; Pielecha, J. *Nanoparticle Emissions from Combustion Engines*; Springer: Berlin/Heidelberg, Germany, 2015; ISBN 3-319-15928-3.
8. Ziółkowski, A.; Fuć, P.; Lijewski, P.; Jagielski, A.; Bednarek, M.; Kusiak, W. Analysis of Exhaust Emissions from Heavy-Duty Vehicles on Different Applications. *Energies* **2022**, *15*, 7886. [CrossRef]
9. Conzen, J.; Lakshmipathy, S.; Kapahi, A.; Kraft, S.; DiDomizio, M. Lithium Ion Battery Energy Storage Systems (BESS) Hazards. *J. Loss Prev. Process Ind.* **2023**, *81*, 104932. [CrossRef]
10. Gao, Z.; Zhang, Y.; Song, N.; Li, X. Biomass-Derived Renewable Carbon Materials for Electrochemical Energy Storage. *Mater. Res. Lett.* **2017**, *5*, 69–88. [CrossRef]
11. Kim, B.K.; Sy, S.; Yu, A.; Zhang, J. Electrochemical Supercapacitors for Energy Storage and Conversion. *Handb. Clean Energy Syst.* **2015**, 1–25. [CrossRef]
12. Ranaweera, C.; Kahol, P.; Ghimire, M.; Mishra, S.; Gupta, R.K. Orange-Peel-Derived Carbon: Designing Sustainable and High-Performance Supercapacitor Electrodes. *C* **2017**, *3*, 25. [CrossRef]
13. Qi, M.; Liu, Y.; Landon, R.S.; Liu, Y.; Moon, I. Assessing and Mitigating Potential Hazards of Emerging Grid-Scale Electrical Energy Storage Systems. *Process Saf. Environ. Prot.* **2021**, *149*, 994–1016. [CrossRef]
14. Rosewater, D.; Williams, A. Analyzing System Safety in Lithium-Ion Grid Energy Storage. *J. Power Sources* **2015**, *300*, 460–471. [CrossRef]
15. Wu, Y.; Zhang, T. Risk Assessment of Offshore Wave-Wind-Solar-Compressed Air Energy Storage Power Plant through Fuzzy Comprehensive Evaluation Model. *Energy* **2021**, *223*, 120057. [CrossRef]
16. Sawczuk, W.; Jüngst, M.; Ulbrich, D.; Kowalczyk, J. Modeling the Depth of Surface Cracks in Brake Disc. *Materials* **2021**, *14*, 3890. [CrossRef]
17. Sawczuk, W.; Merkisz-Guranowska, A.; Ulbrich, D.; Kowalczyk, J.; Cañás, A.-M.R. Investigation and Modelling of the Weight Wear of Friction Pads of a Railway Disc Brake. *Materials* **2022**, *15*, 6312. [CrossRef]
18. Dai, H. Carbon Nanotubes: Opportunities and Challenges. *Surf. Sci.* **2002**, *500*, 218–241. [CrossRef]
19. Frackowiak, E.; Gautier, S.; Gaucher, H.; Bonnamy, S.; Beguin, F. Electrochemical Storage of Lithium in Multiwalled Carbon Nanotubes. *Carbon* **1999**, *37*, 61–69. [CrossRef]
20. Wang, X.; Sun, L.; Susantyoko, R.A.; Fan, Y.; Zhang, Q. Ultrahigh Volumetric Capacity Lithium Ion Battery Anodes with CNT–Si Film. *Nano Energy* **2014**, *8*, 71–77. [CrossRef]
21. Wong, S.I.; Sunarso, J.; Wong, B.T.; Lin, H.; Yu, A.; Jia, B. Towards Enhanced Energy Density of Graphene-Based Supercapacitors: Current Status, Approaches, and Future Directions. *J. Power Sources* **2018**, *396*, 182–206. [CrossRef]
22. Ke, Q.; Wang, J. Graphene-Based Materials for Supercapacitor Electrodes—A Review. *J. Mater.* **2016**, *2*, 37–54. [CrossRef]
23. Chen, C.-M.; Zhang, Q.; Yang, M.-G.; Huang, C.-H.; Yang, Y.-G.; Wang, M.-Z. Structural Evolution during Annealing of Thermally Reduced Graphene Nanosheets for Application in Supercapacitors. *Carbon* **2012**, *50*, 3572–3584. [CrossRef]
24. Fan, Z.; Zhao, Q.; Li, T.; Yan, J.; Ren, Y.; Feng, J.; Wei, T. Easy Synthesis of Porous Graphene Nanosheets and Their Use in Supercapacitors. *Carbon* **2012**, *50*, 1699–1703. [CrossRef]
25. Li, C.; Zhang, X.; Wang, K.; Zhang, H.; Sun, X.; Ma, Y. Three Dimensional Graphene Networks for Supercapacitor Electrode Materials. *New Carbon Mater.* **2015**, *30*, 193–206. [CrossRef]
26. Rudge, A.; Davey, J.; Raistrick, I.; Gottesfeld, S.; Ferraris, J.P. Conducting Polymers as Active Materials in Electrochemical Capacitors. *J. Power Sources* **1994**, *47*, 89–107. [CrossRef]

27. Pigłowska, M.; Kurc, B.; Galiński, M.; Fuć, P.; Kamińska, M.; Szymlet, N.; Daszkiewicz, P. Challenges for Safe Electrolytes Applied in Lithium-Ion Cells—A Review. *Materials* **2021**, *14*, 6783. [CrossRef]
28. Kurc, B.; Pigłowska, M.; Rymaniak, Ł.; Fuć, P. Modern Nanocomposites and Hybrids as Electrode Materials Used in Energy Carriers. *Nanomaterials* **2021**, *11*, 538. [CrossRef]
29. Zhang, J.; Gu, M.; Chen, X. Supercapacitors for Renewable Energy Applications: A Review. *Micro Nano Eng.* **2023**, *21*, 100229. [CrossRef]
30. Gifford, P.; Adams, J.; Corrigan, D.; Venkatesan, S. Development of Advanced Nickel/Metal Hydride Batteries for Electric and Hybrid Vehicles. *J. Power Sources* **1999**, *80*, 157–163. [CrossRef]
31. Kouchachvili, L.; Yaïci, W.; Entchev, E. Hybrid Battery/Supercapacitor Energy Storage System for the Electric Vehicles. *J. Power Sources* **2018**, *374*, 237–248. [CrossRef]
32. Muhiuddin, M.; Khan, A.Z.; Devi, N.A.; Bharadishettar, N.; Meti, S.; Siddique, A.B.; Bhat, K.U.; Akhtar, W.; Rahman, M.R. Cost Effective Synthesis of Sulfur and Nitrogen Co-Doped Graphene Aerogel and Application in Binder Free Supercapacitor. *J. Appl. Phys.* **2024**, *136*, 034901. [CrossRef]
33. Barić, T.; Glavaš, H.; Barukčić, M. Impact of Balance Resistor Uncertainty on Voltages across Supercapacitors. *J. Energy Storage* **2019**, *22*, 131–136. [CrossRef]
34. Xiong, G.; Kundu, A.; Fisher, T.S. Influence of Temperature on Supercapacitor Performance. In *Thermal Effects in Supercapacitors*; Xiong, G., Kundu, A., Fisher, T.S., Eds.; Springer International Publishing: Cham, Switzerland, 2015; pp. 71–114. ISBN 978-3-319-20242-6.
35. Zou, Z.; Cao, J.; Cao, B.; Chen, W. Evaluation Strategy of Regenerative Braking Energy for Supercapacitor Vehicle. *ISA Trans.* **2015**, *55*, 234–240. [CrossRef] [PubMed]
36. Adeyemo, A.A.; Alves, E.; Marra, F.; Brandao, D.; Tedeschi, E. Suitability Assessment of High-Power Energy Storage Technologies for Offshore Oil and Gas Platforms: A Life Cycle Cost Perspective. *J. Energy Storage* **2023**, *61*, 106643. [CrossRef]
37. Kurc, B.; Lijewski, P.; Rymaniak, Ł.; Fuć, P.; Pigłowska, M.; Urbaniak, R.; Ciupek, B. High-Energy Solid Fuel Obtained from Carbonized Rice Starch. *Energies* **2020**, *13*, 4096. [CrossRef]
38. Liu, Y.; Chen, J.; Cui, B.; Yin, P.; Zhang, C. Design and Preparation of Biomass-Derived Carbon Materials for Supercapacitors: A Review. *C* **2018**, *4*, 53. [CrossRef]
39. Wang, X.; Xie, X.; Xie, K.; Liu, Q.; Li, Y.; Tan, X.; Dong, H.; Li, X.; Dong, Z.; Xia, Q. Chitin and Cuticle Proteins Form the Cuticular Layer in the Spinning Duct of Silkworm. *Acta Biomater.* **2022**, *145*, 260–271. [CrossRef]
40. Yang, X.; Fei, B.; Ma, J.; Liu, X.; Yang, S.; Tian, G.; Jiang, Z. Porous Nanoplatelets Wrapped Carbon Aerogels by Pyrolysis of Regenerated Bamboo Cellulose Aerogels as Supercapacitor Electrodes. *Carbohydr. Polym.* **2018**, *180*, 385–392. [CrossRef]
41. Zhuo, H.; Hu, Y.; Chen, Z.; Zhong, L. Cellulose Carbon Aerogel/PPy Composites for High-Performance Supercapacitor. *Carbohydr. Polym.* **2019**, *215*, 322–329. [CrossRef]
42. Cáceres, G.; Fullenkamp, K.; Montané, M.; Naplocha, K.; Dmitruk, A. Encapsulated Nitrates Phase Change Material Selection for Use as Thermal Storage and Heat Transfer Materials at High Temperature in Concentrated Solar Power Plants. *Energies* **2017**, *10*, 1318. [CrossRef]
43. Nazir, H.; Batool, M.; Osorio, F.J.B.; Isaza-Ruiz, M.; Xu, X.; Vignarooban, K.; Phelan, P.; Kannan, A.M. Recent Developments in Phase Change Materials for Energy Storage Applications: A Review. *Int. J. Heat Mass Transf.* **2019**, *129*, 491–523. [CrossRef]
44. Wang, Z.; Sun, S.; Lin, X.; Liu, C.; Tong, N.; Sui, Q.; Li, Z. A Remote Integrated Energy System Based on Cogeneration of a Concentrating Solar Power Plant and Buildings with Phase Change Materials. *Energy Convers. Manag.* **2019**, *187*, 472–485. [CrossRef]
45. Zhao, B.; Cheng, M.; Liu, C.; Dai, Z. Thermal Performance and Cost Analysis of a Multi-Layered Solid-PCM Thermocline Thermal Energy Storage for CSP Tower Plants. *Appl. Energy* **2016**, *178*, 784–799. [CrossRef]

energies

MDPI

Article

The Impact of Mechanical Failure of 18650 Batteries on the Safety of Electric Transport Operations

Henryk Bąkowski, Iga Przytuła *, Wioletta Cebulska, Damian Hadryś and Janusz Ćwiek

Faculty of Transport and Aviation Engineering, Silesian University of Technology, Krasińskiego 8, 40-019 Katowice, Poland; henryk.bakowski@polsl.pl (H.B.); wioletta.cebulska@polsl.pl (W.C.); damian.hadrys@polsl.pl (D.H.); janusz.cwiek@polsl.pl (J.Ć.)
* Correspondence: iga.przytula@polsl.pl

Abstract: The safety of 18650 lithium-ion batteries is critical for the reliability and durability of electric vehicles, especially as interest in sustainable transportation grows. Battery failures, such as fires or explosions, pose significant risks to both users and manufacturers, highlighting the need for advanced power systems. This study used finite element method (FEM) simulations and crash tests to evaluate battery safety in accident scenarios. The results showed that mechanical damage, especially from collisions, can cause internal short circuits, increasing the risk of thermal runaway, especially when combined with high temperatures during normal operation or charging. This can be caused by mechanical damage to the battery causing a change in the distance inside the battery, causing it to short circuit. The results highlight the importance of designing battery systems that prevent internal short circuits, especially under extreme conditions, and the need for continuous monitoring of battery parameters to detect early signs of failure. In the context of improving battery safety, the battery not only saves lives, but also extends vehicle life, reduces electronic waste, and increases energy efficiency, which is consistent with global efforts to minimize the environmental impact of technology and promote safer transportation.

Keywords: electromobility; sustainable transportation; battery safety; finite element method (FEM); crash tests

1. Introduction

Lithium-ion batteries have become the dominant solution for energy storage in electric vehicles (EVs), hybrid electric vehicles (HEVs), and smaller forms of electric transport, such as electric scooters and bicycles. Their widespread adoption can be attributed to their high energy density, which allows for extended driving ranges and more efficient use of vehicle space. This makes lithium-ion batteries particularly suitable for the growing demand in the electromobility sector, where energy efficiency, weight reduction, and compactness are crucial design parameters. However, while the high energy density of these batteries offers significant advantages, it also introduces inherent safety risks. One of the primary concerns is the battery's vulnerability to mechanical stress, which can occur during everyday usage or, more critically, in the event of an accident. Mechanical stress, such as that caused by impacts, punctures, or compression, can lead to internal short circuits within the battery cells. These short circuits, in turn, increase the likelihood of thermal runaway—a dangerous phenomenon where an uncontrolled rise in temperature can cause a chain reaction within the battery. This condition can lead to fires, explosions, and the emission of toxic substances such as gasses, posing serious risks to both vehicle occupants and the surrounding environment. To address these risks, ensuring the safety of lithium-ion batteries in EVs and HEVs is a key priority for manufacturers. A comprehensive approach that integrates rigorous testing and advanced simulations is essential for minimizing the likelihood of battery failures under various operational conditions. One of the most critical stages in the vehicle development process involves simulating potential crash scenarios to understand how battery systems might behave in real-world accidents. A numerical model was developed to simulate the complex conjugate heat transfer in the cell and simulation results were validated. To underscore the practical implications, a case study of a fire accident during a battery cooling experiment was presented in [1].

A widely adopted method for conducting these simulations is the finite element method (FEM). FEM is a numerical modeling technique that enables detailed and precise analysis of the physical responses of battery cells and modules during various types of impact scenarios. By using FEM, engineers can simulate how the battery components deform, fracture, or fail when subjected to different stressors, such as collisions or external forces. This method provides valuable insights into the structural weaknesses of the battery design and helps identify areas that require reinforcement [2].

In practical terms, FEM simulations enable engineers to optimize battery enclosure designs, improving their ability to absorb and dissipate energy during crashes. This includes enhancing the protective casings that surround battery modules, optimizing their placement within the vehicle, and integrating features that prevent mechanical damage from propagating through the system. For example, simulations can help identify the ideal locations for battery mounting, ensuring that the most vulnerable components are shielded from high-impact zones [3,4].

2. Materials and Methods

Lithium-ion batteries represent a relatively modern technology, first introduced in the early 1990s, that has since significantly influenced the energy storage sector. Over time, these batteries have progressively replaced older technologies and are now integral to a wide range of applications. Beyond their common use in consumer electronics such as smartphones and tablets, lithium-ion batteries have become a crucial component in the field of electromobility, powering electric vehicles (EVs), electric bicycles, and electric scooters [3,4].

When operated under normal conditions and in compliance with safety guidelines, lithium-ion batteries are generally considered safe. However, improper handling, storage, or usage can present considerable safety risks. Among the most critical hazards are electrical, thermal, and mechanical stresses, which can lead to dangerous failures. One of the most frequent risks associated with lithium-ion batteries arises during the processes of charging and discharging. Electrical overloads may occur due to several factors, such as the use of inappropriate charging equipment or exceeding the battery's designed capacity. Additionally, deep discharge is another common cause of battery failure, potentially leading to fire hazards. For instance, if lithium-ion batteries are left unused for extended periods, they can become fully discharged, which may damage the cells and increase fire risk [1,5].

Improper storage conditions also pose significant threats. Exposure to temperatures outside the range recommended by manufacturers can result in the degradation of the electrolyte, which releases highly flammable gasses. External heat sources, such as open flames, hot machinery parts, or prolonged exposure to direct sunlight, can heat the batteries, causing thermal overload. Such conditions heighten the risk of internal short circuits or even battery fires [5,6].

Mechanical damage is another critical safety concern when working with lithium-ion batteries. This can occur from collisions, falls, or inappropriate storage that leads to physical deformation of the cells. In the event of such damage, the structural integrity of the battery can be compromised, leading to internal short circuits, which significantly increase the risk of fires or explosions. Given these risks, ensuring the safe handling, charging, and storage of lithium-ion batteries is vital in preventing potentially catastrophic failures. The ongoing development of safety mechanisms, monitoring systems, and robust design strategies remains essential for mitigating these hazards, particularly as the use of lithium-ion batteries continues to expand in both consumer electronics and electromobility applications [5–7].

Mechanical damage can cause a variety of failures. Thermal failure of a single cell can cause a cascading effect, heating neighboring cells in the battery pack and causing a violent chain reaction. Once initiated, this process progresses rapidly and the battery can explode within minutes. Fires caused by lithium-ion batteries are extremely difficult to control due to the rapid spread of flames and the intensity of the combustion process [7,8].

One notable incident occurred in 2017, when a fire broke out in a parking garage in Hannover, attributed to the failure of an electric bicycle battery. Another tragic event took place in 2018, when a man in Hamburg lost his life following an explosion involving a battery charger. More recently, on 8 April 2024, a warehouse fire in Katowice, linked to the storage of Bolt electric scooters, highlighted the serious risks associated with lithium-ion batteries in commercial environments. These incidents demonstrate the potentially catastrophic consequences of battery failures, resulting in significant property damage, financial loss, and, in some cases, the loss of human life [8,9].

Despite these risks, lithium-ion batteries are generally regarded as safe under current manufacturing standards. Before being introduced to the market, lithium-based energy storage systems must undergo rigorous safety testing. One of the most crucial certifications is the UN 38.3 [10], which ensures that batteries can be safely transported. To achieve this certification, batteries are subjected to a range of tests that simulate various transport conditions, including pressure changes, temperature fluctuations, and mechanical stress.

To further enhance the safety of lithium-ion batteries, manufacturers often incorporate a variety of protective measures at the cell level. For instance, if a highly flammable electrolyte is used, flame-retardant additives may be introduced to reduce the risk of combustion. Additionally, robust design solutions, such as crash-resistant casings made from corrosion-resistant materials and insulated with fire-retardant foam, can significantly improve the safety profile of lithium-ion batteries. These measures are essential for mitigating the risks of thermal runaway and ensuring the safe operation of batteries, especially in electric vehicles and other high-stress environments [7–11].

Given the growing prevalence of lithium-ion batteries in everyday applications, including electric vehicles, bicycles, and scooters, the importance of implementing comprehensive safety protocols cannot be overstated. Manufacturers are increasingly focused on refining battery design to improve structural integrity and thermal management, while also investing in safety systems that can detect early signs of failure. For example, advanced monitoring systems are being developed to track temperature, voltage, and pressure fluctuations within the cells in real time, allowing for proactive intervention before a failure occurs [8].

Moreover, fire safety research continues to evolve, with a focus on developing more effective methods for containing and extinguishing lithium-ion battery fires. Given the challenges associated with such fires, including the rapid spread of flames and the potential release of toxic chemicals, fire suppression systems tailored to lithium-ion battery incidents are becoming a key area of focus for both researchers and manufacturers [12–15].

3. Working and Construction of a Li-Ion 18650 Cell

A 18650 lithium-ion battery cell, as illustrated in Figure 1, comprises three primary components: the jelly roll, the outer casing, and terminal parts. The jelly roll, which forms the core of the cell, consists of a tightly wound structure of alternating layers of anode, cathode, and separator materials. This cylindrical spiral configuration enables efficient ion transfer between the electrodes, which is essential for the cell's electrochemical performance. The casing, typically constructed from robust metal, encases the jelly roll to provide protection against external physical impacts and environmental factors that could impair the cell's functionality and reduce its lifespan [16,17].

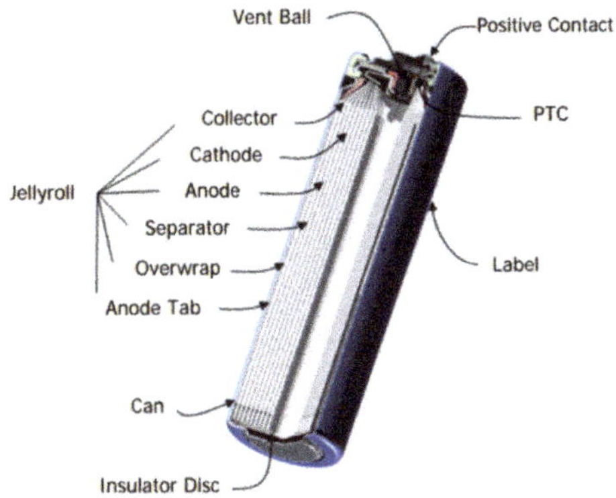

Figure 1. Structure of a cylindrical lithium battery [18].

At the positive terminal, several critical components are located, including a gasket, safety vent, and cathode cap. These features are crucial for maintaining the cell's integrity, especially the safety vent, which serves to release excess pressure in overcharge or fault conditions, thereby preventing potentially catastrophic outcomes. The negative terminal includes an anode cap and insulator, which are designed to stabilize and secure the electrical connections while also providing additional protection for the internal components.

The intricate design of the jelly roll is of particular importance. Its multilayered, compact structure exhibits anisotropic behavior under mechanical stress, meaning that the material properties can differ depending on the direction of the applied force. This anisotropy is a key factor in understanding the cell's mechanical response during operation or under external stress, as it influences heat generation, electrical performance, and overall safety. In high-demand applications, such as electric vehicles and portable electronics, these considerations are critical. The battery's ability to withstand mechanical deformation without compromising safety or efficiency is essential for ensuring both reliability and longevity [4–6,14].

Moreover, the design of 18650 battery cells must address the challenges posed by their operational environments, particularly in electric vehicles, where they are exposed to dynamic mechanical loads, temperature variations, and potential impact forces. The interaction between the jelly roll's layered structure and external forces can lead to uneven stress distributions, which may contribute to localized damage, internal short circuits, or even thermal runaway in severe cases. Understanding these failure mechanisms is essential for improving the safety features of lithium-ion batteries [13–16].

Research into advanced materials and structural optimizations continues to play a pivotal role in addressing these issues. For instance, innovations in separator technology and electrode materials aim to enhance the mechanical and thermal stability of the jelly roll, minimizing the risk of internal damage during high-stress events. Additionally, the development of new casing materials with improved energy absorption and impact resistance can further protect the internal components from external shocks.

Table 1 outlines the technical specifications of the 18650 lithium-ion battery, which will serve as the primary subject of the upcoming research. The battery's nominal capacity is listed at 2200 mAh, with a standard voltage of 3.6 V and a charge voltage of 4.2 ± 0.05 V. The physical dimensions are detailed as 65 mm in length and 18 mm in diameter, while the cell weighs approximately 45.0 ± 1.5 g. This battery demonstrates operational efficiency across a wide temperature range, from −20 °C to 60 °C, highlighting its resilience in various environmental conditions [18].

Table 1. Specifications of the test cell.

Elements	Specifications
Nominal capacity	2200 mAh
Normal voltage	3.6 V
Charge voltage	4.2 ± 0.05 V
Size	65 mm × 18 mm × 18 mm
Weight	45.0 ± 1.5 g
Operating temperature	−20~60 °C
Cathode material	$LiNiMnCoO_2$
Anode material	Graphite

The cathode is composed of lithium nickel manganese cobalt oxide (NMC), while the anode is made from graphite. These materials are specifically chosen for their ability to provide a well-balanced combination of energy density, cycle longevity, and operational stability. The selection of NMC for the cathode is particularly significant as it offers a compromise between energy output, safety, and lifespan, which are crucial parameters for high-demand applications such as electric vehicles and portable electronics. Graphite as the anode material complements these properties by contributing to the cell's overall efficiency and cycle performance [18].

These detailed specifications form the basis for the subsequent experimental studies, which will focus on the battery's performance metrics, such as energy efficiency and durability, particularly under thermal stress conditions. The robustness of the battery materials will also be tested to assess their stability during prolonged use in varied environments. The ability of the battery to withstand thermal fluctuations without significant degradation in performance is essential for its practical application, especially in fields where operational reliability and safety are paramount.

In addition, this research will investigate the cycle life of the battery, evaluating how the anode and cathode materials perform after repeated charge and discharge cycles. Given the increasing demands for energy storage solutions in electric vehicles and consumer electronics, optimizing both energy efficiency and longevity is critical. The data collected will provide valuable insights into the battery's long-term applicability in such contexts, guiding future advancements in lithium-ion battery technology. This foundational technical profiling will serve as a reference for the assessment of both current battery technology and the development of next-generation energy storage systems, aiming to improve safety, efficiency, and durability in various operational scenarios.

Based on the specifications outlined in Table 1, a simulation model of the 18650 lithium-ion battery has been developed, as illustrated in Figure 2. This model incorporates the battery's physical dimensions, electrical parameters, and material characteristics to replicate its behavior under various operational conditions. Accurate representation of these materials is critical to the model's ability to predict chemical reactions, physical interactions, and overall battery performance with a high degree of reliability.

The simulated environment created by this model enables the evaluation of potential degradation mechanisms, such as electrolyte decomposition, electrode wear, and separator failure. It also provides insights into the battery's thermal stability and how it responds to stress factors, including high temperatures and mechanical deformation. By incorporating material properties and thermodynamic behavior, the model can predict the effects of prolonged operational stresses, such as cycling at high charge and discharge rates, which are known to accelerate wear and decrease battery life.

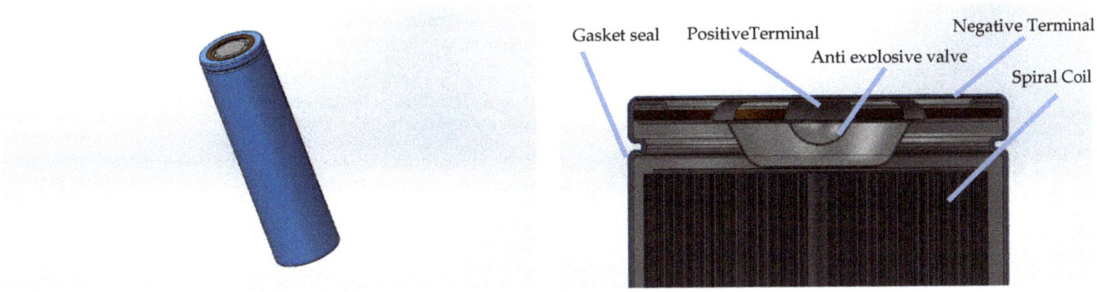

(**a**) (**b**)

Figure 2. Structural model of a 18650 battery: (**a**) frontal view; (**b**) cross-sectional view.

Through such simulations, the impact of thermal stress on the battery's lifespan can be analyzed in detail. The model is particularly useful for assessing thermal runaway risks by simulating scenarios in which local heat buildup within the cell could lead to catastrophic failure. The model allows for visualization of heat dissipation pathways, identifying areas within the battery where thermal management could be improved.

The model also plays a pivotal role in predicting long-term battery behavior through life cycle analysis. By simulating thousands of charge–discharge cycles, it can predict when and how the battery will degrade over time. This type of predictive modeling is essential for understanding the effects of usage patterns on battery efficiency, capacity fade, and overall performance degradation. Additionally, it supports the exploration of new design strategies that could improve battery longevity, such as the use of alternative electrode materials or enhanced cooling systems.

4. Research

The research part of the paper focuses on investigating the effect of dynamic loads on the electrical performance of a 18650 lithium-ion battery. In this study, finite element simulations, specifically a drop test module, were used to investigate how external mechanical stresses affect the battery's capacity, internal structure, and overall operational integrity. Simulation tests using FEM created an impact plane a short distance from the part by giving it a velocity according to Formula (1):

$$V = \mathrm{SQRT}(V0^2 + 2 \times g \times H) \quad (1)$$

where

V0—velocity after release (usually 0);

g—gravity value;

H—height from the release point to the model location.

The analysis parameters were edited and a small enough time step was defined so that the part would complete about 25 time steps before coming into contact with the impact plane. The analyzed model of the 18650 lithium-ion battery contained 10,598 built elements and 6641 nodes. Next, the gravity force was defined. The research was conducted using the drop test module using nonlinear material analysis.

The mesh quality is usually analyzed using two aspects: Aspect Ratio and Jacobian Check. In our model, we used a tetrahedral element with all mid-side nodes located exactly at the middle of the straight edges. The Jacobian ratio is given in the range of 0.8–1. The value of the Aspect Ratio should be kept below 50 in the regions where stresses are of crucial importance. In all other instances, the value should be limited to approximately 500. The conducted research met this criterion.

The simulations were set up with specific boundary conditions designed to reproduce scenarios in which the battery could suffer internal damage leading to short circuits.

This type of damage is critical because it directly affects the battery's ability to operate and may prevent its further use. The analysis provides new insights into how external conditions affect the safety and performance of lithium-ion batteries. This research is particularly important for improving the design and durability of battery systems used in electromobility and portable electronic devices, ensuring that these batteries can withstand and operate safely under various mechanical loads [14].

Tests were conducted at four different speeds—50 km/h, 90 km/h, 140 km/h, and 200 km/h—using a pneumatic valve drain device to launch objects at high speeds. Crashes at different speeds help assess the battery's response to impacts of varying severity, which is key to understanding its structural integrity and safety limits.

Understanding how a battery reacts under these controlled test conditions helps predict how it will perform in real-world scenarios, such as car crashes or handling and shipping incidents. These data serve as a starting point for improving battery designs to withstand physical impacts, improving safety and reliability. By using a tube to launch objects, testing could be conducted in a precise and controlled manner, further increasing the reliability of the results. Figure 3 shows the device used to carry out the crash tests.

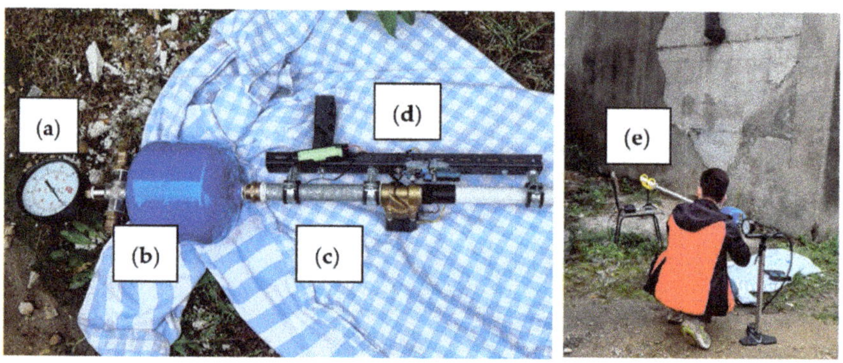

Figure 3. Pneumatic valve drain device: (**a**) pressure gauge, (**b**) pressure tank, (**c**) device barrel, (**d**) control system, and (**e**) recording system.

Figures 4–6 illustrate sample impact test results for a 18650 lithium-ion battery. These figures visually represent the strain, stress, and displacement that the battery experiences under these conditions. On the basis of the experimental studies, simulation studies were carried out which made it possible to determine the distribution of stresses, strains, and displacements. The drop test module allowed the determination of the above-mentioned parameters. Then, three levels of damage were adopted: the minimum, medium, and minimum levels of these parameters.

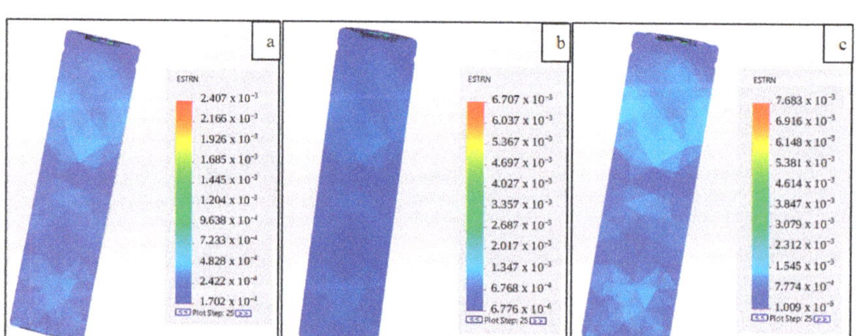

Figure 4. Strains caused by the impact of the battery: (**a**) minimum, (**b**) medium, and (**c**) maximum.

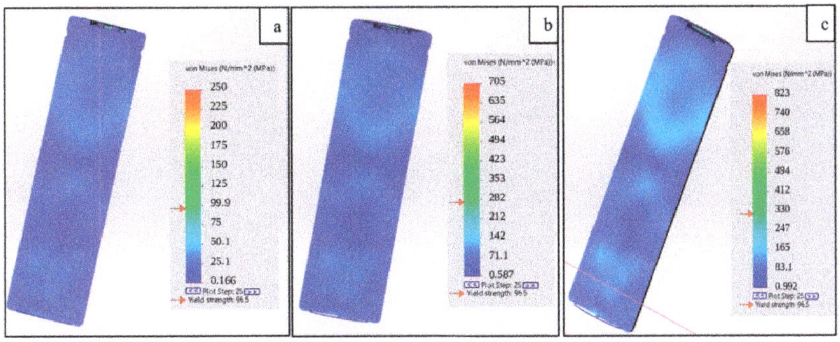

Figure 5. Stresses caused by the impact of the battery: (**a**) minimum, (**b**) medium, and (**c**) maximum.

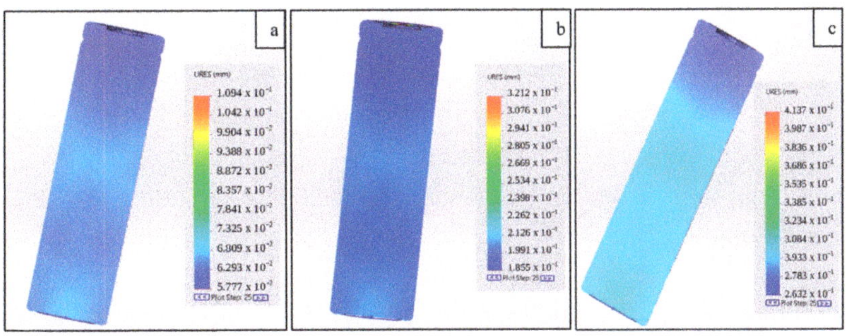

Figure 6. Displacements caused by the impact of the battery: (**a**) minimum, (**b**) medium, and (**c**) maximum.

The simulation tests presented the research results and determined the stress, strain, and displacement distributions and their values. It was found that with the increase in the impact speed, the stresses inside the battery increased. The values of stress, strain, and displacement change depending on the impact speed; therefore, the scale color remains unchanged, but the scale values change. The average stress values at the lowest value (7 J) are about 50 MPA, for the average (15 J) they are about 100 MPa, while for the highest (38 J) they are about 150 MPa.

Figures 7 and 8 show the damage to lithium-ion batteries after the crash tests at different speeds. In the context of battery damage research, the use of a thermal imaging camera can provide valuable information about the condition of the battery. The camera allows the temperature distribution on the cell surface to be monitored, which is crucial in assessing potential problems such as internal short circuits or overheating. After the crash tests, the temperature was monitored during the charging of the cells. The charging rate was 200 mA and the charging time was about 40 min. The results are shown in Figure 9.

Table 2 shows the parameters measured during the impact tests of the 18650 lithium-ion battery at different speeds. It shows the key parameters of the battery performance damage tests, including the speed during impact (km/h) and the parameters during charging of the damaged batteries such as power consumption (J), capacity (mAh), internal resistance (mΩ), charging current (mA), and cell temperature during charging (°C).

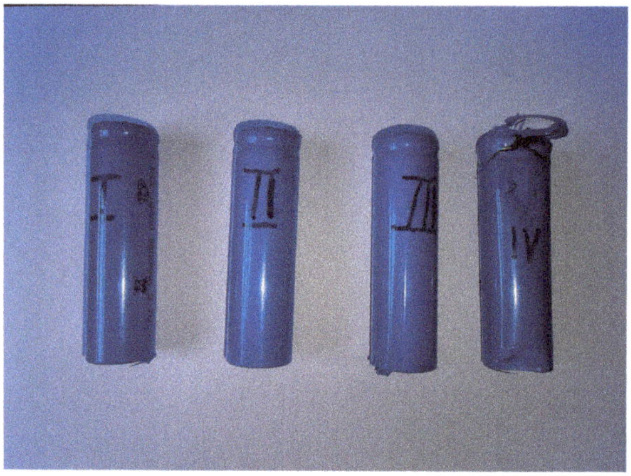

Figure 7. Batteries damaged as a result of collisions at the following speeds: I—50 km/h; II—90 km/h; III—140 km/h; IV—200 km/h.

Figure 8. Extensive battery damage after 200 km/h crash.

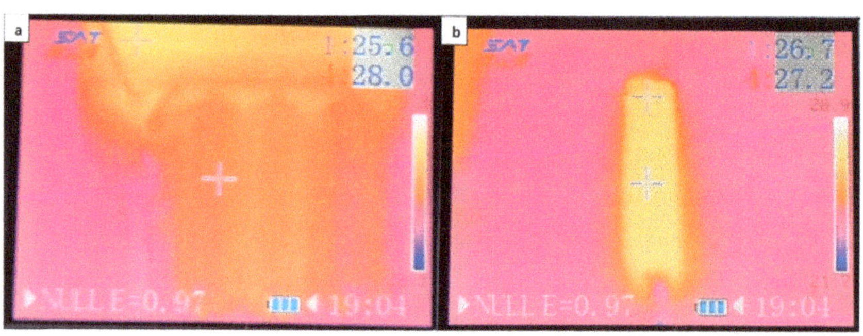

Figure 9. Temperature measurements using a thermal imaging camera: (**a**) battery charging after crash tests; (**b**) battery damaged at a speed of 200 km/h—no charging.

Table 2. Parameters measured after crash tests for the 18650 lithium-ion battery.

Speed (km/h)	Energy (J)	Capacity (mAh)	Internal Resistance (mΩ)	Charging Current (mA)	Cell Temperature during Charging (°C)
50	7	105	113	200	28.2
90	15	122	120	200	28.5
140	38	126	145	200	28.3
200	63	-	255	-	27.2

The impact energy increased with the speed at which the batteries were fired. At 200 km/h, this energy reached 63 J, which suggests that the batteries were seriously damaged under such conditions. The reduction in battery capacity with increasing speed is clear. At 140 km/h, the capacity was 126 mAh, while at 200 km/h it was no longer measured, which may indicate a complete loss of battery function. The battery was seriously damaged in a way that makes it impossible to recharge. This would damage the charger and pose a risk of leakage of electrolyte from the battery. In all cases studied, there was an increase in internal resistance as a result of mechanical damage. The internal resistance increased significantly with speed. At 200 km/h, the resistance was 255 mΩ, suggesting that the battery structure was seriously damaged, leading to significant energy loss and reduced performance. Despite the damage, the cell temperature remained stable during charging, suggesting the absence of internal short circuits, but it is still important to control the temperature with such serious damage.

To calculate the impact energy of a car in a crash at different speeds, we will use the formula for kinetic energy (2):

$$Ek = \frac{1}{2}mv^2 \tag{2}$$

Assuming the mass (m) of the car is 1500 kg and the speed (v) is 50 km/h, we obtain an impact energy of about 144 kJ. In summary, the distribution of impact energy in the car floor area is the result of a complex interaction between materials, the vehicle structure, and safety systems. Proper design of these elements is crucial to minimizing the effects of accidents and protecting occupants.

5. Conclusions

As the speed at which the cell damage occurs increases, the internal resistance increases, which leads to reduced efficiency. The battery capacity was reduced by the damage. In the 200 km/h crash, the damage was so extensive that the battery would not charge.

Despite this damage, the cell temperature remained stable, suggesting that there were no internal short circuits. This is an important aspect, because the absence of short circuits means a lower risk of dangerous situations such as ignition or explosion. However, it is worth noting that serious damage to the battery can occur at an impact energy of 63 J. In the case of electric vehicle crashes, this energy is significantly higher, which can lead to even more serious damage.

Following crash tests, the thermal imaging camera enabled the detection of areas of increased temperature during charging, which could indicate local damage. Real-time temperature measurement allows for faster detection of irregularities, which can help prevent further damage and improve the safety of energy systems. Analyzing the results from the thermal imaging camera in combination with other parameters is important to optimize battery performance and increase reliability. However, the tests did not show any internal short circuits. The temperature level was maintained at an appropriate level, so there was no risk of ignition. Understanding these mechanisms is crucial for optimizing energy systems. Analyzing parameters such as internal resistance, capacity, and temperature makes it possible to identify potential problems and implement appropriate solutions that can increase the efficiency and reliability of batteries. It is therefore important to control battery parameters and temperature. Lithium-ion 18650 batteries should be assembled into packs that are made of impact-absorbing plastic. In this way, the impact energy is dispersed into the structural skeleton of the pack and not directly onto the batteries.

Author Contributions: Conceptualization, I.P., W.C., H.B., D.H. and J.Ć.; methodology, I.P. and W.C.; software, H.B.; validation, I.P., W.C. and H.B.; formal analysis, J.Ć. and D.H.; resources, I.P. and W.C.; writing—original draft preparation, I.P. and W.C.; writing—review and editing, H.B., D.H. and J.Ć.; visualization, H.B., D.H. and J.Ć.; supervision, J.Ć. All authors have read and agreed to the published version of the manuscript.

Funding: The Article Processing Charge was financed under the European Funds for Silesia 2021–2027 Program co-financed by the Just Transition Fund—project entitled "Supporting the staff in intensifying scientific activities in the field of transport transformation towards a green and digital economy". Project number: FESL.10.25-IZ.01-03AF/23-00; Project number at the Silesian University of Technology: 12/010/FSD24/1161.

Data Availability Statement: All data are included in the article.

Conflicts of Interest: The authors declare no conflicts of interest.

References

1. Sutheesh, P.M.; Atul, A.P.; Nichit, R.B.; Bandaru, R. Numerical and experimental investigations on thermal performance of Li-ion battery during explosion. *Int. Commun. Heat Mass Transf.* **2024**, *159*, 107983. [CrossRef]
2. Stan, A.-I.; Swierczynski, M.; Stroe, D.-I.; Teodorescu, R.; Andreasen, S.J. Lithium Ion Battery Chemistries from Renewable Energy Storage to Automotive and Back-up Power Applications—An Overview. In Proceedings of the 2014 International Conference on Optimization of Electrical and Electronic Equipment (OPTIM), Bran, Romania, 22–24 May 2014. [CrossRef]
3. Nitta, N.; Wu, F.; Lee, J.T.; Yushin, G. Li-Ion Battery Materials: Present and Future. *Mater. Today* **2015**, *18*, 252–264. [CrossRef]
4. Dan, D.T. Empowering the Future: Exploring the Construction and Characteristics of Lithium-Ion Batteries. *Adv. Chem. Eng. Sci.* **2024**, *14*, 84–111. [CrossRef]
5. Ezemobi, E.; Silvagni, M.; Mozaffari, A.; Tonoli, A.; Khajepour, A. State of Health Estimation of Lithium-Ion Batteries in Electric Vehicles under Dynamic Load Conditions. *Energies* **2022**, *15*, 1234. [CrossRef]
6. Sakaebe, H.; Yamaki, J. *Lithium Batteries—Secondary Systems—Lithium-Ion Systems | Li-Ion Battery: Overview of Primary and Secondary Systems*; Elsevier eBooks; Elsevier: Amsterdam, The Netherlands, 2024. [CrossRef]
7. Wang, Y.; Xiang, H.; Soo, Y.-Y.; Fan, X. Aging Mechanisms, Prognostics and Management for Lithium-Ion Batteries: Recent Advances. *Renew. Sustain. Energy Rev.* **2025**, *207*, 114915. [CrossRef]
8. United Nations Economic Commission for Europe. *Manual of Tests and Criteria—Eighth Revised Edition*; United Nations: New York, NY, USA, 2023. [CrossRef]
9. Bugryniec, P.J.; Resendiz, E.G.; Nwophoke, S.M.; Khanna, S.; James, C.; Brown, S.F. Review of Gas Emissions from Lithium-Ion Battery Thermal Runaway Failure—Considering Toxic and Flammable Compounds. *J. Energy Storage* **2024**, *87*, 111288. [CrossRef]
10. Maccarelli, M. *Regulation on Machinery: Regulation (EU) 2023/1230*; Certifico S.r.l.: Perugia, Italy, 2023.
11. Yao, X.-Y.; Saxena, S.; Su, L.; Pecht, M.G. The Explosive Nature of Tab Burrs in Li-Ion Batteries. *IEEE Access* **2019**, *7*, 45978–45982. [CrossRef]
12. Zhang, X.; Wierzbicki, T. Characterization of Plasticity and Fracture of Shell Casing of Lithium-Ion Cylindrical Battery. *J. Power Sources* **2015**, *280*, 47–56. [CrossRef]
13. Pfrang, A.; Kersys, A.; Kriston, A.; Sauer, D.U.; Rahe, C.; Käbitz, S.; Figgemeier, E. Long-Term Cycling Induced Jelly Roll Deformation in Commercial 18650 Cells. *J. Power Sources* **2018**, *392*, 168–175. [CrossRef]
14. Babuška, I.; Guo, B.Q. The H, P and H-p Version of the Finite Element Method; Basis Theory and Applications. *Adv. Eng. Softw.* **1992**, *15*, 159–174. [CrossRef]
15. Sahraei, E.; Campbell, J.; Wierzbicki, T. Modeling and short circuit detection of 18650 Li-ion cells under mechanical abuse conditions. *J. Power Sources* **2012**, *220*, 360e37. [CrossRef]
16. Madani, S.S.; Ziebert, C.; Marzband, M. Thermal Behavior Modeling of Lithium-Ion Batteries: A Comprehensive Review. *Symmetry* **2023**, *15*, 1597. [CrossRef]
17. Liu, Y.; Pan, Y.; Wang, H.; Li, H.; Bao, H.; Zhao, Z.; Liu, B. Mechanical Issues of Lithium-Ion Batteries in Road Traffic Conditions: A Review. *Thin-Walled Struct.* **2024**, *201*, 111985. [CrossRef]
18. Energizer Brands, LLC. *Cylindrical Primary Lithium Handbook and Application Manual, Form No. 12004-A*; Energizer Brands, LLC: Saint Louis, MO, USA, 2008.

energies

MDPI

Article

Combined Rough Sets and Rule-Based Expert System to Support Environmentally Oriented Sandwich Pallet Loading Problem

Piotr Sawicki [1,*], Hanna Sawicka [1], Marek Karkula [2] and Krzysztof Zajda [3]

[1] Institute of Transport, Poznan University of Technology, 61-138 Poznań, Poland; hanna.sawicka@put.poznan.pl
[2] Department of Business Informatics and Management Engineering, AGH University of Krakow, 30-067 Kraków, Poland; mkarkula@agh.edu.pl
[3] Grupa Maspex, 34-100 Wadowice, Poland; k.zajda@maspex.com
* Correspondence: piotr.sawicki@put.poznan.pl

Abstract: A sandwich pallet loading problem represents a significant challenge in the logistics of fast-moving consumer goods (FMCG), requiring optimisation of load units (LUs) arrangements to minimise their number in transportation and warehousing processes, leading to an environmental responsibility of organisations. This study introduces an innovative approach combining Dominance-Based Rough Set Theory (DRST) with a rule-based expert system to improve the efficiency of the pallet loading and to provide sustainable development. Key criteria and attributes for the LU assessment, such as weight, height, and fragility, are defined. DRST is utilised to classify these units, leveraging its capability to handle imprecise and vague information. The rule-based system ensures an optimal arrangement of LUs by considering critical control parameters, thereby reducing LU numbers and mitigating the environmental impact of logistics operations, as measured by energy consumption. The proposed approach is validated using real-world data from the FMCG distribution company. Results demonstrate that integrating DRST with an expert system improves decision-making consistency and significantly reduces the number of LUs. This study shows a way to increase the level of environmental responsibility of the organisation by cutting energy consumption and delivering economic and social benefits through fewer shipments. For example, the approach reduces energy consumption for a customer order delivery by 40%, from 0.60 to 0.36 (kWh/pskm).

Keywords: sandwich pallet loading problem; rough sets; machine learning; dominance-based rough sets; learning to optimise; environmental responsibility

1. Introduction

There is extensive research highlighting the negative impact of logistics processes on the environment, including climate change, with substantial evidence pointing to transport as a major contributor to carbon emissions [1,2]. Many activities have been initiated to address this issue, including the adoption of environmentally friendly technologies, such as renewable energy sources, and the reduction in CO_2 emissions in logistics operations within supply chains [3,4]. According to the EU Directive 2023/1791, energy efficiency is a critical area for action, without which the EU's full decarbonisation goal cannot be achieved [5]. Although transport is considered a non-energy sector, it significantly impacts energy consumption and efficiency, accounting for more than 30% of final energy consumption [5]. According to the Sustainability Exchange [6], one litre of diesel fuel for a truck vehicle has an energy content of approximately 10 (kWh), while the Polish Standard (PN) specifies this value at 9.94 (kWh) [7]. Assuming average truck fuel consumption of 40 (dm^3) per 100 (km), the average value of energy consumption (EC) upon the above-mentioned Polish Standard (PN) [7] is 3.976 (kWh/km). Considering the standardised full truckload of 33 pallet spaces (ps) of 1 (EUR) pallet standard and 24 (t) load, the energy content can be estimated at 0.12 (kWh/pskm), that is, 0.12 (kWh) per 1 (ps) per 1 (km). Since total energy consumption in delivery depends on factors such as vehicle type, load weight, empty truck trips, route length, and type, consolidating freight orders and shipping them together rather than separately can yield better environmental and energy outcomes [8].

These findings highlight that the implementation of preventive activities can drastically reduce the harmful effects of logistics operations on the environment and energy consumption [2]. These preventive measures can be applied in three main areas, such as the utilisation of environmentally friendly technologies, the reduction in carbon emissions, and waste reduction [4]. One of the activities that can be applied in the former is the optimal planning of the pallet loading process using advanced tools. This can result in a more efficient use of loading equipment, such as forklifts, by reducing the number of internal transport movements. Furthermore, effective utilisation of vehicle cargo space can reduce the number of vehicles sent to customers, which is directly related to cutting carbon emissions, particularly in transport logistics.

Transport, being integral to various human activities and a key business sector for many companies, is one of the fundamental economic areas. Based on numerous sources [9, 10], the global freight transport sector is expected to grow at a compound annual rate exceeding 11% from 2022 to 2030. Many authors analysing this industry in Poland [11, 12] agree that advanced technologies, particularly Artificial Intelligence (AI), will play a critical role in shaping road transport. Polish transport companies can benefit from AI solutions to manage goods flows, including automated picking in warehouses, autonomous trucking and delivery, traffic control and reduced congestion, smart scheduling, and real-time route adjustment to speed up transport [13]. There are still many barriers in Poland to overcome, including the technology for autonomous fleets, which is still under development. Therefore, digitising transport management systems is recommended for Polish road transport companies [11], especially the large carriers that are the subject of this paper. The expected result is better fleet and goods flow management and improved business planning [11]. Adopting AI technology provides a competitive advantage tool or is at least used to build capabilities to effectively compete between transport companies around the world [12]. In addition, it is perceived as a standard for future activities to make transport operations socially, economically, and environmentally responsible. AI technologies can also be integrated with initiatives related to sustainable development. Kulkov et al. [14] state that these initiatives often require significant changes in business processes and operations, including transport companies, for example, workflow redesign, repetitive tasks automation, and efficiency of the decision-making process, to leverage the full potential of AI technologies. It can result in a positive effect on social, economic, and environmental outcomes [14], including optimised working hours, economic growth through innovation, and resource optimisation to mitigate air pollution.

Artificial Intelligence lacks a single definition due to its evolving nature, and its interpretation depends on the situation, the specific decision problem, and its wider context. Today, it is considered a topic for studying how machines simulate intelligent human activities [15]. Lu [16] defines AI as a collection of theories, methods, and techniques to help machines (especially computers) analyse, simulate, exploit, and explore human thought processes and behaviours. The beginning of this century in the field of AI was dominated by Machine Learning (ML) tools for data processing and analysis. This trend is still relevant [17]; for example, many AI-based language learning tools developed since 2015 utilise ML techniques [18,19]. The application of Machine Learning in road transport includes analysing traffic flow patterns [20], preventing accidents, and increasing user safety through the development of autonomous incident detectors, which is nowadays crucial [21]. Machine Learning techniques are critical in freight transport and logistics, solving value problems such as determining the number of containers to be loaded in transport modes, as well as state problems such as on-time deliveries [22]. Other popular application areas for ML techniques include the Internet of Things and Smart Cities [23–25], e-commerce [26–28], cybersecurity [29–31], and healthcare [32–34], including applications developed during the COVID-19 pandemic [35,36]. Three main approaches in ML can be distinguished, including [37–39]: supervised, unsupervised, and Reinforcement Learning. Some authors [40–42] also include a semi-supervised approach. The supervised approach is based on training the model with the input (training) data and output labels to find the relationship between them. As a result, it is possible to produce accurate labels for new data. In the unsupervised approach, data lack labels. It is assumed that patterns can be discovered. Reinforcement Learning involves an agent learning behaviour through trial-and-error interactions with a dynamic environment [39,43]. In this context, the environment becomes the source of data, replacing training data in the supervised approach. The aim is to maximise the cumulative reward. The semi-supervised learning links labelled and unlabelled data, combining supervised and unsupervised learning. There are many ML techniques, including [44,45]: decision tree, Naïve Bayes classifier, support vector machine, neural network, and k-Nearest Neighbour.

According to Lu [16], AI involves the computation and analysis of data in intelligent ways, focusing on two key streams of development within ML, such as Deep Learning and Big Data. Deep Learning allows computational models composed of multiple processing layers to learn representations of data with multiple levels of abstraction [46]. It is based on the design of deep neural networks, enabling the processing of large amounts of data. The network can automatically learn representations of higher-level features, facilitating the detection of complex patterns in the input data. Exploration of many layers of non-linear information processing leads to feature extraction and transformation in both supervised and unsupervised approaches [47]. Examples of Deep Learning algorithms include [15,48]: Restricted Boltzman Machine, Deep Belief Network, convolutional neural network, Autoencoder, and Long Short-Term Memory. The application of Deep Learning spans a wide range of areas, such as demand forecasting, production, retailing, and quality management issues in supply chain management [49]. Other examples include traffic performance prediction, traffic signal control, lane detection, and identification of objects near vehicles in transportation systems [50]. Furthermore, Deep Learning is applied in autonomous driving, driver behaviour analysis [51], and emergency logistics, such as optimising the distance travelled by truck fleets [52].

Big Data is a term used to describe a huge amount of data, consisting of diverse datasets that are difficult to process and analyse. Some authors [53–55] point out that Big Data can be interpreted in five dimensions: volume—the size of the data, velocity—the speed of incoming and outgoing data, variety—the source and type of data, variability—the inconsistency and trustworthiness of data, and value—the usefulness and importance of data. The aim of Big Data analysis is to reveal patterns and trends and to predict factors such as human behaviours, demand, weather, traffic flow, and traffic congestion [54]. Big Data analysis is usually based on cloud data processing, which requires efficient indexing techniques [56]. There are many different indexing approaches, including [56,57]: bitmap indexes, graph query processing, tree-based indexing, case-based data clustering, the Approximate Nearest Neighbour approach, and hash-based indexing. Big Data is applied in many fields, including public transportation problems—transit users' travel behaviour, transit service quality [58], traffic flow and safety, urban planning [59], crash detection and prediction, e.g., prediction of crash frequency, probability of crash occurrence [60]; manufacturing—real-time processing data, maintenance interval prediction based on machine attributes data [61], quality of raw materials, recycling, identification of labour errors, selection of suppliers, and outsourcing policies [62].

A common characteristic of many real-world problems is imperfect knowledge. There are also well-known AI methods and techniques that address this issue, such as [63, 64]: Fuzzy Set Theory and Rough Set Theory. The Fuzzy set is a mathematical object with a defined membership function that takes values in the range $[0, 1]$. Thanks to this membership function, the degree of belonging of the examined object to a given set is calculated. The function can be represented by various forms, including linear, trapezoidal, triangular, Gaussian, and S-shaped [65]. In Fuzzy Set Theory [66,67], it is assumed that the criteria values are not strictly defined, i.e., they are characterised by vagueness. Zimmermann [63] emphasises that this imprecision of the data does not mean a lack of information or the presence of random variables, but the absence of sharply defined criteria for class membership. Selected application areas of Fuzzy sets include ecological risk [68], safety and reliability systems [69], and production management systems, such as job shop scheduling, quality management, project scheduling, facility location and layout, aggregate planning, production and inventory planning, and forecasting [70,71]. Other applications include deep excavation risk assessment [72].

In Rough Set Theory, that is, a mathematical technique used for the analysis of imprecise and vague descriptions of objects in the classification context [64], approximations are fundamental [73]. According to Pawlak [73], each rough set has a pair of approximations: lower and upper. Objects belonging to the lower approximation are certainly members of this set, while objects classified in the upper approximation of the rough set possibly belong to it. The difference between the upper and lower approximation is called the boundary region [73]. There are many application areas of Rough Set Theory, including image processing; medical diagnosis; intelligent dispatching [74]; business failure prediction; financial risk management; financial investments, e.g., trading rules, financial indicators, and total index of stock exchange forecasting; fault diagnosis, e.g., valve faults and sensor measurements in assembly systems; sunspot recognition and classification; music and acoustics, e.g., violin quality grading and classification of musical instrument sounds [75,76]; collaborative logistics in the manufacturing industry [77]; and supplier selection with capacity of reverse logistics service [78].

Based on the considerations mentioned above, AI techniques are widely applied in many areas to solve different problems. One of them, which is the subject of this study, is the bin packing problem. This is a class of combinatorial problems that attempt to allocate items together into a container. Typically, the goal is either to pack a single container with maximum density or to pack all items using the minimum number of containers. The first paper on the packing problem is credited to Dantzig [79] in the 1950s, who introduced a mathematical model and discussed its applications. Since then, increasingly sophisticated research has been conducted in this area. In the last two decades, however, many review papers have been published on the status quo and future directions of this topic. These reviews concentrate on specific contexts of the packing problem, including two-dimensional strip packing [80,81], two-dimensional packing solved with meta-heuristics [82,83] or solved with swarm intelligence [84], rectangular bin packing [85], irregular, circle, and sphere shape packing [86,87], irregular two-dimensional packing [88], packing under uncertainty [89], and online three- or multidimensional packing [90,91]. Several review papers also focus on fusion packing with other decision problems, such as the cutting stock problem [92] and the vehicle routeing problem [93,94].

Today, real-world industrial and logistics problems focus on practical issues that are constrained by different factors, such as customer positioning requirements, full shipment requirements, load stability, and the minimisation of plastic wrapping [95]. All of these factors are connected to sustainability aspects, including social, economic, and environmental issues. Erbayrak et al. [96] introduce the concept of a product family unit, which focuses on items to be packed in the same bin due to the shared destination, handling requirements, and storage conditions. Most of the elements mentioned above relate to packaging and strive to meet customer expectations at a high standard. Optimising this process can affect product quality, save packaging and transportation costs, reduce warehouse storage space, and decrease the number of transportation instances [97]. This reflects the environmental responsibility of the organisations involved in this process.

Due to the procedural complexity of the loading and/or packaging optimisation problem, it is very often solved with some approximate techniques, mostly heuristics. Despite its proven effectiveness, it is not capable of solving real-world instances in a short amount of time. A promising approach in this area is to employ ML techniques to extend an optimisation procedure and make it more effective in the time domain. This is referred to as the Learn-to-Optimise (L2O) approach, e.g., Chen et al. [98], or the Machine Learning optimisation approach, e.g., Bruni et al. [99]. The idea is to apply an ML algorithm to learn the optimisation process over a set of training problem instances and generalise it to new problems. In the literature, there are examples of studies in various domains that employ ML either to find the optimal solution directly, e.g., Baldo et al. [100] for the knapsack problem and Bruni et al. [99] for the last-mile delivery, or to determine the optimal value of parameters used in solution procedures, e.g., da Costa et al. [101] for routeing problems. The subject of interest in this paper, focused on a Sandwich Pallet Design Problem (SPDP), is to find the optimal value of parameters applied to the sequence of two dependent optimisation procedures.

A detailed state-of-the-art analysis on the subject of this paper has been carried out, and the results obtained, as well as the research gap and objectives, are presented in the following section. The general finding is that the main issue of this paper, i.e., SPDP, becomes evident in the context of reducing packing and transport costs, the high level of customisation, and the growing needs of customers, as well as the environmental responsibility of companies, which is one of the key components of sustainable development. There are many stakeholders involved in the SPDP process, including employees in the company, drivers, customers, and even society, which is interested in environmental protection by reducing shipments, i.e., their harmful effect resulting from high energy consumption, noise pollution, and carbon footprint. However, SPDP has not been analysed in depth, as evidenced by a very limited number of publications, which only consider a poor set of constraints, such as the weight and height of stacked load units, without incorporating specific limitations, including compression strength. Furthermore, there is a lack of procedures supporting the optimisation process, which provides an opportunity for better SPDP solutions. The limitations mentioned above constitute a research gap. In this study, it is addressed by adopting a comprehensive approach to the SPDP solution, including constraints such as the weight and height of load units, the fragility of products, and the different dimensions of pallets. Additionally, this approach provides supporting procedures such as ML, rule generation, and rule-based decision-making, creating possibilities to find better solutions to the problem of stacking palletised load units.

The objective of this paper is to provide a solution to a problem of stacking palletised load units that addresses multiple constraints, such as customer expectations regarding these load units, for example, their weight and height; technical and load safety aspects, e.g., compression strength; product properties, e.g., fragility; and striving to reduce operating costs and energy consumption by planning stacking palletised load units using optimisation methods and supporting techniques. This contributes to the environmental responsibility of companies. The application of the proposed methodology significantly reduces energy consumption due to the decreased number of stacking palletised load units and the reduced number of shipments.

The remaining part of the paper is organised as follows. Section 2 provides a state-of-the-art review on the multi-level pallet problem, specifically related to the Sandwich Pallet Loading Problem and the application of Machine Learning techniques to tackle loading issues. Section 3 presents information on the concept of Dominance-Based Rough Set Theory. In Section 4, the methodology for constructing a Learn-to-Optimise expert system supporting the loading process is proposed, including the procedural scheme and a discussion of its individual phases. Section 5 is dedicated to the experimental verification of the proposed methodology and a discussion of the results. Section 6 concludes the study, addressing the achieved outcomes, advantages, and limitations of the proposed methodology, and indicating directions for future work.

2. Related Work

Aligned with the subject of this paper, the primary focus of this section involves a literature analysis in two key directions. Firstly, a literature review addressing the stacking of elementary load units is presented. Secondly, the application of the Learn-to-Optimise approach is thoroughly examined in the broader context of packaging problems. These areas collectively aid in identifying a research gap, the exploration of which constitutes the central theme of this paper.

2.1. Multi-Level Bin Packaging and Sandwich Pallet Loading Problems

The multi-level pallet problem has been present in the FMCG industry for several years. It involves finding a solution where palletised load units exhibit low values for key parameters, such as permissible weight or height. However, this issue does not arise from inefficiencies in order-picking procedures but rather stems from customer expectations. Consequently, there is a need to stack such units to improve these indicators, and a process commonly referred to as stacking or multi-level pallet design. It is a kind of Multi-Level Bin Packing Problem (MLBP), which represents a combinatorial optimisation challenge where items of varying sizes must be packed into bins arranged hierarchically across multiple levels. Unlike traditional Bin Packing Problems (BPPs), where items are packed into single-level bins, the MLBP involves packing smaller bins into larger ones, forming a hierarchical structure. Each bin level has limited capacity, adding complexity to the optimisation task. The objective is to allocate items efficiently while minimising wasted space or the number of bins used, all while adhering to various constraints.

The MLBP is a type of multi-level pallet decision problem that has not been extensively analysed in the literature. Currently, two specific approaches can be identified, including the hybrid approach [102] and optimisation techniques, both exact [103–105] and approximate [104]. The associated structure is presented in Figure 1.

The MLBP was first discussed by Chen et al. [102] in 2019, introducing a data-driven approach to solve the problem in the manufacturing industry as a type of hybrid approach. Using historical records, there are addressed challenges such as unknown geometric shapes and packing constraints. Their automatic packing framework combined dynamic programming for normal-size problems with a heuristic multi-level Fuzzy-matching algorithm for larger instances, achieving significant acceleration in packing execution time and workload reduction for packing workers.

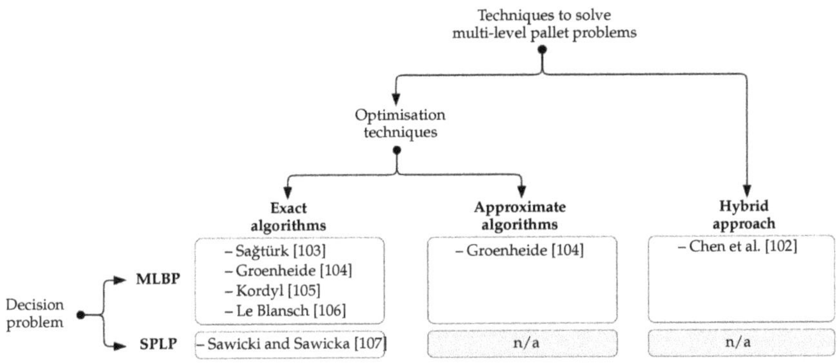

Figure 1. Quantitative techniques used to address multi-level pallet decision problems, including the Multi-Level Bin Packing Problem (MLBP) and the Sandwich Pallet Loading Problem (SPLP).

In contrast to previous work, a typical implementation of optimisation techniques for the MLBP is proposed by Sağtürk [103]. In this study, the MLBP and its variant with partial orders (MLBPPO) are addressed as NP-hard optimisation problems. The research introduces two integer programming models for each problem: one adapted from existing BPP implementations and another incorporating network flow constraints for enhanced linear programming relaxations. Experimental results demonstrate the effectiveness of these models, particularly for solving instances with up to 50 items and five levels in the MLBP. In another study, Groenheide [104] evaluates the performance of solving the MLBP using integer linear programming (ILP) formulations, contrasting this approach with heuristic-based algorithms. It is found that, while simple heuristics produce suboptimal solutions, the computational time required to find optimal solutions with ILP remains impractical for real-world instances. Consequently, the study underscores the need for more specialised algorithms that strike a balance between optimality and computational efficiency, tailored to specific applications. Kordyl [105] explores the MLBP and its variant with class constraints (CCMLBP) using integer programming (IP). These problems aim to minimise the total cost of the bins used while adhering to various constraints. By formulating the problems as IP models and evaluating their performance, the analysis provides insights into effective approaches for solving these complex optimisation issues and assesses the capability of the solution procedures. The research by Le Blansch [106] addresses the MLBP with time windows (MLBPTW), an extension of the classic BPP. This problem involves packing items into bins without exceeding capacity, with additional constraints such as multiple bin levels and time windows for each item. Similarly to Sağtürk [103], Le Blansch proposes IP models to solve both the MLBP and MLBPTW variants, aiming to minimise the cost of used bins while meeting all constraints. By comparing the performance of different IP models across various problem instances, the study identifies the most effective approach to find the optimal solution.

A specific variation of the MLBP involves a scenario in which, instead of multi-level bins, a set of stacked pallets is considered. The key difference lies in the fact that bins provide additional structural strength to the multi-level arrangement and are typically of regular shape, whereas a sandwich consists of pallets stacked on top of one another, with the load primarily supported by the products on the pallets. This problem is referred to as the Sandwich Pallet Loading Problem (SPLP) and is discussed by Sawicki and Sawicka [107].

Their study addresses the optimisation of the construction of Stacked Pallet Load Units (SPLUs), also known as sandwich pallets, from elementary Pallet Load Units (PLUs), focusing on minimising space requirements while adhering to constraints related to the weight, height, and fragility of the products. The paper proposes a single-criterion binary programming model to achieve this optimisation, considering constraints on the fragility of goods and the stacking potential of PLUs. The first constraint addresses the compressive strength of the product at the PLU level, while the second constraint evaluates whether the upper layer of an elementary PLU is robust enough to support additional PLUs. Through experimental verification on a test dataset, the study demonstrates the effectiveness of the optimised SPLU construction in achieving space savings. In general, the paper presents a methodical approach to optimising the design of stacked pallet load units, offering practical benefits in distribution and production systems.

2.2. Machine Learning Techniques Applied to Packaging Problems

In Section 1, the growing interest in the Learn-to-Optimise (L2O) approach for solving decision-making problems across various domains is highlighted. This paper focuses on the Packing Problem (PP), with particular interest in examples of L2O applications in this context. These applications are seen as opportunities to better tailor optimisation procedures to real-world constraints, develop more efficient computational processes, and explore and utilise large datasets. As mentioned earlier, it is a very recent approach. To date, only one comprehensive review on the application of Machine Learning in the multi-dimensional BPP domain has been published by Wu et al. [108]. This study encompasses papers published up to 2023, reviewing a corpus of 21 scholarly contributions. It begins with works by Vinyals et al. [109], Bello et al. [110], and Hu et al. [111] from 2015 and concludes with papers by Puche and Lee [112], Zhao et al. [113,114], and Yang et al. [115] up to 2023. Within this time frame, the research landscape primarily unfolds into two methodological streams: learning-based methods and hybrid methods. Learning-based methodologies solely rely on Machine Learning techniques, tackling a spectrum of BPP tasks, including online, offline, and miscellaneous variations. In contrast, hybrid methodologies amalgamate Machine Learning expertise with conventional heuristics, often incorporating Reinforcement Learning alongside packing heuristics or heuristic search algorithms. These approaches aim to synergistically harness the strengths of both Machine Learning and traditional optimisation methodologies to enhance performance in addressing the complexities of multi-dimensional BPP instances.

As a supplementary perspective, and based on the literature review excluding the problem analysed by Wu et al. [108], this paper identifies two generic areas of L2O application within the PP domain (see Figure 2). Firstly, it intervenes in the computational procedure and the formulation of the mathematical model to enhance the process of finding an optimal solution. Secondly, it supports the analysis and exploration of parameter values to achieve a more favourable optimisation outcome.

Among the instances of the learning phase intervening in the optimisation procedure, there is a group of studies employing partition principles in the PP domain, i.e., classification heuristics [99,116] and Fuzzy clustering [117]. Aylak et al. [116] address the classical pallet loading problem and propose a three-phase procedure. In the first phase, the total number of boxes to be loaded on each horizontal layer is identified using various heuristics, such as one-, three-, five-, hollow-block, and G5 heuristics. In the second phase, the total number of horizontal layers per pallet is calculated based on the available height of the pallet. In the final phase, several ML heuristics are applied for classification, e.g., neural networks, support vector machines, k-Nearest Neighbours, Naïve Bayes classifiers, decision trees, and random forests. ML heuristics are applied to train and then test how to arrange each layer according to a defined type of pallet to minimise the number of pallets used. Bruni et al. [99] consider the Variable Cost and Size Bin Packing Problem with Stochastic Items (VCSBPPSI). This strategic decision problem involves a shipper who needs to secure bin capacity from a carrier to meet uncertain demand. The problem is formulated as a stochastic combinatorial model, where the objective function minimises both the cost of booking the contracted bins (stage 1) and the expected cost associated with extra capacity purchased on the market (stage 2). Since the model reflects a stochastic environment, it is analysed with scenarios and learned, i.e., classified, on various instances. The model is tested with exact algorithms where computation time is a less critical factor, followed by the application of ML heuristics to a large set of instances. A total of 12 different ML heuristics are tested, mostly those applied by Aylak et al. [116], along with others, including the Gaussian process classifier, AdaBoost classifier, Logistic regression, and Linear discriminant analysis.

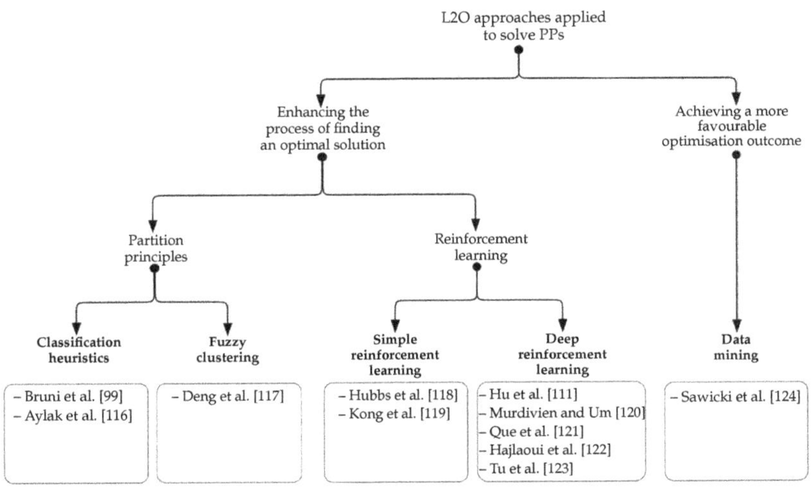

Figure 2. Current Learn-to-Optimise (L2O) approaches applied to address Packing Problems (PPs).

In the optimisation algorithm discussed by Deng [117], clustering is applied using the AMOSA_FCM algorithm. This algorithm employs a Fuzzy C-means (FCM) approach for grouping data objects. The clustering step involves defining complex numbers that represent different clusters and optimising an objective function based on Fuzzy compactness and spatial information. The resulting clustering effect is evaluated against other algorithms, demonstrating improved performance in terms of the objective grouping function value.

Another group of research that combines the learning phase with the optimisation of PP focuses on simple Reinforcement Learning (RL) [118,119] and Deep Reinforcement Learning (DRL) principles [111,120–123]. The study by Hubbs et al. [118] applies RL to packing problems, modelling them as Markov Decision Processes (MDPs) for dynamic decision-making. RL algorithms, such as Proximal Policy Optimisation (PPO), train agents to maximise cumulative rewards by sequentially selecting items based on states such as remaining capacity and item value. For the knapsack problem, RL effectively handles stochastic environments, achieving competitive results with optimisation techniques like integer programming and heuristics, especially in complex, uncertainty-laden scenarios. RL is also explored by Kong et al. [119] in the context of the online knapsack problem. They formulate a novel learning framework for online optimisation problems by introducing the Bang-per-Buck algorithm as a key component of their analysis. This algorithm optimises decision-making by defining a threshold on the value per unit size to determine whether to accept or reject incoming items, effectively maximising the total value of items within the knapsack while adhering to capacity constraints.

To this extent, the DRL principle is explored in the literature in the context of more complex packing problems. The study by Hu et al. [111] primarily aims to minimise surface area in 3D bin packing using the DRL framework, incorporating a heuristic algorithm for sequential item packing decisions. The neural network architecture is inspired by Pointer Net, and the training approach involves memory replay, random sampling during training, and beam search during testing. In contrast, Murdivien and Um [120] focus on training an agent to stack boxes on a designated plate, emphasising optimal placement rather than minimising surface area. This study explicitly defines agents, observations, and rewards and employs the Proximal Policy Optimisation (PPO) algorithm with specific hyperparameter settings for training. The work by Que et al. [121] introduces a distinct approach to the 3D Bin Packing Problem using Deep Reinforcement Learning. This paper addresses a Three-Dimensional Packing Problem (3D-PP) with variable container height. Unlike the previous two investigations [111,120], Que et al. [121] employ an encoder–decoder architecture based on the Transformer for both policy and value networks. They propose a novel state representation method that separates container and box observations, aiming to more effectively mimic human thinking processes during packing. The PPO algorithm is utilised, and the combinatorial action space challenge is addressed by implementing sub-modules in the policy network to effectively reduce the number of action choices.

In the work of Hajlaoui et al. [122], the Single Container Loading Problem (SCLP) with the Full-Support condition (SCLP-FS) is discussed. Full support implies that every loaded box inside the container should be fully supported from below, which is a fundamental requirement to avoid cargo damage. A Mixed Integer Linear Programming (MILP) model is formulated to express the sequence of products in the container and to maximise its volume utilisation. The problem is solved with Deep Reinforcement Learning principles, supported by neural network techniques. The learning procedure involves trial-and-error interaction between the agent and the environment along discrete time steps. During the loading procedure, at every time step, actions are performed by the agent, and the feasibility of new box placement is checked. If the action is incorrect, the loading process is stopped, and a negative reward is assigned. Otherwise, the action updates the list of states and available actions. This process is repeated until the container is fully loaded and no further actions are available. The SCLP-FS problem is experimentally demonstrated, achieving good results with a homogeneous dataset featuring an uncertain number of boxes. DRL is also addressed by Tu et al. [123] to find optimal solutions for two classic packing problems: the 0–1 knapsack problem and the 2D strip packing problem in the online domain. The paper aims to address online packing problems within the request–answer game framework. The core of the proposed approach is a dispatcher sub-algorithm that generates answers at each time step based on the latest request variable and the current solution in the problem environment. A DRL hyper-heuristic is presented, combining end-to-end sequential decision-making with a hyper-heuristic selection. The DRL hyper-heuristic utilises a duelling double deep Q-learning network as the high-level heuristic, trained to make decisions in uncertain online environments. Additionally, a feature fusion method is introduced to represent uncertain environment states, combining three types of features: MDP features, visual features, and histogram information of stochastic attributes. The resulting global representation of the environment state is used as input for the DRL hyper-heuristic.

In contrast, the literature contains a few examples of studies devoted to classification by data mining and applied to the PP domain, e.g., Sawicki et al. [124]. They address a distributor's multi-pallet loading challenge, introducing the stacking of elementary pallet units to create Sandwich Pallet Loading Units (SNDW). The study optimises both elementary units and SNDW, considering weight and height constraints, using knowledge from previous instances. By applying Dominance-Based Rough Sets Theory (DRST) and rule-based decision-making, rules are established to guide the reduction in SNDW by adjusting control parameters in the elementary pallet design. Experimental simulations, augmented by DRST, validate these decisions and contribute to an updated set of decision rules for future order processing. This systematic approach achieves an effectiveness rate of 83% in classifying new orders for potential improvement within the multi-pallet loading process.

Based on the findings mentioned above, key observations are formulated below, serving as the foundation for identifying a research gap and defining the objectives of this study:

- Apart from a few examples, such as the work of Sawicki and Sawicka [107], there is a very limited body of comprehensive studies on the optimisation of SPLP, in contrast to the extensive analyses on the MLBP.
- Existing works on multi-level pallet design consider fundamental constraints, such as the total weight of the constructed pallet load unit and the stacking height, along with specific limitations like the fragility (compression strength) of each elementary unit.
- In the analysed domain of SPLP, advanced procedures supporting the optimisation process of loading are limited. This includes the work of Sawicki et al. [124], where interaction occurs between elementary load units, for example, the MIX unit and the SKU unit, and the design of sandwich-pallet units based on them.
- Few examples of applying advanced Artificial Intelligence techniques in loading and packing problems show limited instances of Machine Learning utilisation to assist in optimising the packing and loading process of sandwich-pallet units. An example is a combination of simulation, Machine Learning, rule generation, and rule-based decision-making (Sawicki et al. [124]).

Given the outlined research gap, the content of this study will be expanded in the following areas:

- Enhancing the existing complex procedure for designing elementary load units and constructing sandwich-pallet units based on them.
- Considering, in the pallet loading process, all previously analysed constraints, including weight, height, and stability of elementary load units.
- Applying previously developed procedures that combine Machine Learning, rule generation, and rule-based decision-making, primarily to acquire comprehensive knowledge about necessary parameter modifications in the design process of sandwich-pallet units.
- Conducting extensive experiments on numerous instances derived from the real processes of designing both elementary and stacked load units, demonstrating the utilitarian benefits of employing such procedures in logistics and supply chains.

3. Preliminaries

In this section, the fundamental assumptions of the Dominance-Based Rough Set Theory are presented, which constitute the basis of the methodology applied to solve the Sandwich Pallet Loading Problem.

3.1. The Concept of Dominance-Based Rough Sets

The Dominance-Based Rough Sets Theory (DRST), also called the Dominance-Based Rough Sets Approach (DRSA), proposed by Greco et al. [125], is an extension of the Rough Sets Theory briefly presented in Section 1. Based on DRST principles, the decision table uses a four-tuple information $DT = \langle U, Q, V, f \rangle$, where U is a finite set of objects, $Q = q_1, \ldots q_m$ is a finite set of characteristics, i.e., columns of the database; V is the domain of characteristics q, expressed as $V = \bigcup_{q \in Q} V_q$, and f is the information function assigned to each pair of an object x and a characteristic q, such that $f\colon U \times Q \to V$ and $f(x,q) \in V_q$, $\forall q \in Q$, $x \in U$.

The set of characteristics Q is composed of both criteria $C^{>}$, i.e., characteristics with preference ordered domains, and attributes $C^{=}$, i.e., characteristics with non-ordered domains. Within the set Q, a conditional part C and a decision part D can be distinguished, such that $C \cup D = Q$. Since $C^{>} \cup C^{=} = C$ and $C^{>} \cap C^{=} = \emptyset$, the decision table is formally expressed as $S = \langle U, C^{>} \cup C^{=} \cup D, V, f \rangle$.

For two objects $x, y \in U$, where x represents reference objects and y represents compared objects, one can specify a binary relation R_P defined in $P \subseteq C$ as described by [125]:

$$yR_Px,\ \ \forall x,y \in U \ \text{if} \begin{cases} yDom_qx & \forall q \in P^{>},\ P^{>} = P \cap C^{>} \\ yInd_qx & \forall q \in P^{=},\ P^{=} = P \cap C^{=} \end{cases} \tag{1}$$

Assuming that P is a subset of conditional characteristics ($P \subseteq C$), the relation yR_Px corresponds to two situations, i.e., object y dominates x from the perspective of all considered criteria $P^{>}$, denoted as $yDom_Px$, and object y is indiscernible with x from all attributes $P^{=}$, denoted as $yInd_Px$. Considering any object $x \in U$ and the relation R_P, one can define a set $R_P^{+}(x)$, which includes the objects y that dominate x from the perspective of all criteria $P^{>}$ and are indiscernible with x for all attributes $P^{=}$. $R_P^{+}(x)$ is called the set of objects dominating object x and is expressed as follows:

$$R_P^{+}(x) = \{y \in U\colon\ yR_Px\}. \tag{2}$$

Similarly, $R_P^{-}(x)$ constitutes the set of those objects y that are dominated by x from all criteria $P^{>}$ and are indiscernible with x from all attributes $P^{=}$. $R_P^{-}(x)$ is called the set of objects dominated by x, and is expressed as follows:

$$R_P^{-}(x) = \{y \in U\colon\ xR_Py\}. \tag{3}$$

In general, the fundamental difference between DRST and the classical RST methodology is that the latter is based solely on attributes $P^{=}$, and the only relationship that can be defined between any two objects is indiscernibility (Ind_P). This significantly limits the scope of application of the classical approach. To illustrate the essence of DRST, a simple example is considered in this section. This example refers to the selection of products according to customer preferences. Three characteristics are considered, namely: product type $C_1^{=}$, as an attribute; supplier category $C_2^{>}$, as a maximised criterion; and product height $C_3^{>}$, also a maximised criterion. The customer's perception of the product is expressed by a maximised criterion D and will be described in the next section. The decision table includes 10 objects (see Table 1).

Table 1. Decision table for an illustrative experiment.

Object	$C_1^{=}$	$C_2^{>}$	$C_3^{>}$	D
1	1002	2	1	A
2	1003	2	1	B
3	1002	3	1	B
4	1001	2	3	A
5	1001	1	2	A
6	1001	2	2	B
7	1003	3	1	C
8	1003	2	2	B
9	1002	2	3	C
10	1003	2	3	C

With respect to the theoretical part, there are several dominating objects, i.e., $R_P^+(1) = \{3, 9\}$, $R_P^+(2) = \{7,8,10\}$, $R_P^+(5) = \{4,6\}$, $R_P^+(6) = \{4\}$, and $R_P^+(8) = \{10\}$. Thus, $R_P^+(1) = \{3,9\}$ means that objects $\{3\}$ and $\{9\}$ dominate $\{1\}$ because object $\{3\}$ is indiscernible with $\{1\}$ from the perspective of $C_1^=$ and dominates $\{1\}$ from the perspective of $C_2^>$, while being equivalent from the perspective of $C_3^>$. Similarly, object $\{9\}$ is indiscernible with $\{1\}$ from the perspective of $C_1^=$, dominates $\{1\}$ from the perspective of $C_3^>$, and is equivalent from the perspective of $C_2^>$, etc. There is also a set of dominated objects, i.e., $R_P^-(3) = \{1\}$, $R_P^-(4) = \{5,6\}$, $R_P^-(6) = \{5\}$, $R_P^-(7) = \{2\}$, $R_P^-(8) = \{2\}$, $R_P^-(9) = \{1\}$, $R_P^-(10) = \{2,8\}$. Thus, $R_P^-(3) = \{1\}$ means that object $\{1\}$ is dominated by $\{3\}$, etc.

3.2. Approximations and Union of Classes in DRST

The decision characteristic D, where $D \subseteq Q$, constitutes the partition of U into a finite number of categories $\mathcal{C\ell}=\{Cl_t,\ t \in T\}$, with $T=\{1,2,\ldots,n\}$. It is also assumed that each object x belongs to only one category $Cl_t \in \mathcal{C\ell}$, and all t-classes are ordered according to an increasing importance of preferences. This implies that objects assigned to class Cl_g are preferred over objects from Cl_h if $g > h$, $\forall g,h \in T$. If an object is assigned to class h according to the characteristic D, and its conditional part C outperforms the conditional part of any other object assigned to class g, it generates a certain inconsistency in DRST. For this reason, two categories of union of classes have been introduced, i.e., upward and downward unions of classes [64,126]. The upward union of classes $Cl_t^{\geq}$ is the set of objects that belong to class Cl_t or to classes more preferred than t, and it is expressed as follows:

$$Cl_t^{\geq} = \bigcup_{h \geq t} Cl_h,\ \forall t = 1,\ldots,n; \tag{4}$$

while downward union of classes $Cl_t^{\leq}$ is the set of objects that belong to class Cl_t or to classes less preferred than t, and it is expressed as follows:

$$Cl_t^{\leq} = \bigcup_{h \leq t} Cl_h,\ \forall t = 1,\ldots,n. \tag{5}$$

Based on [126], from the perspective of $P \subseteq C$ and $t \in T$, the object $x \in U$ undoubtedly belongs to the upward union of classes $Cl_t^{\geq}$ if $x \in Cl_t^{\geq}$ and $y \in Cl_t^{\geq}$ for all objects $y \in U$ dominating x from the perspective of $P^>$ and indiscernible with x from $P^=$. This means that an object $x \in U$ belongs without any doubt to $Cl_t^{\geq}$ if $R_P^+(x) \subseteq Cl_t^{\geq}$, and the set of all such objects constitutes the lower approximation of the upward union of classes $\underline{P}(Cl_t^{\geq})$, denoted by Greco et al. [126] as follows:

$$\underline{P}(Cl_t^{\geq}) = \left\{x \in U_P\colon\ R_P^+(x) \subseteq Cl_t^{\geq}\right\}, \tag{6}$$

where U_P is the set of reference objects from P. At the same time, the set of objects that probably belong to $Cl_t^{\geq}$ constitutes the upper approximation of the upward union of classes $\overline{P}(Cl_t^{\geq})$, denoted as follows:

$$\overline{P}(Cl_t^{\geq}) = \left\{x \in U_P\colon\ R_P^-(x) \cap Cl_t^{\geq} \neq \varnothing\right\} = \bigcup_{x \in Cl_t^{\geq}} R_t^+(x). \tag{7}$$

Similar formulas can be described for the downward unions of classes, i.e., the lower approximation, such as follows:

$$\underline{P}(Cl_t^{\leq}) = \left\{x \in U_P\colon\ R_P^-(x) \subseteq Cl_t^{\leq}\right\}, \tag{8}$$

and the upper approximation, such as the following:

$$\overline{P}(Cl_t^{\leq}) = \left\{x \in U_P\colon\ R_P^{+}(x) \cap Cl_t^{\leq} \neq \varnothing\right\} = \bigcup_{x \in Cl_t^{\leq}} R_t^{-}(x), \tag{9}$$

Comparison of upper and lower approximations of the union of classes results in the definition of the area of uncertainty, called the boundary region Bn_P. The boundary region for the upward union of classes is defined as follows:

$$\mathrm{Bn}_P(Cl_t^{\geq}) = \overline{P}(Cl_t^{\geq}) - \underline{P}(Cl_t^{\geq}), \tag{10}$$

while for the downward union of classes, it is formulated as follows:

$$\mathrm{Bn}_P(Cl_t^{\leq}) = \overline{P}(Cl_t^{\leq}) - \underline{P}(Cl_t^{\leq}). \tag{11}$$

Referring to the example in Table 1 (see Section 3.1), three ascending decision classes have been defined: A, B, and C. The application of DRST allows the formation of four unions of classes, namely, at most A: $Cl_A^{\leq}$, at most B: $Cl_B^{\leq}$, at least B: $Cl_B^{\geq}$, and at least C: $Cl_C^{\geq}$. Considering the dominance relations defined in Section 3.1 and the unions of classes in this section, it is possible to determine the lower and upper approximations of each union of classes and their boundary regions, as follows:

- $\underline{P}(Cl_A^{\leq}) = \{1,5\}$, $\overline{P}(Cl_A^{\leq}) = \{1,4,5,6\}$, and $\mathrm{Bn}_P(Cl_A^{\leq}) = \{4,6\}$.
- $\underline{P}(Cl_B^{\leq}) = \{1,2,3,4,5,6,7,8\}$, $\overline{P}(Cl_B^{\leq}) = \{1,2,3,4,5,6,7,8\}$, and $\mathrm{Bn}_P(Cl_B^{\leq}) = \varnothing$.
- $\underline{P}(Cl_B^{\geq}) = \{2,3,7,8,9,10\}$, $\overline{P}(Cl_B^{\geq}) = \{2,3,4,6,7,8,9,10\}$, and $\mathrm{Bn}_P(Cl_B^{\geq}) = \{4,6\}$.
- $\underline{P}(Cl_C^{\geq}) = \{9,10\}$, $\overline{P}(Cl_C^{\geq}) = \{9,10\}$, and $\mathrm{Bn}_P(Cl_C^{\geq}) = \varnothing$.

The visual interpretation of the unions of classes and their associated objects is presented in Figure 3.

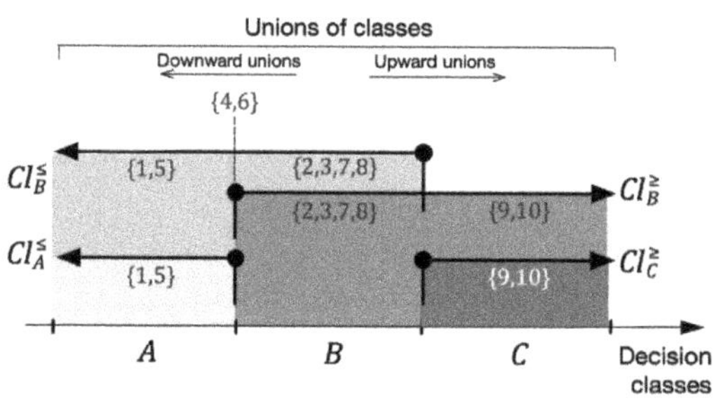

Figure 3. Illustration of downward and upward unions of classes in the considered example.

The membership of $\{4,6\}$ in the boundary region results from the fact that object $\{6\}$ is dominated by $\{4\}$ (see Section 3.1), yet still belongs to the higher class, i.e., B. This inconsistency means that neither object can be precisely assigned to the considered unions of classes, neither to $Cl_A^{\leq}$ nor to $Cl_B^{\geq}$.

3.3. Consistency of Knowledge in DRST

Information about objects in the decision table can be evaluated in terms of consistency and quality of knowledge. The consistency of the available knowledge regarding the approximations of individual classes from the union of classes is described by the accuracy of approximation α_P [126]. It is defined as the ratio of the number of objects appropriately assigned to a specific union of classes to those probably assigned to this union. For upward and downward unions of classes, it is denoted as follows:

$$\alpha_P\left(Cl_t^{\geq}\right) = \frac{|\underline{P}(Cl_t^{\geq})|}{|\overline{P}(Cl_t^{\geq})|}, \tag{12}$$

$$\alpha_P\left(Cl_t^{\leq}\right) = \frac{|\underline{P}(Cl_t^{\leq})|}{|\overline{P}(Cl_t^{\leq})|}. \tag{13}$$

The coherence and quality of information about the objects contained in the decision table are reflected by the quality of sorting $\gamma_P(\mathcal{C}\ell)$ [127]. It is defined as the ratio between the number of P-rightly sorted objects and the total number of objects in U, and it is formulated as follows:

$$\gamma_P(Cl) = \frac{1}{|U|} \cdot \left| U - \left(\left(\bigcup_{t \in T} Bn_P\left(Cl_t^{\geq}\right) \right) \cup \left(\bigcup_{t \in T} Bn_P\left(Cl_t^{\leq}\right) \right) \right) \right| \tag{14}$$

Both consistency measures, i.e., the accuracy of approximation α_P and the quality of sorting γ_P, evaluate the extent to which knowledge about objects and their assignment to individual classes generates conflicts. The accuracy of approximation identifies conflicts occurring separately within each decision class, whereas the quality of sorting refers to all decision classes in total. The higher the values of both indicators, where $0 \leq \alpha_P \leq 1$ and $0 \leq \gamma_P \leq 1$, the more information about the objects can be considered a source of reliable reasoning. From a practical perspective, both measures form the basis for evaluating the validity of using particular parameters that characterise the objects in the decision table.

Referring to the example from Section 3.1 and based on the lower and upper approximations of the union of classes, the accuracy of approximation equals:

- $\underline{P}(Cl_A^{\leq}) = |\{1,5\}|/|\{1,4,5,6\}| = 0.5$;
- $\underline{P}(Cl_B^{\leq}) = |\{1,2,3,4,5,6,7,8\}|/|\{1,2,3,4,5,6,7,8\}| = 1.0$;
- $\underline{P}(Cl_B^{\geq}) = |\{2,3,7,8,9,10\}|/|\{2,3,4,6,7,8,9,10\}| = 0.75$;
- $\underline{P}(Cl_C^{\geq}) = |\{9,10\}|/|\{9,10\}| = 1.0$.

The quality of sorting in this example is $\gamma_P(Cl) = 1/|\{1,2,\ldots,10\}| \cdot |\{1,2,\ldots,10\} - \{4,6\}| = 0.8$.

3.4. Generation of Decision Rules Based on DRST

The Dominance-Based Rough Sets approach represents one of the fundamental sources for generating decision rules that generalise existing knowledge [126,128]. The construction of these rules is based on the upper and lower approximations of the union of object classes. Each decision rule, denoted as r, is a logical formulation composed of a conditional part and a decision part, in the form of "if ..., then ...". The former, known as the predecessor, specifies the value of one or several object characteristics. The latter, referred to as the successor, defines the assignment of the object to one of the unions of classes or individual classes. In the former case, it is called an exact rule, whereas in the latter case, it is termed an approximate rule.

According to [126,129], considering the presence of indiscernibility relations Ind_P with respect to attributes and dominance relations Dom_P with respect to criteria, three types of decision rules are distinguished:

- $DR^{\geq}$, generated on the lower approximation of the upward union of classes $\underline{P}(Cl_t^{\geq})$, as:

$$\begin{aligned} &\text{if } f(x,q_1) \geq r_{q1} \wedge \ldots \wedge f(x,q_w) \geq r_{qw} \wedge f(x,q_{w+1}) = \\ &= r_{qw+1} \wedge \ldots \wedge f(x,q_z) = r_{qz}, \text{ then } x \in Cl_t^{\geq} \end{aligned} \tag{15}$$

where

$$P = \{q_1, \ldots, q_z\} \subseteq C,\ P^{>} = q_1, \ldots, q_w,$$
$$P^{=} = q_{w+1}, \ldots, q_z,\ (r_{q1}, \ldots, r_{qz}) \in V_{q1} \times \ldots \times V_{qz},\ t \in T;$$

- $DR^{\leq}$, generated on the lower approximation of the downward union of classes $\underline{P}(Cl_t^{\leq})$, as

$$\begin{aligned} \text{if } f(x, q_1) \leq r_{q1} \wedge \ldots \wedge\ f(x, q_w) \leq r_{qw} \wedge\ f(x, q_{w+1}) = \\ = r_{qw+1} \wedge \ldots \wedge\ f(x, q_z) = r_{qz},\ \text{ then } x \in Cl_t^{\leq} \end{aligned} \tag{16}$$

where

$$P = \{q_1, \ldots, q_z\} \subseteq C,\ P^{>} = q_1, \ldots, q_w,$$
$$P^{=} = q_{w+1}, \ldots, q_z,\ (r_{q1}, \ldots, r_{qz}) \in V_{q1} \times \ldots \times V_{qz},\ t \in T;$$

- $DR^{\geq\leq}$, generated on the upper approximations of the downward and upward unions of classes $\overline{P}(Cl_t^{\leq}) \cap \overline{P}(Cl_t^{\geq})$, as

$$\begin{aligned} \text{if } f(x, q_1) \geq r_{q1} \wedge \ldots \wedge\ f(x, q_u) \geq r_{qu} \wedge f(x, q_{u+1}) \leq r_{qu+1} \\ \wedge \ldots \wedge\ f(x, q_w) = r_{qw} \wedge f(x, q_{w+1}) = r_{qw+1} \\ \wedge \ldots \wedge\ f(x, q_z) = r_{qz}, \\ \text{then } x \in Cl_s \cup Cl_{s+1} \cup \ldots \cup Cl_t \end{aligned} \tag{17}$$

where

$$P = \{q_1, \ldots, q_z\} \subseteq C,\ P^{>} = q_1, \ldots, q_u \cup q_{u+1}, \ldots, q_w,$$
$$P^{=} = q_{w+1}, \ldots, q_z,\ (r_{q1}, \ldots, r_{qz}) \in V_{q1} \times \ldots \times V_{qz},\ t \in T.$$

Two quantitative measures, called the absolute and relative strength of the generated rules, characterise their importance and classification reliability. The absolute strength AS_r defines the number of objects in the decision table that are consistent with the conditional part of the r-rule. Based on [130], it is denoted for the upper and lower unions of classes as

$$AS_r\left(Cl_t^{\geq}\right) = \left|cond_r \cap Cl_t^{\geq}\right|, \tag{18}$$

$$AS_r\left(Cl_t^{\leq}\right) = \left|cond_r \cap Cl_t^{\leq}\right|. \tag{19}$$

where $cond_r$ denotes the set of objects verifying the conditional part of the r-rule, and $|cond_r \cap Cl_t^{\geq}|$ denotes the cardinality of the set of objects verifying $cond_r$ and belonging to the union of classes $Cl_t^{\geq}$. The relative strength RS_r of the r-rule is the relationship between the absolute strength of the rule and the number of objects in the decision table that constitute the lower approximation of the union of classes suggested by the successor. Based on [130], it is denoted for the upper and lower unions of classes as

$$RS_r\left(Cl_t^{\geq}\right) = \frac{\left|cond_r \cap Cl_t^{\geq}\right|}{\left|Cl_t^{\geq}\right|}, \tag{20}$$

$$RS_r\left(Cl_t^{\leq}\right) = \frac{\left|cond_r \cap Cl_t^{\leq}\right|}{\left|Cl_t^{\leq}\right|}. \tag{21}$$

In the literature, various strategies for generating decision rules based on the decision table are encountered. Among the strategies most frequently mentioned, as proposed by Greco et al. [126,131], Stefanowski [132], and Stefanowski and Vanderpooten [133], are the generations of the following:

- A minimal set of rules, reflecting all objects contained in the decision table.
- An exhaustive set of rules, consisting of all possible decision rules appropriate for the decision table.
- A minimal number of decision rules, covering a relatively large number, but not necessarily all, of the objects contained in the decision table.

Referring to the simple example in Section 3.1, an exemplary minimal set of rules is represented by the following eight items characterised by both absolute and relative strengths of decision rule (see AS_r and RS_r):

- $r = 1$: if $C_1 = 1002$ and $C_3 \geq 3$ then $D = Cl_C^{\geq}$ and $AS_1 = \{9\}$, $RS_1 = 0.50$;
- $r = 2$: if $C_1 = 1003$ and $C_3 \geq 3$ then $D = Cl_C^{\geq}$ and $AS_2 = \{10\}$, $RS_2 = 0.50$;
- $r = 3$: if $C_1 = 1003$ then $D = Cl_B^{\geq}$ and $AS_3 = \{2,7,8,10\}$, $RS_3 = 0.57$;
- $r = 4$: if $C_2 \geq 3$ then $D = Cl_B^{\geq}$ and $AS_4 = \{3,7\}$, $RS_4 = 0.29$;
- $r = 5$: if $C_2 \leq 1$ then $D = Cl_A^{\leq}$ and $AS_5 = \{5\}$, $RS_5 = 0.33$;
- $r = 6$: if $C_1 = 1002$ and $C_2 \leq 2$ and $C_3 \leq 1$ then $D = Cl_A^{\leq}$ and $AS_6 = \{1\}$, $RS_6 = 0.33$;
- $r = 7$: if $C_3 \leq 2$ then $D = Cl_B^{\leq}$ and $AS_7 = \{1,2,3,5,6,7,8\}$, $RS_7 = 0.87$;
- $r = 8$: if $C_1 = 1002$ then $D = Cl_B^{\leq}$ and $AS_8 = \{4,5,6\}$, $RS_8 = 0.37$.

3.5. Classification Based on Decision Rules

The process of generating decision rules based on the objects forming the decision table, as described in Section 3.4, serves as a foundation for the classification of new objects, denoted as z, i.e., objects for which only the conditional characteristics C are known, while the decision characteristic D remains undefined. The assignment of z to the appropriate class is carried out by using the available decision rules, matching the values of the characteristics in the conditional part of the rules with the values of the conditional characteristics C of the new object z. This procedure is referred to as covering z with decision rules. Błaszczyński et al. [130] introduced the concept of cov_z, which denotes the set of decision rules that cover z. Within this framework, they identify three potential cases:

- Case 1: No decision rule covers the object z, i.e., $cov_z = \emptyset$.
- Case 2: Only one decision rule r covers the object z, i.e., $|cov_z| = 1$.
- Case 3: Several decision rules cover the object z, i.e., $|cov_z| = R$,

where $|cov_z|$ represents the cardinality of the set of decision rules covering the object z.

In Case 1, the object z is assigned to all decision classes from $\mathcal{C}\ell$. In Case 2, for the object z and the decision rule r covering it, a value of $score_r(Cl_t, z)$ is calculated for each class Cl_t from $\mathcal{C}\ell$, as follows:

$$score_r(Cl_t, z) = \frac{|cond_r \cap Cl_t|^2}{|cond_r| \cdot |Cl_t|}, \tag{22}$$

where $|cond_r|$ denotes the cardinality of the set of objects verifying $cond_r$, and $|Cl_t|$ represents the set of objects belonging to the decision class Cl_t. In Case 3, for a set of rules $cov_z = R$, a value of $score_R(Cl_t, z)$ is calculated for each decision class Cl_t from $\mathcal{C}\ell$, as follows:

$$score_R(Cl_t, z) = score_R^+(Cl_t, z) - score_R^-(Cl_t, z), \tag{23}$$

where $score_R^+(Cl_t, z)$ and $score_R^-(Cl_t, z)$ represent the positive and negative components of $score_R(Cl_t, z)$.

Thus, the positive component $score_R^+(Cl_t, z)$ is denoted as

$$score_R^+(Cl_t, z) = \frac{\left|(cond_{r_1} \cap Cl_t) \cup \ldots \cup (cond_{r_k} \cap Cl_t)\right|^2}{|cond_{r_1} \cup \ldots \cup cond_{r_k}| \cdot |Cl_t|}, \tag{24}$$

and it takes into account decision rules that are concordant with the assignment of z to Cl_t. These are decision rules $r_i \in cov_z^+(i = 1, \ldots, k)$ that suggest the assignment of the object z to the union of classes $Cl_s^{\geq}$ and $Cl_q^{\geq}$, such that $Cl_t \subseteq Cl_s^{\geq}$ and $Cl_t \subseteq Cl_q^{\leq}$, respectively. In contrast, the negative component $score_R^-(Cl_t, z)$ considers the decision rules discordant with the assignment of z to Cl_t. These are the decision rules $r_i \in cov_z(i = k{+}1, \ldots, h)$ that suggest the assignment of z to the union of classes $Cl_s^{\geq}$ and $Cl_q^{\geq}$, where $Cl_t \cap Cl_q^{\leq} = \emptyset$ and $Cl_t \cap Cl_s^{\geq} = \emptyset$, respectively. This is formally defined as

$$score_R^-(Cl_t, z) = \frac{\left|\left(cond_{r_{k+1}} \cap Cl_{r_{k+1}}^{\geq}\right) \cup \ldots \cup \left(cond_{r_l} \cap Cl_{r_l}^{\geq}\right) \cup \left(cond_{r_{l+1}} \cap Cl_{r_{l+1}}^{\leq}\right) \cup \ldots \cup \left(cond_{r_h} \cap Cl_{r_h}^{\leq}\right)\right|^2}{|cond_{r_{k+1}} \cup \ldots \cup cond_{r_l} \cup cond_{r_{l+1}} \cup \ldots \cup cond_{r_h}| \cdot |Cl_{r_{k+1}}^{\geq} \cup \ldots \cup Cl_{r_l}^{\geq} \cup Cl_{r_{l+1}}^{\geq} \cup \ldots \cup Cl_{r_h}^{\geq}|}, \tag{25}$$

where $Cl_{r_{k+1}}^{\geq}, \ldots, Cl_{r_l}^{\geq}$ are the upward unions of classes suggested for the assignment of z by rules $r_i \in cov_z^-(i = k{+}1, \ldots, l)$, and $Cl_{r_{l+1}}^{\leq}, \ldots, Cl_{r_h}^{\leq}$ are the downward unions of classes suggested for the assignment of z by rules $r_i \in cov_z^-(i = l{+}1, \ldots, h)$. Finally, the object z is assigned to the class Cl_t for which the value of $score_r(Cl_t, z)$, see Case 2, or $score_r(Cl_t, z)$, see Case 3, is the greatest [130].

To illustrate the rule-based decision-making process, reference is made again to the example from Section 3.1 and its analysis in subsequent sections. Assuming an object z, which has not been included in the previous analyses, possesses the following characteristics: $C_1 = 1002$, $C_2 = 3$, and $C_3 = 3$, while the value of the decision class is missing. By comparing object z with the set of rules (see Section 3.4), it can be determined that it corresponds to Rules 2 and 4, as indicated in Figure 4. This scenario aligns with Case 3, where several decision rules cover a new object. Using Equations (23)–(25), it can be calculated to which decision class object z should be assigned, that is, either B or C. For B, the following is calculated: $score_R^+(B,z) = |\{3,7\}|^2/|\{3,7\}| \cdot |\{2,3,6,8\}| = 0.50$ and $score_R^-(B,z) = |\{9\}|^2/|\{9\}| \cdot |\{1,4,5,7,9,10\}| = 0.17$, then $score_R(B,z) = 0.33$. For C, the following is calculated: $score_R^+(C,z) = |\{4\}|^2/|\{4\}| \cdot |\{7,9,10\}| = 0.33$, and $score_R^-(C,z) = |\{3,7\}|^2/|\{3,7\}| \cdot |\{1,2,3,4,5,6,8\}| = 0.29$, then $score_R(C,z) = 0.05$. Since $score_R(B,z) > score_R(C,z)$, object z should therefore be assigned to class B.

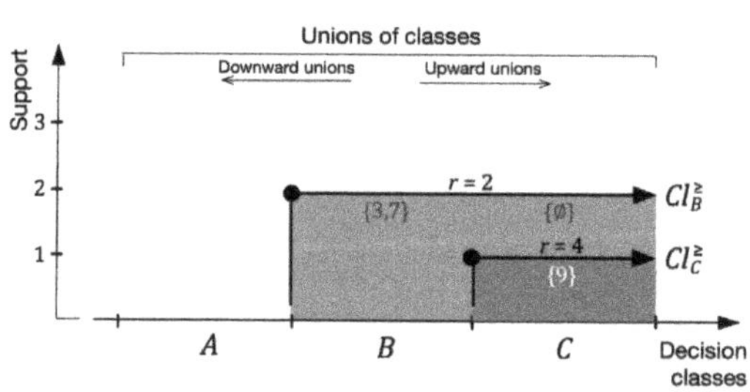

Figure 4. Illustrative example of classification using decision rules—Case 3.

4. Proposed Learning-to-Optimise Expert System

4.1. General Context and Concept of Expert System

In this paper, a methodology for the Learning-to-Optimise expert system is proposed to solve the Sandwich Pallet Loading Problem, leading to environmentally responsible transport operations with social and economic aspects included. The procedure consists of seven steps, as illustrated in Figure 5. All input data are assumed to be available from the Warehouse Management System (WMS). Orders scheduled for daily completion are planned before warehouse operations begin.

In Step 1, based on the default parameters (see P1) of each order and product information retrieved from the WMS, an optimisation algorithm is applied to design a minimum number of MIX pallets (see R1). This result, along with default parameters (see P1), is saved as a single record in the Knowledge Database (KDB). Subsequently, the initial result of the MIX pallet optimisation is analysed in Step 2 to determine the minimum number of SNDW pallets. An optimisation algorithm is used for the design of the SNDW pallets, and the minimum number of SNDW pallets resulting (see R2) is added to the record in the KDB. This record serves as a reference for the optimal number of MIX and SNDW pallets based on P1.

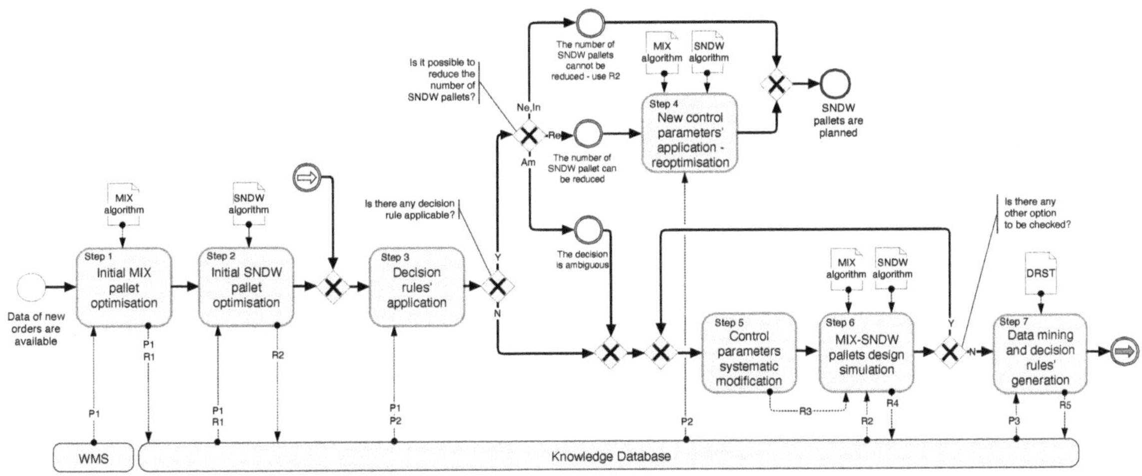

P1 - Default parameters of orders and products; P2 - The current set of decision rules; P3 - The updated KDB (decision table);
R1 - Result of MIX pallets optimisation based on P1; R2 - Result of SNDW pallets optimisation based on P1; R3 - New control parameters; R4 - Result of new MIX-SNDW pallets design simulation;
R5 - Updated set of decision rules

Figure 5. Key steps of the proposed methodology (based on [124]).

The outcome derived from Step 2 is stored in the KDB as R2 and subsequently evaluated in Step 3. This assessment entails the application of the current set of decision rules (see P2) and its comparison with the set of parameters of the order and the products (see P1). The result of Step 3 is to determine whether the existing decision rules allow the conduct of in-depth analyses, that is, if the conditional part of these rules converges with the parameters of the order (see the Y path) or not (see the N path). In the former case, the decision rules facilitate determining whether further enhancement of the optimal number of SNDW pallets is feasible (see the Re path) or not (see the Ne, In path) or if the decision rules do not ensure a definitive decision (see the Am path). If the number of SNDW pallets cannot be reduced (see the Ne, In path), the indicated optimal solution (see R2) should be retained and is considered the final number of SNDW pallets. In contrast, if the number of SNDW pallets can be further reduced (see the Re path), Step 4 is initiated. It is based on the current set of decision rules (see P2) and assumes the application of a new set of control parameters. The MIX and SNDW algorithms are applied, leading to the optimisation of the final result, i.e., a reduced number of SNDW pallets is planned. In the event of an inability to reach a definitive decision (see the Am path), the subsequent procedure follows the same course as if it is not possible to apply a set of decision rules in Step 3 (see the N path). In such cases, the selected control parameters must undergo systematic modification as part of Step 5, followed by further actions. These adjusted parameters, as a set of combinatorial changes, are passed to Step 5 (see R3). The order with adjusted parameters is then simulated in Step 6 using the MIX and SNDW optimisation algorithms. The resulting number of SNDW pallets is compared with the optimal number of SNDW pallets from R2. Steps 5 and 6 are iteratively repeated until all potential changes to the control parameters are simulated and examined, and all new MIX and SNDW pallet design results are recorded in the KDB (see R4). The iterative simulation process ends when no further options need to be checked (see the N path).

Finally, in Step 7, a data mining process based on the updated decision table in the KDB is carried out (see P3), supported by Dominance-Based Rough Sets Theory (DRST). A new set of decision rules is generated as an output (see R5) and recorded in the KDB. The process of applying decision rules repeats for each new case analysed in Step 3. The current set of these rules (see P2), i.e., the updated set of decision rules generated in Step 7 (see R5), is automatically incorporated into the KDB as the current set of decision rules to be applied in Step 3 (see P2).

4.2. Initial Optimisation of MIX and SNDW Pallets—Steps 1 and 2

In Step 1, a dataset containing information about customers and the set of products in their orders is retrieved from the WMS, as indicated by P1 in Figure 5. The customers' preferences are primarily characterised by a set of default parameters, such as the permissible weight and height of the MIX pallet and consent for combining different products on a pallet, i.e., mixing. Information about the products in the order includes a list of all the assortment items, the weight and dimensions of the collective packaging (i.e., its length, width, and height), and the quantity of collective packaging for each item. The fragility class of each product is also specified, i.e., the higher the number of the class, the more fragile the products.

The essence of optimisation at this stage is to construct the smallest possible number of MIX load units (LU) necessary to accommodate all ordered products. This optimisation is preliminary in nature, as it considers the default parameters assigned to the customer and aims to ensure an optimal outcome under these conditions. From a procedural standpoint, the planning of MIX LU entails the following activities:

- The collective packaging of the same product on one pallet unit is planned without dividing this product into different layers separated by another product.
- The planning of subsequent layers of products from the bottom up on a pallet unit is carried out while maintaining the fragility classes, which means that the products on subsequent layers must have the same or greater fragility.
- The arrangement of subsequent layers of products is only possible when the base layer is filled with products to a certain minimum extent.
- The arrangement of different products on one layer is possible when the height difference of the collective packaging falls within the tolerance limits.
- The arrangement of the product is planned with respect to a standard EUR pallet carrier with dimensions of 800 × 1200 × 25 mm.
- The planning of the arrangement of the products must not exceed the outline of the pallet base.

Taking these assumptions into account, a pseudocode of the generic algorithm to build an optimal MIX LU is presented in Algorithm 1.

Following the optimisation of the MIX pallets, each unit is assigned a set of characteristics (see R1 in Figure 5), such as the following:

- The number of products allocated and their location on individual layers.
- The total weight and height of the constructed MIX LU.
- The MIX LU status $St = \{Top, Base\}$, which determines the arrangement of the products on the top layer. The *Top* status indicates that the top layer of the MIX LU is filled with products below the minimum assumed level; otherwise, it is *Base*.

Algorithm 1 Constructing MIX pallet load units

function createMixLUs($\mathcal{O}$, $\mathcal{P}$, *LUTypes*, *params*)

Input: $\mathcal{O}$, $\mathcal{P}$, *LUTypes*, *params* are orders list, list of products in orders, list of pallet carrier types, list of planning parameters, respectively.

Output: A mixed LUs list `mixedLUs_list`.

```
mixedLUs_list ← null                                  //initialize mixed LUs list
for each order ∈ O do
    init_planning_params(params)
    if product ∈ order is the same (LU = SKU) then
        assign_mixedLU_type()                         //determining LU type-Top or Base
    end if
    generate_planning_products(order)
    sort_products()
    //main steps of the algorithm
    create_layers(order)                              //creating a list of layers for each product
    add_normative_layers()                            //adding normative layers
    add_nonnormative_layers()                         //adding non-normative layers
    add_remaining_products()                          //adding remaining products to LU
    assign_mixedLU_type()                             //determining LU type-Top or Base
end for
return mixedLUs_list;
```

In Step 2, the creation of the SNDW LUs is planned by stacking MIX LUs using the optimisation algorithm. The default control parameters P1 are utilised, including the allowed weight and height of the SNDW load unit. The objective of the optimisation is to construct the smallest possible number of SNDW LUs to ensure customer requirements, full shipments, stability of the loading, minimisation of plastic wrapping, and fulfilment of the environmental responsibility of the organisation. From a procedural standpoint, the planning of the SNDW LUs entails actions such as the following:

- MIX LUs with the *Base* status can be stacked, while those with *Top* status can only be stacked as the top unit.
- The fragility class of the MIX LU being placed on the top of the SNDW LU cannot be lower than the fragility class of the products placed below.
- The fragility class of the MIX LU is calculated according to an approach based on the weight shares and fragility classes of the products constituting this unit.

Taking these rules into consideration, the generic algorithm for building optimal SNDW LUs is presented below (see Algorithm 2).

Algorithm 2 Generating of SNDW LUs

function createSNDWLUs($\mathcal{O}$, *mixedLUs_list*, *params*)

Input: $\mathcal{O}$, *mixedLUs_list*, *params* are orders list, list of mixed LUs—result of Algorithm 1, list of planning parameters, respectively.

Output: A stacked and non-stacked mixed LUs list `SNDWLUs_list`.

```
SNDWLUs_list ← null                                  //initialize SNDW LUs list
for each order ∈ O do
    for each mixedLU ∈ mixedLUs_list do
        split_mixedLUs_byType()              //splitting mixed LUs into two subsets
                                                                //- Base and Top
        order_mixed_LUs()                 //sorting LUs according to fragility class
        run_DP_algorithm()              //running dynamic programming algorithm
                                                //to minimise number of SNDWLUs
    end for
end for
return SNDWLUs_list;
```

A dynamic programming algorithm (*run_DP_algorithm()* function) was utilised to optimise the SNDW LUs' volume. The objective of the algorithm was to fill the LU to its maximum height while adhering to the weight limitations.

Information about the number of MIX LUs included is assigned to each optimally planned SNDW LU. After completing Steps 1 and 2, a comprehensive description of the conditional part C of each individual record, representing a single object in the decision table, is generated and stored in the KDB (see Figure 5). This description consists of five subsets, that is, $C = O \cup MCP \cup M \cup SCP \cup S$, namely: the structural parameters of the items included in the order (O); the control parameters applied to the MIX algorithm (MCP); the output of the MIX algorithm along with the characteristics of the other elementary LUs in the order, such as SKU and FULL (M); the control parameters used in the SNDW LU algorithm (SCP); and the output of the SNDW algorithm (S). At this stage of the proposed algorithm, the decision part is not yet defined, i.e., $D = \emptyset$.

4.3. Rule-Based Decision-Making—Steps 3 and 4

As mentioned in Section 4.2, each newly created record in the KDB possesses a complete set of conditional characteristics, while its decision characteristic remains undefined. Therefore, Step 3 aims first to assess whether the available decision rules r are applicable to classify such an object z—decision yes (see Y) or no (see N), as shown in Figure 5. Secondly, if yes, it is necessary to determine to which of the predefined classes the new object z should be allocated based on the MPA classifier. In this case, the following options are distinguished:

- It is not possible to reduce the number of SNDW LUs, which is represented by the following formula (26):

$$cond_r(z) \wedge MPA(r) = \begin{cases} Ne \Longrightarrow MPA(z) = Ne \\ In \Longrightarrow MPA(z) = In \end{cases} \tag{26}$$

- It is probable that the number of SNDW LUs can be decreased, which can be defined as

$$cond_r(z) \wedge MPA(r) = Re \Longrightarrow MPA(z) = Re \tag{27}$$

- There are insufficient arguments to assess the possibility of reducing the number of SNDW LUs, which means that a decision is ambiguous (Am) and can be formulated as

$$cond_r(z) \wedge MPA(r) = Re \cup Ne \cup In \Longrightarrow MPA(z) = Am, \tag{28}$$

where $cond_r(z)$ denotes the correspondence of the conditional part of the rule r with the description of the object z, $MPA(r)$ denotes the decision class assignment indicated by the rule r, and $MPA(z)$ denotes the resulting decision class assignment of z to the decision class of MPA. A definition of MPA is presented in Section 4.4. The decision to allocate the new object z to the decision class, using the MPA classifier, is made based on the strength of the decision rules r applicable to the classification, i.e., based on the $score_r(MPA_t, z)$ for $|cov_z| = 1$ or $score_R(MPA_t, z)$ for $|cov_z| = R$, as discussed in Section 3.5.

In the case where the number of SNDW LUs cannot be reduced, i.e., $MPA(z) = In \cup Ne$, the solution obtained in Step 2, i.e., the optimisation result based on the default parameters R2, is considered final and passed on for implementation. When there is an indication that the number of SNDW LUs can be reduced, i.e., $MPA(z) = Re$, Step 4 is executed. In this step, the recommended values of the control parameters are implemented based on the previously applied set of decision rules, i.e., P2 (see Figure 5), allowing for the reduction in the number of SNDW LUs. The procedure for generating new optimal solutions is repeated using the MIX and SNDW optimisation algorithms, and the obtained result is passed on for implementation. In the third scenario, when $MPA(z) = Am$, the decision-making process requires additional actions, i.e., executing Steps 5–7 and repeating Step 3. This process involves modifying control parameters (Step 5), simulating pallet designs with MIX and SNDW optimisation algorithms (Step 6), and updating the set of decision rules (Step 7) to be reapplied in Step 3. The purpose of this sequence is to determine the assignment to a single decision class, $MPA(z)$. This process is discussed in the following section.

4.4. Modification of Control Parameters—Step 5, and Simulation Design of MIX-SNDW Pallets—Step 6

In cases where the available decision rules cannot be applied in Step 3 to assess the potential improvement of the obtained solution, an experimental evaluation of improvement possibilities should be performed. For this purpose, in Step 5, a systematic modification of selected characteristics q_i for a single object x, that is, $f(x, q_i)$, is carried out. A modified characteristic $f(x, q_i)^{mod}$ satisfies the following condition:

$$\begin{gathered} f(x,q_i)^- \leq f(x,q_i)^{mod} \leq f(x,q)^+; \\ q_i \in MCP \cup SCP; \;\; f(x,q_i)^{mod} \in V_{q_i}, \end{gathered} \tag{29}$$

where $f(x, q_i)^-$ denotes the minimum value that characteristic q_i can achieve; $f(x, q_i)^+$ represents its maximum attainable value; MCP refers to a characteristic being a subset of control parameters for the MIX algorithm, and SCP denotes a characteristic being a subset of control parameters for the SNDW algorithm. An iterative change in the control parameter value is based on a unit increment size, which can be formulated as follows:

$$f(x,q_i)^{mod} = f(x,q_i)^- + n \cdot \Delta f(x,q_i), \tag{30}$$

and

$$\Delta f(x,q_i) = \frac{1}{k}\left(f(x,q_i)^+ - f(x,q_i)^-\right), \;\; k \in Z, \;\; \forall k > 0, \tag{31}$$

where $f(x, q_i)^{mod}$ indicates a modified value of characteristic q_i for an object x, n denotes the multiplicity, and k denotes the number of intervals ($n = 1, 2, \ldots, k$).

Any modification of selected control parameters initiates the simulation of the sequence of both the MIX and SNDW optimisation algorithms in Step 6. Iteratively conducted Steps 5 and 6 allow for an assessment of the impact of control parameter modifications on the change in the number of constructed SNDW LUs. Based on the results of $n = 1, 2, \ldots, k$

iterations, the *MPA* classifier, reflecting the most probable allocation of a given object to one of the classes, is formulated as follows:

$$MPA = \begin{cases} In & \text{if} \quad a_{In}^{-} \leq \min_n(S_0 - S_n) \leq a_{In}^{+} \wedge b_{In}^{-} \leq \max_n(S_0 - S_n) \leq b_{In}^{+} \\ Ne & \text{if} \quad a_{Ne}^{-} \leq \min_n(S_0 - S_n) \leq a_{Ne}^{+} \wedge b_{Ne}^{-} \leq \max_n(S_0 - S_n) \leq b_{Ne}^{+} \\ Re & \text{if} \quad a_{Re}^{-} \leq \min_n(S_0 - S_n) \leq a_{Ne}^{+} \wedge b_{Re}^{-} \leq \max_n(S_0 - S_n) \leq b_{Re}^{+} \end{cases} \tag{32}$$

$$a_{MPA}^{-} \leq a_{MPA}^{+}, \; b_{MPA}^{-} \leq b_{MPA}^{+},$$
$$((a_{In}^{-}, a_{In}^{+}) \times (b_{In}^{-}, b_{In}^{+}) \cap (a_{Ne}^{-}, a_{Ne}^{+}) \times (b_{Ne}^{-}, b_{Ne}^{+}) \cap (a_{Re}^{-}, a_{Re}^{+}) \times (b_{Re}^{-}, b_{Re}^{+})) = \varnothing,$$

where S_0 denotes the quantity of the SNDW LUs constructed under default control parameters, whereas S_n represents the quantity of the SNDW LUs constructed under control parameters in the $n = 1, 2, \ldots, k$ iterations, $(a_{MPA}^{-}, a_{MPA}^{+})$ are the lower and upper limits of $\min(S_0 - Sn)$ for the *MPA* classifier, and $(b_{MPA}^{-}, b_{MPA}^{+})$ are the lower and upper limits of $\max(S_0 - Sn)$ for the *MPA* classifier, where *MPA* denotes the decision class *In*, *Ne*, *Re*, i.e., the possibility of increase, no change (neutrality), and reduction in the number of constructed SNDW LUs, respectively.

For object x classified into the *Re* class, it is possible to determine the minimum level of modified control parameters based on the simulations carried out. The application of these values forms the basis for reducing the number of SNDW LUs. This minimum level of the control parameter is determined as

$$MPA(x) = Re \rightarrow \begin{cases} f(x, q_i)^{mod} \geq f(x, q_i)^{Re}: \; \max_n (S_0 - S_n) \geq b_{Re}^{-}, \; \forall q_i \in \max(C^{>}) \\ f(x, q_i)^{mod} \leq f(x, q_i)^{Re}: \; \min_n (S_0 - S_n) \leq a_{Re}^{-}, \; \forall q_i \in \min(C^{>}) \end{cases} \tag{33}$$

for the parameter being maximised, i.e., $\max(C^{>})$, or minimised, i.e., $\min(C^{>})$, where $f(x, q_i)^{Re}$ denotes the recommended value for object x to achieve a reduction in SNDW LUs. Other notations are as previously defined.

The recommended value is assigned to the record for object x in the KDB as a supplementary value and is later used in the decision-making process in Step 4. For object x being assigned to *In* or *Ne*, it is null. From a practical perspective, the modification of control parameters is possible within an acceptable range, i.e., without exceeding the boundary values indicated by safety reasons, such as the stability of the products on the pallet, or organisational constraints, such as the ability to place the pallet in a rack slot of a specific height. All parameters, together with decision variables, are applied to the design of pallets in an optimal way. The result is equally important from an economic and environmental point of view. However, it is not the subject of this analysis to judge which step of the decision process relates to the aforementioned aspects. The result can be expressed not only in the number of SNDW LUs but also in economic and environmental measures. From an economic perspective, each SNDW LU reduction translates into savings and greater flexibility in matching different orders into a single delivery for customers. At the same time, a lower number of SNDW LUs reduces energy consumption. The number of products sent to each customer remains unchanged, but their reorganisation on LUs with new parameters after repeated optimisation allows for the design of stacked palletised load units more effectively.

4.5. Data Mining and Decision Rule Generation—Step 7

After performing Steps 5 and 6, a new object z, for which the decision class of *MPA* has been determined, is added to the KDB, becoming another object within the finite set of objects U. This signifies that the decision table gains an additional instance, enabling comprehensive data mining, which constitutes the essence of Step 7.

Data mining in Step 7 encompasses a set of routine calculations that form the core of DRST. At the beginning of this step, the consistency of information contained in the expanded decision table is established by determining the upper and lower approximations of the union of classes (see Section 3.2). Since there are three defined decision classes, i.e., In, Ne, and Re, four unions of classes are considered, i.e., at most In denoted as $In^{\leq}$, at most Ne denoted as $Ne^{\leq}$, at least Ne denoted as $Ne^{\geq}$, and at least Re denoted as $Re^{\geq}$ (see Figure 6). Subsequently, based on this assessment, the consistency and quality of classification are evaluated (see Section 3.3).

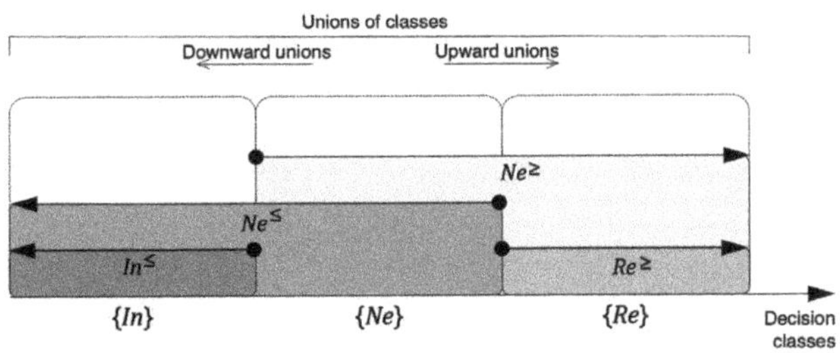

Figure 6. Unions of classes defined for decision classes In, Ne, and Re.

The greater the consistency and quality of the classification within the decision table, the higher the probability of obtaining credible decision rules. Rule generation, as a generalisation of knowledge contained in the decision table, is performed in the summary of Step 7. The principles of rule generation are presented in Section 3.4. However, in the proposed procedure, it is assumed that classification is conducted solely based on exact rules, i.e., those based exclusively on the lower approximation of classes. This approach reduces the number of rules and allows the generation of stable rules supported by unambiguous examples. A possible action to further reduce the number and complexity of the decision rules is to limit the number of parameters analysed. Such an analysis can be conducted by deleting redundant characteristics in the reducts and the core of classification. However, this is beyond the scope of this paper.

Ultimately, a new set of decision rules is updated, replacing the existing set of rules contained in the KDB (see R5 in Figure 5). This allows for current decision-making regarding each new object subjected to evaluation in Step 3.

5. Experimental Application of Expert System

5.1. Context of Decision-Making and Experiments

The proposed methodology has been verified with one of the largest market players in the Central and Eastern European food industry, the Maspex Group. This company manufactures products and distributes them to customers, mainly retail chains and wholesalers. Since sustainable development has become one of the key competitive advantages in the market, environmental responsibility, as well as social and economic considerations, are important aspects of the company's operations. Customer orders are characterised by a high diversity of products and order volumes. A significant phenomenon in these orders is quantitative fragmentation, i.e., small quantities of a large variety of products, necessitating the formation of heterogeneous pallet load units, referred to as MIX LUs.

The paper focuses on applying the developed methodology to process orders for which the customer allows the creation of load units:

- Composed of multiple products, i.e., heterogeneous MIX units (see Figure 7a).

- Composed of a single product, where the quantity of product units constitutes a full pallet load unit, i.e., FULL (see Figure 7b).
- Composed of a single product, where the quantity of product units is less than a full pallet load unit, i.e., SKU (see Figure 7c).
- Stacked, where the SNDW unit is formed from other MIX and/or SKU units (see Figure 7d).

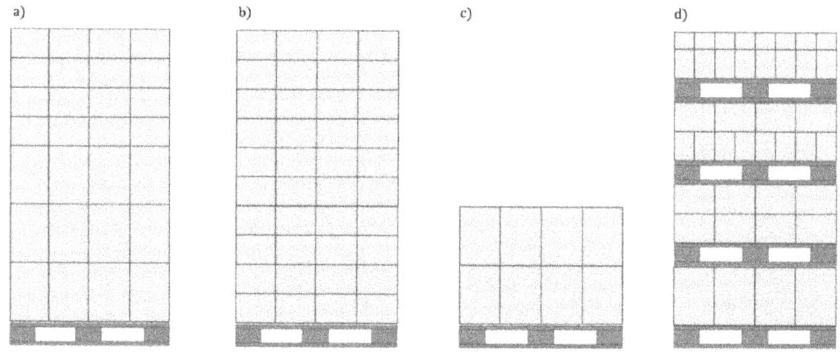

Figure 7. Types of load units applied in the research, i.e., elementary units, including (**a**) MIX, (**b**) FULL, (**c**) SKU load, and a set of elementary units (**d**) SNDW.

When shaping heterogeneous LUs, i.e., MIX and SNDW, a significant limitation is the strength of individual products. Each product belongs to one of six fragility classes (FC), i.e., $FC = 0, 1, \ldots, 5$, where the lower the value of the fragility class, the stronger the packaging and the product itself. For example, in the FMCG sector considered, a typical product with the lowest fragility class, i.e., $FC = 0$, is a multipack of juices in carton–aluminium packaging or a multipack of water in plastic or glass containers. In contrast, $FC = 5$ refers to multipacks of tea in cardboard packaging or instant products in foil packaging. The FC parameter is assigned once to a specific product configuration and is based on laboratory strength tests.

In the subsequent parts of this section, the key application context of the proposed methodology is presented. It begins with information on the construction of the KDB, followed by descriptions of the learning and testing phases, and concludes with an exemplary application to classify new orders.

5.2. Knowledge Database

The KDB, presented in Section 3.1 as a decision table, serves in this methodology as a matrix to collect comprehensive information on individual orders. Each order represents a single object in the KDB, i.e., a single record composed of conditional characteristics C (both criteria and attributes), decision characteristics D, and supplementary information. It is assumed that the structure of an individual record includes a set of characteristics, starting with the conditional ones C, in the following order:

- Structural parameters of the items, i.e., products, included in the order (O_i).
- Control parameters used in the MIX algorithm and SKU LUs (MCP_i).
- Output of the MIX algorithm, as well as characteristics of other elementary LUs in the order, i.e., SKU and FULL LUs (M_i).
- Control parameters used in the SNDW algorithm (SCP_i).
- Output of the SNDW algorithm (S_i).

The decision characteristic (D) entails the allocation of each order to a specific class of objects, directly resulting from the MPA. Supplementary to the characteristics contained in a single KDB record is information about the unique order number and the ordinal number of the KDB record (No), as well as the recommended value of the control parameters (SU_i). The detailed structure of the record is presented in Figure 8, and the definition of the name and domain of each characteristic is included in Table 2.

(a) **No:** $O_1, O_2, O_3, O_4, O_5, O_6, O_7, O_8, O_9, O_{10}, O_{11}, O_{12}, O_{13}, O_{14}, O_{15}, O_{16}, O_{17}, O_{18}, O_{19}, O_{20}, O_{21}, O_{22}, MCP_{23}, MCP_{24}, MCP_{25}, MCP_{26}, MCP_{27}, M_{28}, M_{29}, M_{30}, M_{31}, M_{32}, M_{33}, M_{34}, M_{35}, M_{36}, M_{37}, M_{38}, SCP_{39}, SCP_{40}, S_{41}, MPA_{42}, SU_{43}$

(b) **7845719:** 1, 0, 0, 0, 0, 1, 0, 0, 0, 0, 0, 1, 0, 0, 0, 0, 1, 0, 0, 0, 0, 0, 1450, 900, 10, 0.7, 1, 4, 0, 0, 0, 0, 0, 0, 0, 0, 0, 0, 1800, 750, 4, Ne, (0, 0)

Figure 8. Structure of the KDB, including (**a**) names and arrangement of KDB record characteristics and (**b**) an exemplary record of the KDB.

Table 2. Names and domains of individual KDB record characteristics.

Symbol	Name (Unit)	Domain	Q Type [1], Pref. [2]
No	Record ID (–)	Integer	Su, n/a
O_1	Number of items in order (pcs.)	Integer	$C^=$, n/a
O_2	Diversity of FC [3] in order (pcs.)	Integer	$C^=$, n/a
O_3	Number of FC in order (pcs.)	Integer	$C^=$, n/a
O_4	Weight share of products with FC = 0 (–)	Real	$C^=$, n/a
O_5	Weight share of products with FC = 1 (–)	Real	$C^=$, n/a
O_6	Weight share of products with FC = 2 (–)	Real	$C^=$, n/a
O_7	Weight share of products with FC = 3 (–)	Real	$C^=$, n/a
O_8	Weight share of products with FC = 4 (–)	Real	$C^=$, n/a
O_9	Weight share of products with FC = 5 (–)	Real	$C^=$, n/a
O_{10}	Quantitative share of products with FC = 0 (–)	Real	$C^=$, n/a
O_{11}	Quantitative share of products with FC = 1 (–)	Real	$C^=$, n/a
O_{12}	Quantitative share of products with FC = 2 (–)	Real	$C^=$, n/a
O_{13}	Quantitative share of products with FC = 3 (–)	Real	$C^=$, n/a
O_{14}	Quantitative share of products with FC = 4 (–)	Real	$C^=$, n/a
O_{15}	Quantitative share of products with FC = 5 (–)	Real	$C^=$, n/a
O_{16}	Quantitative share of full-layer products (–)	Real	$C^>$, max
O_{17}	Maximum height difference of products with FC = 0 (mm)	Integer	$C^>$, min
O_{18}	Maximum height difference of products with FC = 1 (mm)	Integer	$C^>$, min
O_{19}	Maximum height difference of products with FC = 2 (mm)	Integer	$C^>$, min
O_{20}	Maximum height difference of products with FC = 3 (mm)	Integer	$C^>$, min
O_{21}	Maximum height difference of products with FC = 4 (mm)	Integer	$C^>$, min
O_{22}	Maximum height difference of products with FC = 5 (mm)	Integer	$C^>$, min
MCP_{23}	Maximum height of MIX or SKU LU in order (mm)	Integer	$C^>$, max
MCP_{24}	Maximum weight of MIX or SKU LU in order (kg)	Integer	$C^>$, max
MCP_{25}	Tolerance of height difference of products on the same layer of MIX (mm)	Integer	$C^>$, max
MCP_{26}	Lower limit of filling with products the upper layer of MIX/SKU LU (–)	Real	$C^>$, min
MCP_{27}	Upper limit of filling with products the upper layer of MIX/SKU LU (–)	Real	$C^>$, max
M_{28}	Number of MIX LU built upon order's parameters (pcs.)	Integer	$C^>$, min
M_{29}	Number of SKU LU in order (pcs.)	Integer	$C^>$, min
M_{30}	Number of FULL LU in order (pcs.)	Integer	$C^>$, min
M_{31}	Height difference between the highest and the lowest MIX LU in order (mm)	Integer	$C^=$, n/a
M_{32}	Height difference between the highest and the lowest SKU LU in order (mm)	Integer	$C^=$, n/a
M_{33}	Height difference between the highest and the lowest FULL LU in order (mm)	Integer	$C^=$, n/a
M_{34}	Weight difference between the heaviest and the lightest MIX LU in order (kg)	Real	$C^=$, n/a
M_{35}	Weight difference between the heaviest and the lightest SKU LU in order (kg)	Real	$C^=$, n/a
M_{36}	Weight difference between the heaviest and the lightest FULL LU in order (kg)	Real	$C^=$, n/a
M_{37}	Quantitative share of MIX LU with TOP status (pcs.)	Real	$C^>$, min
M_{38}	Quantitative share of SKU LU with TOP status (pcs.)	Real	$C^>$, min
SCP_{39}	Maximum height of SNDW LU in order (mm)	Integer	$C^>$, max
SCP_{40}	Maximum weight of SNDW LU in order (kg)	Integer	$C^>$, max
S_{41}	Number of SNDW LU built upon control parameters in the order (pcs.)	Integer	$C^>$, min
MPA_{43}	The most probable assignment $\{In, Ne, Re\}$	String	D, max
SU_{43}	Supplementary information, i.e., the recommended value of control parameter	String	Su, n/a

[1] Type of characteristics: Su—supplementary; C—conditional, i.e., $C^>$—criterion or $C^=$—attribute; D—decision;
[2] Preference direction: n/a—not applicable, min—minimised, max—maximised; [3] FC = fragility class.

5.3. Learning Phase

The application of the developed methodology requires an initial learning phase, which aims to develop and gather the necessary knowledge to make current operational decisions, i.e., to classify incoming customer orders. Referring to Figure 5 (see Section 4.1), the course of the learning phase is as follows:

- A set of historical orders, meeting the basic characteristics discussed in Section 5.1, is retrieved from the WMS, and this set serves as the testing set of orders.

- For each order, Steps 1 and 2 are executed using default parameters, and their effects are recorded as successive data in the KDB, according to the structure described in Section 5.2.
- In Step 3, there are no decision rules that can be applied to classify objects—all orders in the KDB are directed to the N path for the execution of Steps 5–7.
- In Step 5, the variability range of selected control parameters from MCP_i and/or SCP_i is determined for each order, as described in Section 4.4 (see Equations (29)–(31)).
- In Step 6, for each modified control parameter (see Step 5), a simulation of the MIX and SNDW algorithms is conducted, resulting in the determination of the number of SNDW LUs (S_{41} parameter). Subsequently, classification with MPA is carried out based on Equation (32).
- In Step 7, the principles of DRST are applied, including the approximation of the union of classes (see Section 3.2), the evaluation of knowledge consistency (see Section 3.3), and the generation of decision rules (see Section 3.4).

The learning phase enables the creation of a set of decision rules and their recording in the KDB. Based on these rules, it is possible to classify new orders (see Step 3 in Figure 5). In the learning phase, the following assumptions and settings are adopted:

- The set of orders is retrieved from the WMS and consists of 709 items.
- Two control parameters are modified, namely MCP_{23}, i.e., maximum height of MIX and SKU LUs, in (mm), and MCP_{24}, i.e., maximum weight of MIX and SKU LUs, in (kg).
- The range of variations of both parameters is $f(x,q_i)^- = 0.7{\cdot}f(x,q_i)^{\text{def}}$ and $f(x,q_i)^+ = 1.1{\cdot}f(x,q_i)^{\text{def}}$, where $q_i = (MCP_{23}, MCP_{24})$, and the number of intervals is $k = 9$.
- The complete set of solutions is explored, i.e., MIX-SNDW results are generated for combinations of $V_{MCP_{23}} \times V_{MCP_{24}}$, i.e., totalling 81 combinations $(9 \cdot 9)$ and resulting in 57,429 simulations.
- The input file with the empty value of MPA_{42} is available at the Mendeley Data repository: https://doi.org/10.17632/4dwry6rhk4.1. Supplementary information (see SU_{43}) differentiates the results of simulations obtained using default (d) and modified (m) values of parameters MCP_{23} and MCP_{24}.

For these assumptions, a series of experiments are carried out, and the results for selected orders are depicted in Figure 9. Changes in control parameters MCP_{23} and MCP_{24} can lead to no change in the number of SNDW LUs (see Figure 9a), or only an increase in the number of SNDW LUs (see Figure 9c), or both an increase and a reduction in the number of SNDW load units (see Figure 9b,d).

The results of all simulations for this set of orders are initially assigned to one of the classes, i.e., In, Ne, or Re (see Section 4.4, Equation (32)). The solutions are grouped according to the increase in the maximum value of the parameter S_{41}, as presented in Figure 10. Establishing the boundary between solution classes for the MPA classifier is not a trivial task. Therefore, three decision strategies are created based on the difference between the values of S_0 and S_n, that is, $(S_0 - S_n)$, resulting from changes in control parameters, namely the following:

- Strategy 1 (see Figure 10a and Table 3)—orders belonging to class In are those for which the difference $(S_0 - S_n)$ indicates a significant increase in the SNDW LUs; orders in the Ne class indicate a situation with no change or a small change in comparison with the S_0 value. The Re class includes orders with the maximum value of the difference $(S_0 - S_n)$ that is greater than or equal to 1, regardless of the minimum value of $(S_0 - S_n)$.

- Strategy 2 (see Figure 10b and Table 3)—this strategy differs from Strategy 1 by a slight modification to the boundary between the In and Ne classes. In Strategy 2, the orders assigned to the Ne class represent solutions where the difference $(S_0 - S_n)$ equals zero only. Consequently, the difference $(S_0 - S_n)$ of minus one (-1) is assigned to the In class. The assignment of the remaining orders is the same as in Strategy 1.
- Strategy 3 (see Figure 10c and Table 3)—orders belonging to class In are those indicating a significant increase in SNDW LUs along with changes in control parameters, and a slight reduction in SNDW LUs is also possible. The Ne class tends to compensate for the difference $(S_0 - S_n)$ between the resulting minimum and maximum values, i.e., on one side, there are orders with a difference $(S_0 - S_n)$ of minus six (-6), and on the other side, orders with a difference of plus four $(+4)$. The class Re is oriented towards solutions aimed at reducing SNDW LUs while allowing for the inclusion of solutions that result in a slight increase in the number of SNDW LUs.

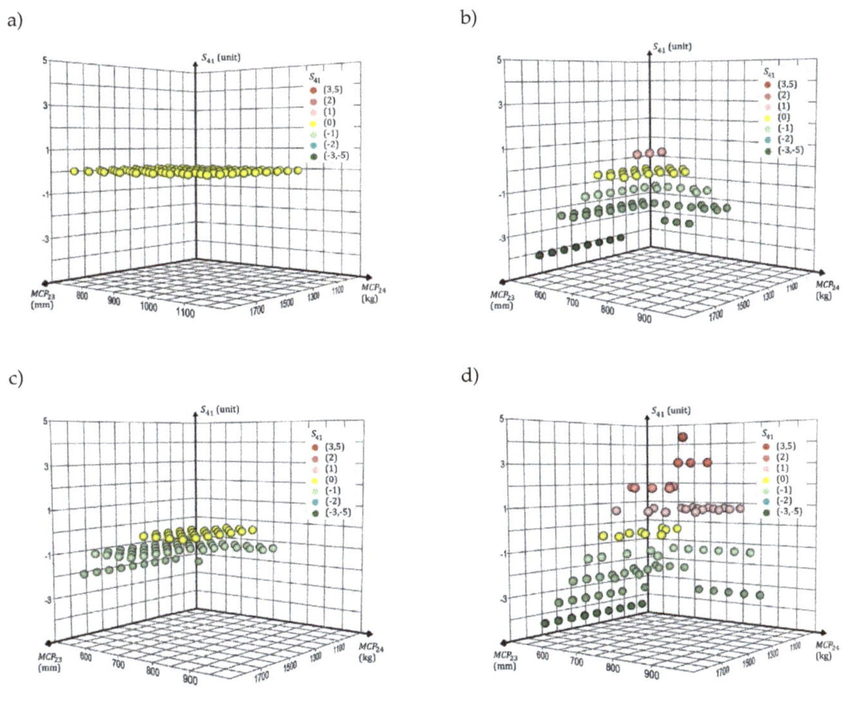

Figure 9. Relationship between control parameters, namely maximum height (MCP_{23}) and maximum weight (MCP_{24}) of SNDW pallets, and the number of created SNDW pallets (S_{41}). Specifically, red, yellow, and green dots represent a reduction, no change, and an increase in the number of SNDW pallets, respectively. The resulting number of SNDW pallets is as follows: (**a**) remains unchanged; (**b**) significantly increases and occasionally shows a slight decrease; (**c**) remains constant or increases; (**d**) demonstrates a tendency for significant increases and reductions.

Table 3. Results of the analysis of the set of objects under DRST principles.

Decision Class	Limits	Strategies		
		Strategy 1	**Strategy 2**	**Strategy 3**
In	$(a_{In}^{-}, a_{In}^{+}) \cap (b_{In}^{-}, b_{In}^{+})$	$(-5,-2) \cap (0,0)$	$(-5,-1) \cap (0,0)$	$(-7,-2) \cap (0,1)$
Ne	$(a_{Ne}^{-}, a_{Ne}^{+}) \cap (b_{Ne}^{-}, b_{Ne}^{+})$	$(-1,0) \cap (0,0)$	$(0,0) \cap (0,0)$	$(-1,0) \cap (0,1)$ OR $(-6,-4) \cap (2,4)$
Re	$(a_{Re}^{-}, a_{Re}^{+}) \cap (b_{Re}^{-}, b_{Re}^{+})$	$(-7,0) \cap (1,4)$	$(-7,0) \cap (1,4)$	$(0,0) \cap (1,1)$ OR $(-1,-1) \cap (2,2)$

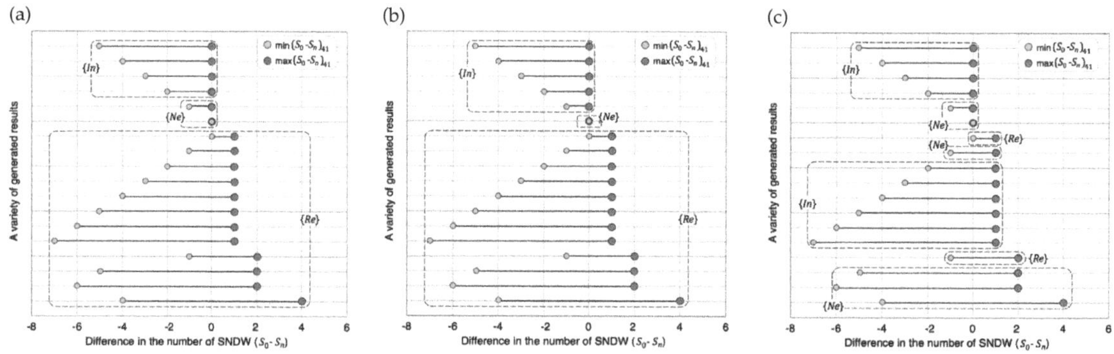

Figure 10. Three strategies of MPA value assignment, i.e., In, Ne, and Re, based on the minimum and maximum value of the difference $(S_o - S_n)$ in the number of SNDW LUs, including (**a**) Strategy 1, (**b**) Strategy 2, and (**c**) Strategy 3.

Based on the assumed classification of objects, i.e., the assignment of each order to the class resulting from MPA, a series of calculations are conducted according to the principles of DRST, separately for each strategy. These calculations include determining the approximations of the unions of classes, assessing knowledge consistency, and generating decision rules. For this purpose, the jMAF software [134] has been used, and a summary of the results is presented in Table 4. The following section discusses the practical utility of each strategy and its consistency with the set of orders under Strategies 1–3.

Table 4. Results of the analysis of the set of orders under Strategies 1–3 using DRST principles.

Profile of Analysis		Strategies		
		Strategy 1	**Strategy 2**	**Strategy 3**
Number of objects assigned to MPA	Class In	48	144	74
	Class Ne	581	486	603
	Class Re	80	79	32
	Total	709	709	709
Accuracy (and cardinality) of union of classes	At most In	1.000 (48)	0.959 (144)	1.000 (74)
	At least Ne	1.000 (661)	0.989 (564)	1.000 (635)
	At most Re	0.995 (629)	0.995 (629)	0.996 (677)
	At least Re	0.963 (80)	0.963 (79)	0.912 (32)
Quality of approximations	–	0.996	0.987	0.996
Number of certain decision rules	Standard unions	290	354	252
	Monotonic unions	284	351	247

A comparison of the results presented in Table 4 indicates that each of the applied strategies leads to a different allocation of 709 orders to the respective classes, i.e., In, Ne, and Re. For instance, class In is represented by 48 orders in Strategy 1, 74 orders in Strategy 3, and 144 orders in Strategy 2. From the perspective of information consistency within these sets of orders, Strategies 1 and 3 exhibit higher approximation accuracy indices compared with Strategy 2. However, in Strategy 3, the approximation of the union of classes at least Re performs relatively worse compared with Strategies 1 and 2, with a score of 0.912. Globally, the quality of the approximations based on Strategies 1 and 3 is at a high level of 0.996, making both strategies the most attractive. The final selection, however, is determined after performing the tests described in the following section.

5.4. Testing Phase

The outcome of the conducted learning phase results in the assignment of all objects in the set of orders to MPA according to Strategies 1–3. This forms the foundation for assessing knowledge consistency and generating decision rules. The testing phase, which is the subject of this section, aims to determine the final strategy (one of 1–3) for classifying new orders. The essence of the testing phase, conducted separately for the set of orders under Strategies 1–3, is as follows:

- Test 1 (Reclassification): This test involves subjecting the generated set of decision rules to a procedure of reclassifying all training objects. That is, all orders that served as the basis for generating the rules are reclassified. The difference between the original classification, i.e., the MPA value, and the classification suggested by the decision rules determines the effectiveness of the rule set.
- Test 2 (Cross-validation): This test involves dividing the testing set into p-subsets, where $(p-1)$ subsets constitute the training set (used to generate decision rules), and a single subset p constitutes the test set, which serves as the basis for reclassification, similar to Test 1. Test 2 is repeated p times so that, in each iteration, the subset p appears only once as the test set.

These two tests serve distinct purposes: the reclassification test evaluates the accuracy of rule generation on known examples (see Test 1), while the cross-validation test evaluates the performance of decision rules when applied to previously unseen objects (see Test 2). In the reclassification test, the same dataset previously used to generate the rules serves as the training set. Consequently, this test evaluates only the quality of decisions that can be made based on one or more decision rules aligned with the reclassified objects. The handling of negative examples in the cross-validation test involves two distinct approaches: (1) Covers only inconsistency: Negative examples are classified as negative only when they violate rule consistency. This approach focuses on detecting errors caused by rule imprecision; (2) covers none: this stricter approach assumes that the rules do not classify any negative examples as belonging to a class, thereby eliminating the risk of false positives.

In the testing phase, the following assumptions and settings are adopted:

- The reclassification test (Test 1): This test is performed alternatively for standard and monotonic unions of classes. Alternative classification principles are also considered, namely, the Dominance-Based Rough Sets Approach (DRSA) and the Variable Consistency Dominance-Based Rough Sets Approach (VCDRSA) [126].
- The cross-validation test (Test 2): This test is conducted at a consistency level of 1.0, with the testing set of orders divided into $p = 10$ subsets. The classification algorithms considered include DRSA and VCDRSA. An interpretation of classification errors is also carried out, considering two distinct cases: it covers only inconsistent (ambiguous) assignments, or it covers orders with no assignment, i.e., no rule matches the order.

For these assumptions, a series of experiments is conducted using jMAF software (downloadable at: http://www.cs.put.poznan.pl/jblaszczynski/Site/jRS.html, version-08.03.2019). The obtained results of the reclassification test are presented in Table 5, while the results of the cross-validation test are shown in Table 6.

To refine the assessment of classification efficiency for new objects within cross-validation, a confusion matrix as a heat map is created for each test profile (see Table 7).

Table 5. Results of the reclassification test.

Test Profile			Testing Set of Orders		
Union Type	**Classification Method**	**Classification Result**	**Strategy 1**	**Strategy 2**	**Strategy 3**
Standard	DRSA	Correct Incorrect Unknown	708 (99.86%) 1 (0.14%) 0 (0.00%)	704 (99.43%) 4 (0.56%) 0 (0.00%)	708 (99.86%) 1 (0.14%) 0 (0.00%)
	VCDRSA	Correct Incorrect Unknown	700 (98.73%) 9 (1.27%) 0 (0.00%)	677 (95.62%) 31 (4.38%) 0 (0.00%)	679 (95.77%) 30 (4.23%) 0 (0.00%)
Monotonic	DRSA	Correct Incorrect Unknown	708 (99.86%) 1 (0.14%) 0 (0.00%)	704 (99.43%) 4 (0.56%) 0 (0.00%)	708 (99.86%) 1 (0.14%) 0 (0.00%)
	VCDRSA	Correct Incorrect Unknown	695 (98.02%) 14 (1.97%) 0 (0.00%)	676 (95.48%) 32 (4.52%) 0 (0.00%)	685 (96.62%) 24 (3.38%) 0 (0.00%)

Table 6. Results of the cross-validation test.

Test Profile			Testing Set of Orders		
Classification Algorithm	**Negative Example Treatment**	**Classification Result**	**Strategy 1**	**Strategy 2**	**Strategy 3**
DRSA	Covers only inconsistency	Correct Incorrect Unknown	569 (80.25%) 140 (19.75%) 0 (0.00%)	550 (77.68%) 158 (22.32%) 0 (0.00%)	610 (86.04%) 99 (13.96%) 0 (0.00%)
	Covers none	Correct Incorrect Unknown	572 (80.68%) 137 (19.32%) 0 (0.00%)	550 (77.68%) 158 (22.32%) 0 (0.00%)	596 (84.06%) 113 (15.94%) 0 (0.00%)
VCDRSA	Covers only inconsistency	Correct Incorrect Unknown	534 (75.32%) 94 (13.26%) 81 (11.42%)	448 (63.28%) 147 (20.76%) 113 (15.96%)	556 (78.42%) 91 (12.83%) 62 (8.74%)
	Covers none	Correct Incorrect Unknown	528 (74.47%) 99 (13.96%) 82 (11.56%)	452 (63.84%) 137 (19.35%) 119 (16.81%)	547 (77.15%) 98 (13.82%) 64 (9.03%)

Table 7. Confusion matrix of the cross-validation test (normalised).

Test Profile		Testing Set of Orders											
Classification Algorithm	**Negative Example Treatment**	**Strategy 1**				**Strategy 2**				**Strategy 3**			
DRSA	Covers only inconsistency		*In*	*Ne*	*Re*		*In*	*Ne*	*Re*		*In*	*Ne*	*Re*
		In	0.40	0.35	0.25	*In*	0.41	0.43	0.16	*In*	0.74	0.24	0.01
		Ne	0.05	0.92	0.03	*Ne*	0.02	0.96	0.02	*Ne*	0.08	0.91	0.01
		Re	0.36	0.41	0.23	*Re*	0.44	0.22	0.34	*Re*	0.09	0.63	0.28
	Covers none		*In*	*Ne*	*Re*		*In*	*Ne*	*Re*		*In*	*Ne*	*Re*
		In	0.38	0.4	0.23	*In*	0.4	0.42	0.18	*In*	0.7	0.28	0.01
		Ne	0.05	0.91	0.04	*Ne*	0.02	0.96	0.02	*Ne*	0.1	0.89	0.01
		Re	0.28	0.41	0.31	*Re*	0.47	0.2	0.33	*Re*	0.09	0.63	0.28
VCDRSA	Covers only inconsistency		*In*	*Ne*	*Re*		*In*	*Ne*	*Re*		*In*	*Ne*	*Re*
		In	0.14	0.81	0.06	*In*	0.22	0.69	0.08	*In*	0.13	0.87	0
		Ne	0.01	0.97	0.02	*Ne*	0.06	0.93	0.01	*Ne*	0.03	0.97	0
		Re	0.05	0.78	0.17	*Re*	0.22	0.51	0.27	*Re*	0	0.66	0.34
	Covers none		*In*	*Ne*	*Re*		*In*	*Ne*	*Re*		*In*	*Ne*	*Re*
		In	0.07	0.89	0.04	*In*	0.28	0.68	0.04	*In*	0.14	0.86	0
		Ne	0.02	0.96	0.03	*Ne*	0.06	0.94	0	*Ne*	0.04	0.96	0
		Re	0.07	0.79	0.14	*Re*	0.24	0.55	0.2	*Re*	0	0.69	0.31

In each matrix, the rows indicate the cardinality of objects based on their original classification, i.e., In, Ne, and Re, according to the testing set of orders, while the columns indicate the cardinality of objects based on the predicted classification derived from decision rules. Thus, the diagonal of the confusion matrix shows the cardinality of objects for which the original classification and its prediction are consistent. The diagonal-off values highlight the discrepancies between the classes.

Since the cardinality of objects according to the original classification varies across Strategies 1–3 (see Table 4), each matrix is normalised to the range $(0, 1)$, ensuring that the sum of each row in each matrix equals 1. In Figure 7, consistency between the original and predicted classifications is distinguished by colour, using a scale where 0.01 corresponds to light blue and 1.0 to dark blue. The table containing matrices expressed in absolute numbers of objects within each class, i.e., before normalisation, is provided in Appendix B (see Table A2).

The interpretation of the results presented in Table 7 is as follows:

- In all test profiles, the highest consistency between the original classification and the prediction is achieved for Ne, ranging from 0.89 to 0.97. For In, the consistency ranges from 0.07 to 0.74, and for Re, it ranges from 0.14 to 0.34.
- For Strategy 1, the highest consistency between the original classification and the prediction of classes In, Ne, and Re is achieved with the application of the DRSA algorithm, assuming the lack of matching rules, namely, "covers only inconsistency", with respective values of 0.40, 0.92, and 0.23. A relatively high number of errors occurs for Re, where the predictions fall into Ne (the consistency value is 0.78; see the VCDRSA classifier and the "covers only inconsistency" test). The lowest confusion value concerns Ne in the VCDRSA test—"covers only inconsistency"—with a value of 0.01 for predictions to In classes.
- For Strategy 2, the highest consistency between the original classification and the prediction of classes In, Ne, and Re is achieved for the DRSA test—"covers only inconsistency"—with values of 0.41, 0.96, and 0.34, respectively. Relatively high confusion occurs for In, where the predictions fall into Ne (consistency value is 0.69, using the VCDRSA test—"covers only inconsistency"). The lowest confusion value concerns Ne in the VCDRSA test—"covers only inconsistency"—and the VCDRSA test—"covers none", with a value of 0.01 for predictions to the Re classes.
- For Strategy 3, the highest consistency is achieved with the application of the DRSA test—"covers only inconsistency"—with values of 0.74, 0.91, and 0.28 for classes In, Ne, and Re, respectively. Relatively high confusion occurs for In, where predictions fall into Ne, with a consistency value of 0.87 using the VCDRSA test—"covers only inconsistency". The lowest confusion value concerns In, Ne, and Re classes in the VCDRSA test—"covers only inconsistency", with a value of 0 for predictions to Ne, Re, and In classes, respectively. The same result is achieved for the VCDRSA test—"covers none".

These observations confirm the highest predictive usefulness of the set of orders designed under Strategy 3. The reasons primarily lie in the method of assigning simulation results to one of the decision classes, i.e., through the use of the MPA classifier (see Figure 10). In the case of Strategy 3, only objects for which the interval between $max(S_0 - S_n)_{41}$ and 0 is greater than the interval between $min(S_0 - S_n)_{41}$ and 0 are considered to belong to the class Re. This scenario can be described as a case of positive asymmetry. In contrast, for Strategies 1 and 2, the Re class also includes cases of negative asymmetry or symmetry. This leads to greater inconsistency in the test results based on Strategies 1 and 2 compared with those derived from Strategy 3. Therefore, Strategy 3 is adopted as the baseline for the next phase, i.e., the classification of new objects.

5.5. Application Phase

Based on the actions undertaken during the learning and testing phases (see Sections 5.3 and 5.4), the set of orders under Strategy 3 is recommended as the most suitable. This strategy, and parametrisation of the considered set in the form of classification of objects into *In*, *Ne*, or *Re* classes, is adopted as the standard and used in the application phase for making operational decisions, i.e., classifying new orders. Referring to Figure 5 (see Section 4.1), the course of the application phase is as follows:

- New orders, meeting the basic characteristics discussed in Section 5.1, are individually, i.e., one by one, retrieved from the WMS.
- For each order, optimisation in Steps 1 and 2 is performed, and their results are recorded in the KDB as subsequent data according to the adopted structure (see Section 5.3).
- In Step 3, each order is classified into *In*, *Ne*, or *Re* classes using the available decision rules. The assessment determines whether the available decision rules are applicable to the specific order. If applicable, the order is directed along the Y path; otherwise, the N path is selected. If there is a match between the order characteristics and the rules, the next question is whether a reduction in the number of SNDW LUs is possible. In this case, the following options can be considered: (a) If the rules classify the order into the *In* or *Ne* classes, the solution developed in Steps 1 and 2 is final, and the order is directed for execution accordingly. (b) If the rules classify the order into the *Re* class, Step 4 is executed, and the key actions are discussed below. (c) If the classification result is ambiguous, the order follows the sequence of Steps 5–7.
- In Step 4, new parameters obtained in Steps 1 and 2 that are necessary to achieve a reduction in SNDW LUs are applied. The decision rule classifying the order into the *Re* class, contained in the set P2 (see Figure 5), specifies in its descriptive part the minimum values of the control parameters required to achieve the reduction in SNDW LUs (see $f(x,\ q_i)^{mod}$ in Equation (29), i.e., $q_i = MCP_{23} \wedge MCP_{24}$). Based on the modified control parameters, the MIX and SNDW algorithms are rerun, generating a new optimal solution, which is then directed for execution.
- If Step 3 results indicate that no rules are consistent with the order characteristics (see the N path in Figure 5), or if the available rules suggest an ambiguous classification (see the Am path), the sequence of Steps 5–7 is executed, following the procedure described in the learning phase (see Section 5.3). The KDB, augmented with each new record, serves as the basis for performing all operations according to the relevant DRST, and new decision rules are generated for the updated set. Upon completion of Step 7, the order is redirected to Step 3, where a decision is made following the procedure described above.

Considering all procedural steps in the application phase, the following assumptions are made in the experiments:

- The KDB is populated with a set of 709 orders, and their assignment to one of the *In*, *Ne*, or *Re* classes is performed under Strategy 3.
- The KDB initially contains a set of 252 certain decision rules generated based on the standard union of classes (see Table 4).
- The set of new objects to be classified in the application phase comprises 20 orders; the input file is available in the Mendeley Data repository: https://doi.org/10.17632/4dwry6rhk4.1.

A series of experiments has been conducted under these assumptions, and successive new orders have been iteratively retrieved. The detailed compilation of results for each iteration, i.e., for each of the 20 new orders, is presented in Table A3, in Appendix C. These results are divided into four parts:

- Part 1 presents the decisions made.
- Part 2 describes the changes in the KDB content.
- Part 3 shows the results of the cross-validation for the variable KDB.
- Part 4 includes the diagonal part of the confusion matrix.

In Part 1 of Table A3 (see Appendix C), the classification results of the new orders according to the *MPA* and based on decision rules are recorded. If a decision is correct, it is marked as "correct"; otherwise, it is marked as "incorrect". If no decision is made, it is marked as "unknown". The evaluation of the accuracy of decisions is possible because each of the 20 orders is classified in parallel according to Strategy 3, as described in Section 5.4 during the learning phase. The outcome of the iterative classification of these 20 orders is summarised in Figure 11a. In Step 3, if there is no classification based on the rules, i.e., if the decision is marked as "unknown" or "incorrect", the order is directed along the path for Steps 5–7, as presented in Figure 5. In this case, the control parameters are modified (see Step 5), and based on the MIX and SNDW simulations, the classification according to *MPA* is made (see Step 6). In Step 7, analyses using DRST are repeated to evaluate the consistency of information, new decision rules are generated, and the KDB is updated. The results of this stage are included in Part 2 of Table A3, which shows the number of objects in the database, the quality of classification in each iteration, and the number of generated decision rules. These results are visualised in Figure 11b.

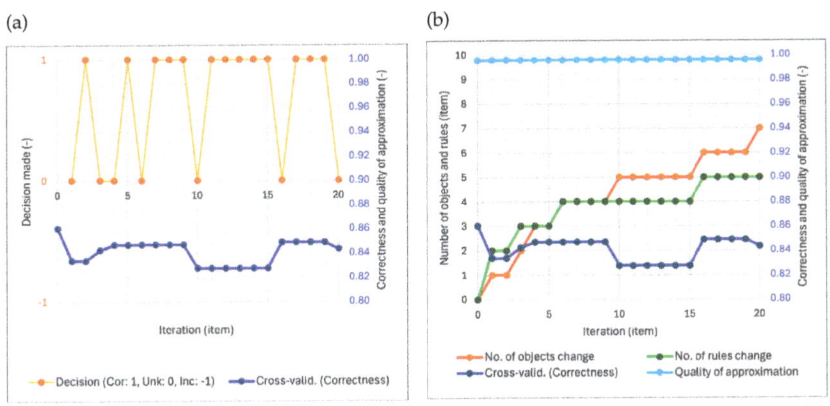

Figure 11. Graphical interpretation of application phase results, considering its iterative nature, including (**a**) correctness of cross-validation test results against the type of decision made, (**b**) changes in the total number of objects (*In*, *Ne*, and *Re* in total) against the increase in the number of objects in the KDB, the number of generated rules, and the quality of approximation.

Part 3 of Table A3 presents the cross-validation results for the iteratively updated KDB. With the addition of new orders to the KDB, cross-validation is repeated, allowing for tracking iterative changes in the number of correctly classified objects. These results are depicted in Figure 11a,b as reference points in both analyses. Part 4 of Table A3 indicates the number of objects correctly classified into *In*, *Ne*, or *Re* classes, representing the diagonal of the confusion matrix after each iteration. Both summaries presented in Figure 11a,b demonstrate the following:

- Sixty-five percent of the decisions, i.e., 13 out of 20, are made correctly and are based on the decision rules available in the KDB. There are no incorrect decisions recorded at this stage, as shown in the chart with decision values 0 and 1 in Figure 11a.
- Thirty-five percent of the decisions are marked as "unknown", i.e., they are represented by the value 0 in the chart of decisions made in Figure 11a. In these cases, there are no matching rules in the available set of decision rules. Consequently, the procedure for expanding the KDB and increasing the set of decision rules is initiated.
- The result of the cross-validation test, marked as the navy-blue line in Figure 11a, indicates that as the database expands with additional orders, the accuracy of the classification evolves. This evolution is characterised by deterioration (iterations 1, 10, and 20), stability (iteration 6), or improvement (iterations 3, 4, and 16). In general, classification accuracy remains at an average level of 84% over 20 iterations.
- Increasing the number of orders in the KDB leads to a proportional increase in the number of decision rules, as illustrated in Figure 11b. After 20 iterations, the expansion of the KDB by five objects resulted in an increase of the rule set by 7.
- Changes in the size of the KDB do not result in deterioration of key metrics, namely: (a) The classification accuracy remains consistently high at 0.996 (see the blue line in Figure 11b); (b) The quality of rule-based decisions, after iterative updates, consistently enables correct object classification, maintaining a range of 0.83–0.86 (see the navy-blue lines in Figure 11a,b).

The analysis of each new object leads to its classification according to MPA, namely into In, Ne, or Re classes. In the experiment, an iterative classification of 20 new orders has been performed, resulting in three orders assigned to the In class, sixteen orders to the Ne class, and one order to the Re class (see Part 1 of Table A3). In the latter case, the classification of an object into the Re class indicates the potential for reducing the number of SNDW units, which requires the application of recommended values. These values are specified as supplementary attributes in the description of objects that support the decision rule on which the classification into the Re class is based. This value is determined according to Equation (33). Table 8 presents the default values of the control parameters MCP_{23} and MCP_{24}, along with their recommended values, which facilitate the reduction in the SNDW LUs. The default parameter values are derived from the WMS and serve as the basis for optimisation using the MIX and SNDW algorithms in Steps 1 and 2. Meanwhile, the recommended values are implemented in Step 4.

Table 8. Default versus recommended values of control parameters.

Iteration	Order Type	Value of Control Parameters				SNDW LUs		Energy Consumption	
		MCP_{23} (mm)		MCP_{24} (kg)		S_{41} (item)		EC (kWh/pskm)	
		default	recommended	default	recommended	default	recommended	default	recommended
20	new	1450	1595	900	810	5	3	0.601	0.360

As shown in the example presented in Table 8, the reduction in the number of SNDW LUs can be achieved by increasing the default value of the control parameter MCP_{23} and decreasing the default value of MCP_{24}. An increase in MCP_{23} by 145 mm and a decrease in MCP_{24} by 90 kg can result in the reduction in the SNDW LUs by 2. This outcome translates into reduced energy consumption during delivery. Based on the calculation presented earlier in this paper (see Section 1—Introduction), one LU can generate 0.12 (kWh) per 1 (km), i.e., 0.12 (kWh/pskm). Therefore, a reduction from 5 to 3 LUs results in a 40% reduction in energy consumption, i.e., from 0.60 (kWh/pskm) to 0.36 (kWh/pskm).

5.6. Discussion of Results

The methodology presented in Section 5 aims to classify customer orders through the exploration of historical knowledge, optimisation, simulation experiments, and learning and testing procedures. These orders are divided into three classes using the proposed classifier *MPA* within three alternative strategies. The results are compared, and an evaluation of the consistency of these three classification strategies is conducted using the DRST. This enables the selection of the most advantageous classifier, which forms the basis for generating decision rules that are subsequently applied in the decision-making process, i.e., the classification of new orders. It is crucial to assess whether new orders have the potential to reduce the number of SNDW LUs by adjusting the values of the control parameters and applying the MIX and SNDW optimisation algorithms. Detailed planning of these LUs typically leads to improved organisation of loading activities, such as optimising the number of internal transport movements. Fewer stacked palletised LUs result in fewer shipments organised for customers due to more efficient use of the vehicle's cargo space. The outcome of minimising internal transport operations and customer shipments is a reduction in the company's transport costs. A smaller number of palletised LUs has a positive impact on energy consumption, with a reduction of 0.12 (kWh/km) per palletised LU, as presented in Section 1 of this paper. Society also benefits from the proposed approach: fewer transport operations result in lower noise emissions and a reduced carbon footprint. These aspects, along with reduced energy consumption, are identified as key factors of environmental protection. Thus, the proposed approach supports the company's environmental responsibility in the context of social, economic, and environmental benefits.

The developed methodology enables the classification of new orders with an accuracy of 83–86%, distinguishing those that lack the potential to improve the solution (assigned to the *In* or *Ne* classes) from those with potential for reducing the number of SNDW LUs (assigned to the *Re* class). In the former case, further analysis time is reduced. In the latter case, necessary changes in parameter values can be identified, and applying the optimisation algorithm leads to a reduction in the number of SNDW LUs.

The proposed methodology is designed to ensure that all cases where correct classification is not immediately possible (representing 14% of orders) can be addressed by additional simulation experiments. This approach facilitates decision-making and fills the gap that previously hindered classification using decision rules. Consequently, correct decisions can be made for 100% of cases, with a portion of classification based on decision rules and the remainder using supplementary simulations. In a broader context, within both the training set and the application set, the proportion of orders identified with potential for improvement (i.e., $MPA = Re$) is 4.5% (32 out of 709 orders; see Strategy 3 in Table 4) and 5.0% (1 out of 20 orders; see Part 1 in Table A3), respectively. The proposed methodology, verified using a small research dataset, demonstrates the effectiveness of the decision-making process in real-world scenarios and provides a foundation for further studies in a broader context. This aspect is discussed in the following section.

6. Conclusions and Future Work

This study presented a novel approach to optimise the Sandwich Pallet Loading Problem (SPLP) in FMCG logistics, aiming to minimise the number of LUs required for warehousing and transportation. The proposed methodology integrates the strengths of Dominance-Based Rough Set Theory (DRST) and a rule-based expert system to enhance the efficiency, automation, and reliability of the pallet loading planning process while reducing energy consumption. Moreover, the newly introduced *MPA* classifier for customer orders has demonstrated effective results in numerical experiments using a medium-scale dataset.

The combination of decision rules and simulation accurately determines whether it is possible to improve the solution of SPLP.

The experimental application of this methodology in a real-world company involved the classification of customers' orders into three classes using the proposed classifier under three alternative strategies. Following a comparative analysis, the most advantageous classifier was selected, facilitating the classification of new orders. The results demonstrated that 86% of the orders were correctly classified. Additionally, 4.5% of the orders in the training set (32 out of 709 orders, see Strategy 3 in Table 4) and 5% of the orders in the application set (1 out of 20 orders, see Part 1 in Table A3) had the potential for reducing SNDW LUs. These orders were assigned to the *Re* class. The parameter values for SNDW LUs revealed that, after applying the optimisation algorithm, it was possible to reduce the number of stacked palletised load units. For example, an analysed order from the application set demonstrated a potential reduction in two SNDW LUs. This improvement was achieved by modifying two control parameters, i.e., increasing the LU height by 145 (mm) and decreasing the LU weight by 90 (kg). These adjustments provided precise input data for the optimisation algorithm. The reduction in the number of LUs positively impacted energy consumption, which decreased by 0.241 (kWh/pskm). This improvement benefitted the environment. In a broader context, reducing the number of SNDW LUs can lead to fewer shipments to customers. For instance, it may allow consolidation into one transport instead of two due to more efficient utilisation of vehicle cargo space. The benefits of fewer shipments extend beyond environmental factors—such as reduced energy consumption, a lower carbon footprint, and decreased noise pollution—to include economic advantages (lower company costs) and social improvements (enhanced quality of life).

The proposed SPLP solution builds the company's environmental responsibility while simultaneously addressing current industry challenges, including the efficient management of goods flow through the application of advanced technology. These aspects are significant for the company as they serve as a tool for gaining a competitive advantage in the market and align with numerous business initiatives related to sustainable development. The methodology proposed in this paper meets the company's expectations by reducing costs and improving the efficiency of warehouse and transport operations. It also responds to the demands of increasingly environmentally conscious customers and society at large. An additional advantage of the DRST is its capability to effectively handle imprecise and ambiguous data, which is critical for addressing the inherent uncertainties of the SPLP. These uncertainties stem from the diverse nature of orders, which may consist of various products differing in type, fragility, and dimensions. This diversity affects how products are arranged on pallet units and ultimately how they are positioned on the sandwich LUs. Another key benefit of the proposed methodology lies in its ability to classify LUs based on essential attributes, such as weight and height. The classification provides valuable insights into the potential for reducing the number of sandwich pallets in an efficient manner, as it leverages previously acquired knowledge about orders and palletised load units. Consequently, there is no need to manually test new combinations of weight and height, conduct optimisation experiments, or analyse results to identify opportunities for reducing the number of LUs. Instead, the data-driven approach suggests optimal values for SNDW LUs that deliver improved outcomes. Moreover, every new order contributes to the knowledge database (KDB), enriching the dataset and enhancing its classification capabilities over time. The methodology also accounts for important constraints, such as fragility and pallet dimensions, which contribute to its flexibility and practical applicability. The significance of a company's environmental responsibility is emphasised as a critical driver of competitive advantage and a future standard for industry practices. Verification of the proposed methodology using real FMCG distribution data underscores the effectiveness of integrating DRST with a rule-based expert system, ensuring both operational efficiency and environmental sustainability.

In addition to the many advantages of the proposed solution, certain limitations of the methodology should be highlighted. One limitation is its restricted learning capability—DRST relies on pre-defined attributes and cannot automatically discover new patterns from the dataset as effectively as Machine Learning (ML) or Artificial Intelligence (AI) models. This paper does not address the scalability of research, particularly in terms of defining the maximum number of characteristics and objects in the decision table that would make this methodology fully applicable in practice. Scalability could primarily be achieved through the analysis of reducts and the core of the classification. The parameters of the SNDW LUs considered in the presented approach are strict, meaning they do not account for a permissible tolerance limit regarding the height differences of products placed on the same LU layer. Additionally, there is no flexibility in defining the lower and upper degrees of LU layer filling with products. Introducing tolerances for these parameters would allow the layer to be considered homogeneous in terms of height and fully filled with products. This adjustment could create more options for constructing sandwich LUs. The positive impact of the applied methodology on the environment is currently expressed through a single quantitative measure, i.e., energy consumption. Expanding this analysis to include tools for calculating the carbon footprint (e.g., life-cycle assessment, LCA) or noise pollution (e.g., noise mapping) could provide a more comprehensive understanding of the organisation's level of environmental responsibility.

This study paves the way for further advances in FMCG logistics optimisation. Future studies could explore the integration of Machine Learning algorithms for even more dynamic and adaptable decision-making in SPLP scenarios. Although the proposed methodology provides a structured approach, other techniques such as ML/AI, feature engineering, optimisation algorithms, and data augmentation could offer additional benefits. Future work plans to expand the studies by incorporating additional planning parameters, including the parameter for the permissible tolerance limit regarding the difference in the height of products placed on the same LU layer; parameters defining the lower and upper limits of the degree of filling of the LU layer; parameters specifying product dimensions (length, width, shape) to enable tighter packing and reduce wasted space; and a parameter addressing the centre of gravity of the LU to eliminate uneven weight distribution and improve the pallet stability.

This paper contributes to the advancement of the environmental responsibility of companies. The application of the proposed methodology provides a perspective on the reduction in the number of SNDW LUs while simultaneously decreasing warehousing operation costs as well as delivery costs. It also serves as a tool to strengthen the decision-making process regarding the execution of orders. The development of this approach enables companies to anticipate improvements in the social, economic, and environmental aspects of their activities.

Author Contributions: P.S.: Conceptualization, methodology, investigations, formal analysis, software (implementation of jMAF software), writing—original draft, visualization; H.S.: Conceptualization, methodology, formal analysis, writing—original draft, validation, writing—review and editing, software (support in MIX and SNDW algorithm designs); M.K.: Writing—original draft, writing—review and editing, validation, software, project administration, K.Z.: Data curation, software (supervision of MIX and SNDW code design). All authors have read and agreed to the published version of the manuscript.

Funding: This work was supported by the National Centre for Research and Development (NCBR), Poland, under the POIR.01.01.01-00-0482/18 programme—R&D project titled "R&D on automation of logistics processes using intelligent algorithms and prototypes of equipment in a distributed structure of distribution centres". Piotr Sawicki and Hanna Sawicka also acknowledge financial support from the scientific research project funded by the Ministry of Science and Higher Education in the Republic of Poland—project 0416/SBAD/0006. Marek Karkula also acknowledges financial support from a scientific research project funded by the Ministry of Science and Higher Education in the Republic of Poland—project 16.16.200.396/B310.

Data Availability Statement: Anonymized data summaries are available in the repository, accessible via Mendeley Data: https://doi.org/10.17632/4dwry6rhk4.1.

Conflicts of Interest: Author Krzysztof Zajda is employed by the company Grupa Maspex. The remaining authors declare that the research was conducted in the absence of any commercial or financial relationships that could be construed as a potential conflict of interest.

Appendix A

Below is a compilation of all abbreviations, indices, and symbols used throughout the paper.

Abbreviations	
DRSA	Dominance-Based Rough Sets Approach, also known as DRST;
DRST	Dominance-Based Rough Sets Theory;
EC	Energy consumption, expressed in (kWh/pskm);
FC	Fragility class;
KDB	Knowledge database;
MIX	A pallet load unit composed of different types of products, i.e., different SKUs;
SKU	Stock keeping unit;
SNDW	A pallet load unit called a sandwich;
VCDRSA	Variable Consistency Dominance-Based Rough Sets Approach.
Indices	
i	An index of characteristics q contained in the KDB, $i = 1, 2, \ldots, m$; (applicable for defining a finite set of characteristics, see Section 3.1);
n	An index of intervals into which the range of variability is divided, $n = 1, 2, \ldots, k$ (applicable to defining modifications to control parameters, see Section 4.4);
t	An index of an object's categories, $t \in T$; where g and h represent distinct categories, i.e., $g, h \in T$ (applicable to defining object categories, see Section 3.2);
Notations	
a^-_{MPA}	The lower limit of $min(S_0 - Sn)$ for the MPA classifier, $MPA = \{In, Ne, Re\}$;
a^+_{MPA}	The upper limit of $min(S_0 - Sn)$ for the MPA classifier, $MPA = \{In, Ne, Re\}$;
$\alpha_P()$	Accuracy of approximation of the union of classes defined on P;
b^-_{MPA}	The lower limit of $max(S_0 - Sn)$ for the MPA classifier, $MPA = \{In, Ne, Re\}$;
b^+_{MPA}	The upper limit of $max(S_0 - Sn)$ for the MPA classifier, $MPA = \{In, Ne, Re\}$;
$Bn_P()$	Boundary region defined on P;
C	Conditional part of characteristics Q, being a criterion $C^>$ or attribute $C^=$;
$C\ell$	Set of object categories;
Cl_t	Object category of t-class;
D	Decision part of characteristics Q, $C \cup D = Q$;
Dom_P	Dominance relation defined on P;
$DR^{\geq}$	Set of decision rules generated on $\underline{P}(Cl_t^{\geq})$;
$DR^{\leq}$	Set of decision rules generated on $\underline{P}(Cl_t^{\leq})$;
$DR^{\geq\leq}$	Set of decision rules generated on $\overline{P}(Cl_t^{\leq}) \cap \overline{P}(Cl_t^{\geq})$;
f	Information function assigned to the pair of object x and characteristic q;
$\gamma_P(Cl)$	Quality of sorting to object categories Cl defined on P;

In	Assignment to the decision class indicating that an increase in the number of SNDW pallets is most probable; $MPA = \{In\}$;
Ind_P	Indiscernible relation defined on P;
M_i	A kind of characteristic that is an output of MIX algorithm;
MCP_i	A kind of characteristic used as a control parameter applied for MIX algorithm;
MPA	the most probable assignment; it is a classifier that assigns an order to one of the decision classes, i.e., $MPA = \{In, Ne, Re\}$;
Ne	Assignment to the decision class indicating that a neutral result in the number of SNDW pallets is most probable; $MPA = \{Ne\}$;
n	Assumed number of variable intervals;
P	Subset of conditional characteristics, $P \subseteq C$;
$\underline{P}(Cl_t^{\geq})$	Lower approximation of the upward union of classes upon P;
$\underline{P}(Cl_t^{\leq})$	Lower approximation of the downward union of classes upon P;
$\overline{P}(Cl_t^{\geq})$	Upper approximation of the upward union of classes upon P;
$\overline{P}(Cl_t^{\leq})$	an upper approximation of the downward union of classes upon P;
Q	A finite set of characteristics;
q	A single characteristic, $Q = \{q_1, \ldots q_m\}$;
r	A decision rule;
r_q	The value of conditional part of decision rule referred to a characteristic q;
Re	Assignment to the decision class indicating that a reduced number of SNDW pallets is most probable; $MPA = \{Re\}$;
R_P	A binary relation defined on a subset of characteristics P;
S_0	The quantity of SNDW LUs constructed under default control parameters;
S_n	The quantity of SNDW LUs constructed under control parameters in the $n = 1, 2, \ldots, k$ iterations;
SCP_i	A kind of control parameter applied for SNDW pallet design algorithm;
U	A finite set of objects;
V	The domain of characteristics q;
x	A reference object, $x \in U$;
y	A compared object, $y \in U$;
z	A new object for which $D = \emptyset$;

Appendix B

Table A2 shows confusion matrices derived from cross-validation during the testing phase (see Section 5.4). The reported values are expressed in absolute numbers prior to normalisation, while the results after normalisation are shown in Table 7. Strategies 1–3 differ in their eligibility thresholds for assigning objects to one of three classes, i.e., In, Ne, and Re. The class In indicates that changes in control parameters are expected to increase the number of SNDW LUs. The class Ne represents neutrality, where no significant increase or decrease in the number of LUs is anticipated. Finally, the class Re corresponds to an expected reduction in the number of SNDW LUs.

Table A2. Confusion matrix of the cross-validation test (prior to normalisation).

Test Profile		Testing Set of Orders											
Classification Algorithm	**Negative Example Treatment**	**Strategy 1**				**Strategy 2**				**Strategy 3**			
DRSA	Covers only inconsistency		*In*	*Ne*	*Re*		*In*	*Ne*	*Re*		*In*	*Ne*	*Re*
		In	19	17	12	*In*	59	62	23	*In*	55	18	1
		Ne	29	532	20	*Ne*	11	464	10	*Ne*	50	546	7
		Re	29	33	18	*Re*	35	17	27	*Re*	3	20	9
	Covers none		*In*	*Ne*	*Re*		*In*	*Ne*	*Re*		*In*	*Ne*	*Re*
		In	18	19	11	*In*	58	60	26	*In*	52	21	1
		Ne	29	529	23	*Ne*	10	466	9	*Ne*	61	535	7
		Re	22	33	25	*Re*	37	16	26	*Re*	3	20	9
VCDRSA	Covers only inconsistency		*In*	*Ne*	*Re*		*In*	*Ne*	*Re*		*In*	*Ne*	*Re*
		In	5	29	2	*In*	22	68	8	*In*	8	54	0
		Ne	6	519	8	*Ne*	27	411	4	*Ne*	15	537	1
		Re	3	46	10	*Re*	12	28	15	*Re*	0	21	11
	Covers none		*In*	*Ne*	*Re*		*In*	*Ne*	*Re*		*In*	*Ne*	*Re*
		In	2	24	1	*In*	27	66	4	*In*	9	54	0
		Ne	10	518	14	*Ne*	25	415	3	*Ne*	21	528	1
		Re	4	46	8	*Re*	12	27	10	*Re*	0	22	10

Appendix C

In the application phase of the proposed procedure (see Section 5.5), a series of experiments was conducted, during which successive new orders were iteratively retrieved. A detailed compilation of the results for each iteration, i.e., for each of the 20 new orders, is presented in Table A3. In this phase, the following assumptions were made for the experiments:

- The KDB was populated with a dataset of 709 orders, and their assignment to one of the *In*, *Ne*, or *Re* classes was carried out using Strategy 3.
- The KDB initially contained a set of 252 certain decision rules generated based on the standard union of classes (see Table 4).
- The set of new objects to be classified in the application phase consisted of 20 orders; the input file is available on the Mendeley Data repository: https://doi.org/10.17632/4dwry6rhk4.1.

The results in Table A3 are divided into four parts, including:

- Decisions made (see Part 1 of Table A3);
- Changes in the KDB content (see Part 2 of Table A3);
- Results of cross-validation for the variable KDB (see Part 3 of Table A3);
- Diagonal part of the confusion matrix (see Part 4 of Table A3).

Table A3. Detailed results of the application phase.

Iteration	0	1	2	3	4	5	6	7	8	9	10	11	12	13	14	15	16	17	18	19	20
1. Result																					
MPA	–	*In*	*Ne*	*In*	*In*	*Ne*	*Ne*	*Ne*	*Ne*	*Ne*	*Ne*	*Ne*	*Ne*	*Ne*	*Ne*	*Ne*	*Ne*	*Ne*	*Ne*	*Ne*	*Re*
Dec. [1]	–	Unk	Cor	Unk	Unk	Cor	Unk	Cor	Cor	Cor	Unk	Cor	Cor	Cor	Cor	Cor	Unk	Cor	Cor	Cor	Unk
2. KDB structure																					
No of objects	709	710	710	711	712	712	713	713	713	713	714	714	714	714	714	714	715	715	715	715	716
Quality of app.	0.996	0.996	0.996	0.996	0.996	0.996	0.996	0.996	0.996	0.996	0.996	0.996	0.996	0.996	0.996	0.996	0.996	0.996	0.996	0.996	0.996
No. of rules	252	254	254	255	255	255	256	256	256	256	256	256	256	256	256	256	257	257	257	257	257
3. Cross-validation																					
Corect (card)	610	592	592	599	603	603	604	604	604	604	591	591	591	591	591	591	607	607	607	607	603
(%)	86.04	83.38	83.38	84.25	84.69	84.69	84.71	84.71	84.71	84.71	82.77	82.77	82.77	82.77	82.77	82.77	84.89	84.89	84.89	84.89	84.34
Incorect (card)	99	118	118	112	109	109	109	109	109	109	123	123	123	123	123	123	108	108	108	108	112
(%)	13.96	16.62	16.62	15.75	15.31	15.31	15.29	15.29	15.29	15.29	12.23	12.23	12.23	12.23	12.23	12.23	15.11	15.11	15.11	15.11	15.66
Unknown (card)	0	0	0	0	0	0	0	0	0	0	0	0	0	0	0	0	0	0	0	0	0
(%)	0	0	0	0	0	0	0	0	0	0	0	0	0	0	0	0	0	0	0	0	0
4. Confusion																					
In (card)	55	51	51	47	54	54	53	53	53	53	48	48	48	48	48	48	51	51	51	51	50
(%)	0.74	0.68	0.68	0.62	0.7	0.7	0.69	0.69	0.69	0.69	0.62	0.62	0.62	0.62	0.62	0.62	0.66	0.66	0.66	0.66	0.65
Ne (card)	546	532	532	543	540	540	542	542	542	542	535	535	535	535	535	535	547	547	547	547	545
(%)	0.91	0.88	0.88	0.9	0.9	0.9	0.9	0.9	0.9	0.9	0.88	0.88	0.88	0.88	0.88	0.88	0.9	0.9	0.9	0.9	0.9
Re (card)	9	9	9	9	9	9	9	9	9	9	8	8	8	8	8	8	9	9	9	9	8
(%)	0.28	0.28	0.28	0.28	0.28	0.28	0.28	0.28	0.28	0.28	0.25	0.25	0.25	0.25	0.25	0.25	0.28	0.28	0.28	0.28	0.25

[1] Decision: Cor—correct, Inc—incorrect, Unk—unknown.

References

1. Dekker, R.; Bloemhof, J.; Mallidis, I. Operations Research for green logistics—An overview of aspects, issues, contributions and challenges. *Eur. J. Oper. Res.* **2012**, *219*, 671–679. [CrossRef]
2. Pal, P.; Gopal, P.; Ramkumar, M. Impact of transportation on climate change: An ecological modernization theoretical perspective. *Transp. Policy* **2023**, *130*, 167–183. [CrossRef]
3. Jorsfeldt, L.; Hvolby, H.H.; Nguyen, V. Implementing environmental sustainability in logistics operations: A case study. *Strateg. Outsourcing Int. J.* **2016**, *9*, 98–125. [CrossRef]
4. Younus, M.; Purnomo, E.; Husein, R.; Khairunnisa, T. Towards a Greener Future: Promoting Green and Sustainable Development in Transportation Operation. *E3S Web Conf.* **2023**, *440*, 01004. [CrossRef]
5. *Directive (EU) 2023/1791 of the European Parliament and of the Council of 13 September 2023 on Energy Efficiency and Amending Regulation (EU) 2023/955*; FAO: Rome, Italy, 2023.
6. Sustainability Exchange. Available online: https://www.sustainabilityexchange.ac.uk (accessed on 6 November 2024).
7. *PN-EN 590:2022-08*; Paliwa dla Pojazdów Samochodowych (Fuels for Vehicles). Polish Committee for Standardization: Warsaw, Poland, 2022.
8. Kirschstein, T.; Heinold, A.; Behnke, M.; Meisel, F.; Bierwirth, C. Eco-labeling of freight transport services: Design, evaluation, and research directions. *J. Ind. Ecol.* **2022**, *23*, 801–814. [CrossRef]
9. Zion Market Research. Available online: https://www.zionmarketresearch.com (accessed on 6 November 2024).
10. Grand View Research. Available online: https://www.grandviewresearch.com (accessed on 6 November 2024).
11. PricewaterhouseCoopers. *Transport of the Future, Report on Prospects for the Development of Road Transport in Poland in 2020–2030*; PwC: London, UK, 2019.
12. Hitchcock, L.; Mehboob, Y.; Mullapudi, M.; Rauch, M.; Naik, U. *The Future of Freight: Transforming the Movement of Goods*; Deloitte Insights; Deloitte: Baltimore, MD, USA, 2022.
13. PricewaterhouseCoopers. *What's the Real Value of AI for Your Business and How Can You Capitalise?* Report; PwC: London, UK, 2017.
14. Kulkov, I.; Kulkova, J.; Rohrbeck, R.; Menvielle, L.; Kaartemo, V.; Makkonen, H. Artificial intelligence–driven sustainable development: Examining organizational, technical, and processing approaches to achieving global goals. *Sustain. Dev.* **2024**, *32*, 2253–2267. [CrossRef]
15. Zhang, C.; Lu, Y. Study of artificial intelligence: The state of the art and future prospects. *J. Ind. Inf. Integr.* **2021**, *23*, 100224. [CrossRef]
16. Lu, Y. Artificial Intelligence: A survey on evolution, models, application and future trends. *J. Manag. Anal.* **2019**, *6*, 1–29. [CrossRef]
17. Lundberg, L.; Boldt, M.; Borg, A.; Grahn, H. Bibliometric Mining of Research Trends in Machine Learning. *AI* **2024**, *5*, 208–236. [CrossRef]
18. Sharadgah, T.A.; Sa'di, R.A. A Systematic Review of Research on the Use of Artificial Intelligence in English Language Teaching and Learning (2015–2021): What are the Current Effects? *J. Inf. Technol. Educ. Res.* **2022**, *21*, 337–377. [CrossRef]
19. Thinh, N.T.; Hai, N.D.X.; Tho, T.P. The Influential Role of Robot in Second Language Classes Based on Artificial Intelligence. *Int. J. Mech. Eng. Robot. Res.* **2020**, *9*, 1306–1311. [CrossRef]
20. Gangwani, D.; Gangwani, P. Applications of Machine Learning and Artificial Intelligence in Intelligent Transportation System: A Review. In *Applications of Artificial Intelligence and Machine Learning*; Lecture Notes in Electrical Engineering; Choudhary, A., Agrawal, A.P., Logeswaran, R., Unhelkar, B., Eds.; Springer: Singapore, 2021; Volume 778, pp. 203–216. [CrossRef]
21. Olugbade, S.; Ojo, S.; Imoize, A.L.; Isabona, J.; Alaba, M.O. A Review of Artificial Intelligence and Machine Learning for Incident Detectors in Road Transport Systems. *Math. Comput. Appl.* **2022**, *27*, 77. [CrossRef]
22. Tsolaki, K.; Vafeiadis, T.; Nizamis, A.; Ioannidis, D.; Tzovaras, D. Utilizing machine learning on freight transportation and logistics applications: A review. *ICT Express* **2023**, *9*, 284–295. [CrossRef]
23. Ullah, A.; Anwar, S.; Li, J.; Nadeem, L.; Mahmood, T.; Rehman, A.; Saba, T. Smart cities: The role of Internet of Things and machine learning in realizing a data-centric smart environment. *Complex Intell. Syst.* **2024**, *10*, 1607–1637. [CrossRef]
24. Din, I.U.; Guizani, M.; Rodrigues, J.J.P.C.; Hassan, S.; Korotaev, V.V. Machine learning in the Internet of Things: Designed techniques for smart cities. *Future Gener. Comput. Syst.* **2019**, *100*, 826–843. [CrossRef]
25. Alaoui, E.A.A.; Tekouabou, S.C.K. Intelligent management of bike sharing in smart cities using machine learning and Internet of Things. *Sustain. Cities Soc.* **2021**, *67*, 102702. [CrossRef]
26. Zhang, X.; Guo, F.; Chen, T.; Pan, L.; Beliakov, G.; Wu, J. A Brief Survey of Machine Learning and Deep Learning Techniques for E-Commerce Research. *J. Theor. Appl. Electron. Commer. Res.* **2023**, *18*, 2188–2216. [CrossRef]
27. Policarpo, L.M.; Silveira, D.E.; Rosa Righi, R.; Stoffel, R.A.; Costa, C.A.; Barbosa, J.L.V.; Scorsatto, R.; Arcot, T. Machine learning through the lens of e-commerce initiatives: An up-to-date systematic literature review. *Comput. Sci. Rev.* **2021**, *41*, 100414. [CrossRef]

28. Ballestar, M.; Grau-Carles, P.; Sainz, J. Predicting customer quality in e-commerce social networks: A machine learning approach. *Rev. Manag. Sci.* **2019**, *13*, 589–603. [CrossRef]
29. Sarker, I.H.; Kayes, A.S.M.; Badsha, S.; Alqahtani, H.; Watters, P.; Ng, A. Cybersecurity data science: An overview from machine learning perspective. *J. Big Data* **2020**, *7*, 1–29. [CrossRef]
30. Shaukat, K.; Luo, S.; Varadharajan, V.; Hameed, I.; Chen, S.; Liu, D.; Li, J. Performance Comparison and Current Challenges of Using Machine Learning Techniques in Cybersecurity. *Energies* **2020**, *13*, 2509. [CrossRef]
31. Apruzzese, G.; Laskov, P.; Oca, E.M.; Mallouli, W.; Rapa, L.B.; Grammatopoulos, A.V.; Franco, F. The Role of Machine Learning in Cybersecurity. *Digit. Threat. Res. Pract.* **2023**, *4*, 8–38. [CrossRef]
32. Alanazi, A. Using machine learning for healthcare challenges and opportunities. *Inform. Med. Unlocked* **2020**, *30*, 100924. [CrossRef]
33. Jayatilake, S.M.D.A.C.; Ganegoda, G.U. Involvement of Machine Learning Tools in Healthcare Decision Making. *J. Healthc. Eng.* **2021**, *2021*, 6679512. [CrossRef]
34. Habehh, H.; Gohel, S. Machine Learning in Healthcare. *Curr. Genom.* **2021**, *22*, 291–300. [CrossRef]
35. Lalmuanawma, S.; Hussain, J.; Chhakchhuak, L. Applications of machine learning and artificial intelligence for COVID-19 (SARS-CoV-2) pandemic: A review. *Chaos Solitons Fractals* **2020**, *139*, 110059. [CrossRef] [PubMed]
36. Malki, Z.; Atlam, E.S.; Ewis, A.; Dagnew, G.; Ghoneim, O.; Mohamed, A.; Abdel-Daim, M.; Gad, I. The COVID-19 pandemic: Prediction study based on machine learning models. *Environ. Sci. Pollut. Res.* **2021**, *28*, 40496–40506. [CrossRef] [PubMed]
37. Muhammad, I.; Yan, Z. Supervised Machine Learning Approaches: A Survey. *ICTACT J. Soft Comput.* **2015**, *5*, 946–952. [CrossRef]
38. Usama, M.; Qadir, J.; Raza, A.; Arif, H.; Yau, K.L.; Elkhatib, Y.; Hussain, A.; Al-Fuqaha, A. Unsupervised Machine Learning for Networking Techniques, Applications and Research Challenges. *IEEE Access* **2019**, *7*, 65579. [CrossRef]
39. Kaelbling, L.; Littman, M.; Moore, A. Reinforcement Learning: A Survey. *J. Artif. Intell. Res.* **1999**, *4*, 237–285. [CrossRef]
40. Sarker, I. Machine Learning: Algorithms, Real-World Applications and Research Directions. *SN Comput. Sci.* **2021**, *2*, 21. [CrossRef] [PubMed]
41. Reddy, Y.; Viswanath, P.; Reddy, B.E. Semi-supervised learning: A brief review. *Int. J. Eng. Technol.* **2018**, *7*, 81–85. [CrossRef]
42. Bhowmick, K.; Narvekar, M.; Khatkhatay, M.A. A Comprehensive Study and Analysis of Semi Supervised Learning Techniques. *Int. J. Eng. Res. Technol.* **2019**, *8*, 810–816.
43. Zuccotto, M.; Castellini, A.; Torre, D.; Mola, L.; Farinelli, A. Reinforcement learning applications in environmental sustainability: A review. *Artif. Intell. Rev.* **2024**, *57*, 88. [CrossRef]
44. Soofi, A.A.; Awan, A. Classification Techniques in Machine Learning: Applications and Issues. *J. Basic Appl. Sci.* **2017**, *13*, 459–465. [CrossRef]
45. Mahesh, B. Machine Learning Algorithms - A Review. *Int. J. Sci. Res.* **2020**, *9*, 381–386. [CrossRef]
46. LeCun, Y.; Bengio, Y.; Hinton, G. Deep learning. *Nature* **2015**, *521*, 436–444. [CrossRef]
47. Woschank, M.; Rauch, E.; Zsifkovits, H. A Review of Further Directions for Artificial Intelligence, Machine Learning, and Deep Learning in Smart Logistics. *Sustainability* **2020**, *12*, 3760. [CrossRef]
48. Shrestha, A.; Mahmood, A. Review of Deep Learning Algorithms and Architectures. *IEEE Access* **2019**, *7*, 53040–53065. [CrossRef]
49. Shavaki, F.; Ghahnavieh, A.E. Applications of deep learning into supply chain management: A systematic literature review and a framework for future research. *Artif. Intell. Rev.* **2023**, *56*, 4447–4489. [CrossRef] [PubMed]
50. Guerrero-Ibañez, J.; Contreras-Castillo, J.; Zeadally, S. Deep learning support for intelligent transportation systems. *Trans. Emerg. Telecommun. Technol.* **2020**, *32*, e4169. [CrossRef]
51. Nguyen, H.; Kieu, L.M.; Wen, T.; Cai, C. Deep learning methods in transportation domain: A review. *IET Intell. Transp. Syst.* **2018**, *12*, 998–1004. [CrossRef]
52. Song, A.; Yang, X.; Ni, L.; Liu, C.; Yao, Y.; Pan, L. Optimization Analysis of the Emergency Logistics Identification Method Based on the Deep Learning Model under the Background of Big Data. *Wirel. Commun. Mob. Comput.* **2022**, *2022*, 463035. [CrossRef]
53. Yaqoob, I.; Hashem, I.A.T.; Gani, A.; Mokhtar, S.; Ahmed, E.; Anuar, N.B.; Vasilakos, A.V. Big data: From beginning to future. *Int. J. Inf. Manag.* **2016**, *36*, 1231–1247. [CrossRef]
54. Tang, L.; Li, J.; Du, H.; Li, L.; Wu, J.; Wang, S. Big Data in Forecasting Research: A Literature Review. *Big Data Res.* **2022**, *27*, 100289. [CrossRef]
55. Berisha, B.; Mëziu, E.; Shabani, I. Big data analytics in Cloud computing: An overview. *J. Cloud Comput.* **2022**, *11*, 10. [CrossRef]
56. Gani, A.; Siddiqa, A.; Shamshirband, S.; Hanum, F. A survey on indexing techniques for big data: Taxonomy and performance evaluation. *Knowl. Inf. Syst.* **2016**, *46*, 241–284. [CrossRef]
57. Othman, A.U.; Moses, T.; Aisha, U.Y.; Gital, A.Y.; Souley, B.; Adeleke, B.T. Big Data Indexing: Taxonomy, Performance Evaluation, Challenges and Research Opportunities. *J. Comput. Sci. Eng.* **2022**, *3*, 71–94. [CrossRef]
58. Welch, T.F.; Widita, A. Big data in public transportation: A review of sources and methods. *Transp. Rev.* **2019**, *39*, 795–818. [CrossRef]

59. Neilson, A.; Indratmo, B.; Tjandra, S. Systematic Review of the Literature on Big Data in the Transportation Domain: Concepts and Applications. *Big Data Res.* **2019**, *17*, 35–44. [CrossRef]
60. Lian, Y.; Zhang, G.; Lee, J.; Huang, H. Review on big data applications in safety research of intelligent transportation systems and connected/automated vehicles. *Accid. Anal. Prev.* **2020**, *146*, 105711. [CrossRef]
61. Cui, Y.; Kara, S.; Chan, K.C. Manufacturing big data ecosystem: A systematic literature review. *Robot. Comput. Integr. Manuf.* **2020**, *62*, 101861. [CrossRef]
62. Fahmideh, M.; Beydoun, G. Big data analytics architecture design—An application in manufacturing systems. *Comput. Ind. Eng.* **2019**, *128*, 948–963. [CrossRef]
63. Zimmermann, H.J. Fuzzy set theory. *Adv. Rev.* **2010**, *2*, 317–332. [CrossRef]
64. Pawlak, Z. Rough sets. *Int. J. Comput. Inf. Sci.* **1982**, *11*, 341–356. [CrossRef]
65. Ali, O.A.M.; Ali, A.Y.; Sumait, B.S. Comparison between the Effects of Different Types of Membership Functions on Fuzzy Logic Controller Performance. *Int. J. Emerg. Eng. Res. Technol.* **2015**, *3*, 76–83.
66. Zadeh, L.A. Fuzzy Sets. *Inf. Control.* **1965**, *8*, 338–353. [CrossRef]
67. Dutta, P.; Konwar, A. Quintic fuzzy sets: A new class of fuzzy sets for solving multi-criteria decision-making problems under uncertainty. *Decis. Anal. J.* **2024**, *11*, 100449. [CrossRef]
68. Užga-Rebrovs, O.; Kulešova, G. Comparative Analysis of Fuzzy Set Defuzzification Methods in the Context of Ecological Risk Assessment. *Inf. Technol. Manag. Sci.* **2017**, *20*, 25–29. [CrossRef]
69. Kabir, S.; Papadopoulos, Y. A review of applications of fuzzy sets to safety and reliability engineering. *Int. J. Approx. Reason.* **2018**, *100*, 29–55. [CrossRef]
70. Guiffrida, A.L.; Nagi, R. Fuzzy set theory applications in production management research: A literature survey. *J. Intell. Manuf.* **1998**, *9*, 39–56. [CrossRef]
71. Huang, Z.; Zhang, H.; Wang, D.; Yu, H.; Wang, L.; Yu, D.; Peng, Y. Preference-based multi-attribute decision-making method with spherical-Z fuzzy sets for green product design. *Eng. Appl. Artif. Intell.* **2023**, *126*, 106767. [CrossRef]
72. Lin, S.S.; Shen, S.L.; Zhou, A.; Xu, Y.S. Risk assessment and management of excavation system based on fuzzy set theory and machine learning methods. *Autom. Constr.* **2021**, *122*, 103490. [CrossRef]
73. Pawlak, Z. Rough set theory and its applications. *J. Telecommun. Inf. Technol.* **2002**, *3*, 7–10. [CrossRef]
74. Zhang, Q.; Xie, Q.; Wang, G. A survey on rough set theory and its applications. *CAAI Trans. Intell. Technol.* **2016**, *1*, 323–333. [CrossRef]
75. Shen, Q.; Jensen, R. Rough Sets, Their Extensions and Applications. *Int. J. Autom. Comput.* **2007**, *4*, 217–228. [CrossRef]
76. Liu, T.; Yang, L. Financial Risk Early Warning Model for Listed Companies Using BP Neural Network and Rough Set Theory. *IEEE Access* **2024**, *12*, 27456–27464. [CrossRef]
77. Gao, J. Performance evaluation of manufacturing collaborative logistics based on BP neural network and rough set. *Neural Comput. Appl.* **2021**, *33*, 739–754. [CrossRef]
78. Huang, C.C.; Liang, W.Y.; Tseng, T.L.; Chen, P.H. The rough set based approach to generic routing problems: Case of reverse logistics supplier selection. *J. Intell. Manuf.* **2016**, *27*, 781–795. [CrossRef]
79. Danzig, G. *Notes on Linear Programming: Part XXXV—Discrete-Variable Extremum Problems*; U.S. Air Force, Research Memorandum; Department of the Air Force: Washington, DC, USA, 1956.
80. Bezerra, V.M.R.; Leao, A.A.S.; Oliveira, J.F.; Santos, M.O. Models for the two-dimensional level strip packing problem—A review and a computational evaluation. *J. Oper. Res. Soc.* **2020**, *71*, 606–627. [CrossRef]
81. Oliveira, J.F.; Júnior, A.N.; Silva, E.; Carravilla, M.A. A survey on heuristics for the two-dimensional rectangular strip packing problem. *Pesqui. Oper.* **2016**, *36*, 197–226. [CrossRef]
82. Hopper, E.; Turton, B.C.H. A review of the application of meta-heuristic algorithms to 2D strip packing problems. *Artif. Intell. Rev.* **2001**, *16*, 257–300. [CrossRef]
83. Franklin, J.; Rosete, A.; Sosa-Gómez, G.; Rojas, O. A Metaheuristic Approach for a Two-dimensional Fuzzy Version of the Variable Size and Cost Bin Packing Problem. *Int. J. Comput. Intell. Syst.* **2024**, *17*, 281. [CrossRef]
84. Xu, Y.; Yang, G.; Bai, J.; Pan, C. A review of the application of swarm intelligence algorithms to 2D cutting and packing problem. *Lect. Not. Comp. Sci* **2011**, *6728*, 64–70. [CrossRef]
85. Mezghani, S.; Haddar, B.; Chabchoub, H. The evolution of rectangular bin packing problem—A review of research topics, applications, and cited papers. *J. Ind. Manag. Optim.* **2023**, *19*, 3329–3361. [CrossRef]
86. Leao, A.; Toledo, F.; Oliveira, J.; Carravilla, M.; Alvarez-Valdés, R. Irregular packing problems: A review of mathematical models. *Eur. J. Oper. Res* **2020**, *282*, 803–822. [CrossRef]
87. Hifi, M.; M'Hallah, R. A literature review on circle and sphere packing problems: Models and methodologies. *Adv. Oper. Res.* **2009**, *2009*, 150624. [CrossRef]
88. Guo, B.; Zhang, Y.; Hu, J.; Li, J.; Wu, F.; Peng, Q.; Zhang, Q. Two-dimensional irregular packing problems: A review. *Front. Mech. Eng.* **2022**, *8*, 966691. [CrossRef]

89. Salem, K.H.; Silva, E.; Oliveira, J.F. Cutting and packing problems under uncertainty: Literature review and classification framework. *Int. Trans. Oper. Res.* **2023**, *30*, 3329–3360. [CrossRef]
90. Ali, S.; Ramos, A.G.; Carravilla, M.A.; Oliveira, J.F. On-line three-dimensional packing problems: A review of off-line and on-line solution approaches. *Comput. Ind. Eng.* **2022**, *168*, 108122. [CrossRef]
91. Christensen, H.I.; Khan, A.; Pokutta, S.; Tetali, P. Approximation and online algorithms for multidimensional bin packing: A survey. *Comput. Sci. Rev.* **2017**, *24*, 63–79. [CrossRef]
92. Delorme, M.; Iori, M.; Martello, S. Bin packing and cutting stock problems: Mathematical models and exact algorithms. *Eur. J. Oper. Res.* **2016**, *255*, 1–20. [CrossRef]
93. Göçmen, E.; Erol, R. Transportation problems for intermodal networks: Mathematical models, exact and heuristic algorithms, and machine learning. *Expert Syst. Appl.* **2019**, *135*, 374–387. [CrossRef]
94. Zhang, Q.; Wei, L.R.; Hu, R.; Yan, R.; Li, L.H.; Zhu, X.N. A review on the bin packing capacitated vehicle routing problem. *Adv. Mater. Res.* **2013**, *853*, 668–673. [CrossRef]
95. Ananno, A.A.; Ribeiro, L. A Multi-Heuristic Algorithm for Multi-Container 3-D Bin Packing Problem Optimization Using Real World Constraints. *IEEE Access* **2024**, *12*, 42105–42130. [CrossRef]
96. Erbayrak, S.; Özkir, V.; Yildirim, U.M. Multi-objective 3D bin packing problem with load balance and product. *Comput. Ind. Eng.* **2021**, *159*, 107518. [CrossRef]
97. Wang, G.; Zhu, J.; Cai, W.; Liu, B.; Tian, Y.; Meng, F. Research on Packaging Optimization in Customized Panel Furniture Enterprises. *BioResources* **2021**, *16*, 1186–1206. [CrossRef]
98. Chen, T.; Chen, X.; Chen, W.; Heaton, H.; Liu, J.; Wang, Z.; Yin, W. Learning to optimize: A primer and a benchmark. *J. Mach. Learn. Res* **2022**, *23*, 8562–8620. [CrossRef]
99. Bruni, M.E.; Fadda, E.; Fedorov, S.; Perboli, G. A machine learning optimization approach for last-mile delivery and third-party logistics. *Comput. Oper. Res.* **2023**, *157*. [CrossRef]
100. Baldo, A.; Boffa, M.; Cascioli, L.; Fadda, E.; Lanza, C.; Ravera, A. The polynomial robust knapsack problem. *Eur. J. Oper. Res.* **2023**, *305*, 1424–1434. [CrossRef]
101. de Costa, P.; Rhuggenaath, J.; Zhang, Y.; Akcay, A.; Kaymak, U. Learning 2-opt heuristics for routing problems via deep reinforcement learning. *SN Comput. Sci.* **2021**, *2*, 388. [CrossRef]
102. Chen, L.; Tong, X.; Yuan, M.; Zeng, J.; Chen, L. A data-driven approach for multi-level packing problems in manufacturing industry. In Proceedings of the 25th ACM SIGKDD International Conference on Knowledge Discovery & Data Mining, Anchorage, AK, USA, 4–8 August 2019; pp. 1762–1770. [CrossRef]
103. Sağtürk, C. An Integer Programming Approach to the Multi-Level Bin Packing Problem with Partial Orders. Bachelor's Thesis, Delft University of Technology, Delft, The Netherlands, 19 June 2022.
104. Groenheide, J.J. Comparing Approximate and Optimal Solution Algorithms for the Multi-Level Bin Packing Problem. Bachelor's Thesis, Delft University of Technology, Delft, The Netherlands, 17 June 2022.
105. Kordyl, A.N. Integer Programming Models for the Class Constrained Multi-Level Bin Packing Problem. Bachelor's Thesis, Delft University of Technology, Delft, The Netherlands, 19 June 2022.
106. Le Blansch, M. Solving the Multi-Level Bin Packing Problem with Time Windows Using Integer Programming. Bachelor's Thesis, Delft University of Technology, Delft, The Netherlands, 19 June 2022.
107. Sawicki, P.; Sawicka, H. Design optimization of stacked pallet load units. *Appl. Sci.* **2023**, *13*, 2153. [CrossRef]
108. Wu, W.; Fan, C.; Huang, J.; Liu, Z.; Yan, J. Machine learning for the multi-dimensional bin packing problem: Literature review and empirical evaluation. *arXiv* **2023**, arXiv:2312.08103. [CrossRef]
109. Vinyals, O.; Fortunato, M.; Jaitly, N. Pointer Networks. In *Proceedings of the Advances in Neural Information Processing Systems 28 (NIPS 2015)*; Cortes, C., Lawrence, N., Lee, D., Sugiyama, M., Garnett, R., Eds.; Curran Associates, Inc.: New York, NY, USA, 2015; Volume 28, pp. 2692–2700. [CrossRef]
110. Bello, I.; Pham, H.; Le, Q.; Norouzi, M.; Bengio, S. Neural combinatorial optimization with reinforcement learning. *arXiv* **2016**, arXiv:1611.09940. [CrossRef]
111. Hu, H.; Zhang, X.; Yan, X.; Wang, L.; Xu, Y. Solving a new 3D bin packing problem with deep reinforcement learning method. *arXiv* **2017**, arXiv:1708.05930. [CrossRef]
112. Puche, A.V.; Lee, S. Online 3D Bin Packing Reinforcement Learning Solution with Buffer. In Proceedings of the 2022 IEEE/RSJ International Conference on Intelligent Robots and Systems (IROS), Kyoto, Japan, 23–27 October 2022; pp. 8902–8909. [CrossRef]
113. Zhao, H.; Zhu, C.; Xu, X.; Huang, H.; Xu, K. Learning practically feasible policies for online 3D bin packing. *Sci. China Infor. Sci.* **2022**, *65*, 112105. [CrossRef]
114. Zhao, H.; Yu, Y.; Xu, K. Learning efficient online 3D bin packing on packing configuration trees. In Proceedings of the International Conference on Learning Representations, Virtual, 25–29 April 2022; pp. 1–18.
115. Yang, S.; Song, S.; Chu, S.; Song, R.; Cheng, J.; Li, Y.; Zhang, W. Heuristics Integrated Deep Reinforcement Learning for Online 3D Bin Packing. *IEEE Trans. Autom. Sci. Eng.* **2024**, *21*, 939–950. [CrossRef]

116. Aylak, B.L.; İnce, M.; Oral, O.; Süer, G.; Almasarwah, N.; Singh, M.; Salah, B. Application of Machine Learning Methods for Pallet Loading Problem. *Appl. Sci.* **2021**, *11*, 8304. [CrossRef]
117. Deng, D. Optimization Analysis of Clustering Algorithm under the Background of Big Data Data Mining. In Proceedings of the 2022 7th International Conference on Intelligent Computing and Signal Processing (ICSP), Xi'an, China, 15–17 April 2022; pp. 652–655. [CrossRef]
118. Hubbs, C.D.; Perez, H.D.; Sarwar, O.; Sahinidis, N.V.; Grossmann, I.E.; Wassick, J.M. OR-Gym: A Reinforcement Learning Library for Operations Research Problems. *arXiv* **2020**, arXiv:2008.06319. [CrossRef]
119. Kong, W.; Liaw, C.; Mehta, A.; Sivakumar, D. A new dog learns old tricks: RL finds classic optimization algorithms. In Proceedings of the International Conference on Learning Representations, Vancouver, BC, USA, 30 April–3 May 2018; pp. 1–25.
120. Murdivien, S.; Um, J. BoxStacker: Deep reinforcement learning for 3D bin packing problem in virtual environment of logistics systems. *Sensors* **2023**, *23*, 6928. [CrossRef]
121. Que, Q.; Yang, F.; Zhang, D. Solving 3D packing problem using Transformer network and reinforcement learning. *Expert Syst. Appl.* **2023**, *214*, 119153. [CrossRef]
122. Hajlaoui, Y.; Jaoua, A.; Layeb, S.B. Deep reinforcement learning for solving the single container loading problem. *Eng. Optim.* **2023**, *55*, 668–684. [CrossRef]
123. Tu, C.; Bai, R.; Aickelin, U.; Zhang, Y.; Du, H. A deep reinforcement learning hyper-heuristic with feature fusion for online packing problems. *Expert Syst. Appl.* **2023**, *230*, 120568. [CrossRef]
124. Sawicki, P.; Sawicka, H.; Karkula, M.; Zajda, K. Expert system based support to sandwich pallet loading problem in the FMCG distribution company. In Proceedings of the 11th Carpathian Logistics Congress, Prague, Czech Republic, 8–10 November 2023.
125. Greco, S.; Matarazzo, B.; Słowiński, R. The use of rough sets and fuzzy sets in MCDM. In *Multicriteria Decision Making. Advances in MCDM Models, Algorithms, Theory, and Applications*; Gel, T.; Stewart, T.J.; Hanne, T., Eds.; Springer: Dordrecht, The Netherlands, 1999; pp. 397–455. [CrossRef]
126. Greco, S.; Matarazzo, B.; Słowiński, R. Rough set theory for multicriteria decision analysis. *Eur. J. Oper. Res.* **2001**, *129*, 1–47. [CrossRef]
127. Greco, S.; Matarazzo, B.; Słowiński, R. A New Rough Set Approach to Multicriteria and Multiattribute Classification. In *Rough Sets and Current Trends in Computing*; Lecture Notes in Artificial Intelligence; Polkowski, L., Skowron, A., Eds.; Springer: Berlin/Heidelberg, Germany, 1998; Volume 1424, pp. 60–67. [CrossRef]
128. Azibi, R.; Vanderpooten, D. Construction of rule-based assignment models. *Eur. J. Oper. Res.* **2002**, *138*, 274–293. [CrossRef]
129. Słowiński, R.; Vanderpooten, D. A generalized definition of rough approximations based on similarity. *IEEE Trans. Knowl. Data Eng.* **2000**, *12*, 331–336. [CrossRef]
130. Błaszczyński, J.; Greco, S.; Słowiński, R. Multi-criteria classification—A new scheme for application of dominance-based decision rules. *Eur. J. Oper. Res.* **2007**, *181*, 1030–1044. [CrossRef]
131. Greco, S.; Matarazzo, B.; Słowiński, R. Multicriteria classification by dominance-based rough set approach. Methodological basis of the 4eMka system. In *Handbook of Data Mining and Knowledge Discovery*; Kloesgen, W., Zytkow, J., Eds.; Oxford University Press: New York, NY, USA, 2002; pp. 318–328.
132. Stefanowski, J. On rough set based approaches to induction of decision rules. In *Rough Sets in Data Mining and Knowledge Discovery*; Skowron, A., Polkowski, L., Eds.; Physica: Berlin/Heidelberg, Germany, 1998; Volume 1, pp. 530–553.
133. Stefanowski, J.; Vanderpooten, D. A general two-stage approach to inducing rules from examples. In *Rough Sets, Fuzzy Sets and Knowledge Discovery*; Ziarko, W., Ed.; Springer: London, UK, 1994; pp. 317–325.
134. Błaszczyński, J.; Greco, S.; Matarazzo, B.; Słowiński, R.; Szeląg, M. jMAF—Dominance-Based Rough Set Data Analysis Framework. In *Rough Sets and Intelligent Systems—Professor Zdzisław Pawlak in Memoriam: Volume 1*; Skowron, A., Suraj, Z., Eds.; Springer Berlin Heidelberg: Berlin/Heidelberg, Germany, 2013; pp. 185–209. [CrossRef]

energies

MDPI

Article

Deep Learning vs. Gradient Boosting: Optimizing Transport Energy Forecasts in Thailand Through LSTM and XGBoost

Thanapong Champahom [1], Chinnakrit Banyong [2], Thananya Janhuaton [2], Chamroeun Se [3], Fareeda Watcharamaisakul [3], Vatanavongs Ratanavaraha [2] and Sajjakaj Jomnonkwao [2,*]

1 Department of Management, Faculty of Business Administration, Rajamangala University of Technology Isan, Nakhon Ratchasima 30000, Thailand; thanapong.ch@rmuti.ac.th

2 School of Transportation Engineering, Institute of Engineering, Suranaree University of Technology, Nakhon Ratchasima 30000, Thailand; d6500733@g.sut.ac.th (C.B.); d6700942@g.sut.ac.th (T.J.); vatanavongs@g.sut.ac.th (V.R.)

3 Institute of Research and Development, Suranaree University of Technology, Nakhon Ratchasima 30000, Thailand; chamroeun.s@g.sut.ac.th (C.S.); fareeda@sut.ac.th (F.W.)

* Correspondence: sajjakaj@g.sut.ac.th

Abstract: Thailand's transport sector faces critical challenges in energy management amid rapid economic growth, with transport accounting for approximately 30% of total energy consumption. This study addresses research gaps in transport energy forecasting by comparing Long Short-Term Memory (LSTM) neural networks and XGBoost models for predicting transport energy consumption in Thailand. Utilizing a comprehensive dataset spanning 1993–2022 that includes vehicle registration data by size category, vehicle kilometers traveled, and macroeconomic indicators, this research evaluates both modeling approaches through multiple performance metrics. The results demonstrate that XGBoost consistently outperforms LSTM, achieving an R-squared value of 0.9508 for test data compared to LSTM's 0.2005. Feature importance analysis reveals that medium vehicles contribute 36.6% to energy consumption predictions, followed by truck VKT (20.5%), with economic and demographic factors accounting for a combined 15.2%. This research contributes to both methodological understanding and practical application by establishing XGBoost's superior performance for transport energy forecasting, quantifying the differential impact of various vehicle categories on energy consumption, and demonstrating the value of integrating vehicle registration and usage data in predictive models. The findings provide evidence-based guidance for prioritizing policy interventions in Thailand's transport sector to enhance energy efficiency and sustainability.

Keywords: transportation energy consumption; XGBoost; LSTM; vehicle fleet composition; energy forecasting

1. Introduction

Thailand's transport sector faces critical challenges in energy management and sustainable development amid rapid economic growth and urbanization. As the country's second-largest energy-consuming sector, transport accounts for approximately 30% of total final energy consumption, with a consistent upward trend over the past three decades. This escalating energy demand poses significant challenges to energy security, environmental sustainability, and the nation's commitment to reducing greenhouse gas emissions under international agreements [1,2]. The complexity of Thailand's transport energy consumption patterns is uniquely characterized by several factors. First, the country has a distinctive vehicle fleet composition, with an unusually high proportion of motorcycles alongside a growing number of private cars and commercial vehicles [3,4]. Second, there are significant disparities in transportation needs and energy consumption patterns between urban and rural areas, reflecting Thailand's diverse geographical and economic landscape [5,6]. Third, the rapid expansion of e-commerce and logistics services has led to increased energy demand in the freight transport sector [7,8]. In this context, the ability to accurately predict transport energy consumption becomes crucial for effective policy planning and resource allocation [9,10].

Recent advancements in artificial intelligence and machine learning, particularly Long Short-Term Memory (LSTM) neural networks and XGBoost, offer promising alternatives for more accurate transport energy consumption predictions. These methods have demonstrated superior capabilities in handling complex, time-dependent data and capturing non-linear relationships between variables. LSTM networks excel in modeling sequential data with temporal dependencies, making them particularly suitable for forecasting energy consumption patterns over time [11,12]. XGBoost, as an ensemble learning method, has gained significant traction due to its computational efficiency and ability to handle diverse datasets with minimal preprocessing [13,14]. Both approaches can effectively incorporate multiple variables and identify complex interactions that traditional statistical methods often struggle to capture. Recent studies by Çınarer et al. [15] and Hoxha et al. [16] have demonstrated that these advanced methods significantly outperform conventional approaches in predictive accuracy for transportation-related energy consumption and emissions. However, their application in the specific context of Thailand's transport sector remains limited, particularly in incorporating the unique characteristics of the country's transport system and vehicle usage patterns [17,18].

Regarding a comprehensive review of relevant literature, several key trends emerge in the field of transport energy consumption prediction. Recent research by Pongthanaisawan et al. [1] has highlighted the significant growth in fuel consumption and greenhouse gas emissions in Thailand's transport sector, emphasizing the need for accurate predictive models. Studies by Ağbulut [2] and Ji et al. [13] have demonstrated the efficacy of machine learning approaches in forecasting transportation energy demand and emissions, with their models showing high prediction accuracies ranging from 90.8% to 95.2%. The application of LSTM networks has been particularly noteworthy, with researchers such as Duan et al. [12] and Ghanbari and Borna [19] demonstrating their effectiveness in capturing temporal dependencies in time-series data related to energy consumption. In parallel, XGBoost has emerged as a powerful alternative, with studies by Çınarer et al. [15] showcasing its superior performance in predicting carbon emissions and energy needs in transportation contexts. These advanced modeling techniques have significantly outperformed traditional statistical approaches, as demonstrated by Rahman et al. [20] and Shams Amiri et al. [21] in their comparative analyses. When considering this body of research alongside the available dataset (1993–2022), several specific research gaps become apparent. First, while previous studies have incorporated vehicle-related variables, none have comprehensively analyzed the combined impact of differentiated vehicle categories (small, medium, and large) alongside their corresponding vehicle kilometers traveled (VKT) data for motorcycles, passengers, and trucks. This unique combination of variables presents an opportunity for more nuanced prediction modeling that has not been explored in existing literature. Second, although studies like Mohsin et al. [22] and Limanond et al. [23] have used GDP and population data, they have not specifically examined the long-term relationship (30-year period) between these macroeconomic indicators and transport energy consumption in Thailand's context. The extensive temporal coverage of the dataset (1993–2022) offers a unique opportunity to analyze long-term patterns that most existing studies, typically using 5–10-year periods, cannot address. Furthermore, while existing research has employed LSTM and XGBoost models, there is a limited exploration of how these methods perform with segregated vehicle data (both registration numbers and VKT). The granularity of the vehicle-related variables provides an opportunity to develop more sophisticated prediction models that can account for the distinct impacts of different vehicle types and their usage patterns on energy consumption. Additionally, there is a notable gap in understanding how the relationship between vehicle registration numbers and actual vehicle usage (VKT) influences energy consumption predictions. This aspect is particularly relevant for Thailand's transport policy planning but remains unexplored in the current literature.

This study utilizes a comprehensive dataset spanning from 1993 to 2022, which provides a unique opportunity for detailed analysis and modeling. The dataset includes the following: (1) vehicle registration data categorized by size (small, medium, and large vehicles), (2) vehicle kilometers traveled (VKT) for different vehicle types (motorcycles, passenger vehicles, and trucks), (3) macroeconomic indicators (GDP and population), and (4) historical transport energy consumption data. The length and granularity of this dataset enable a more nuanced understanding of the relationships between vehicle ownership, usage patterns, economic factors, and energy consumption. This rich data environment allows for the development of more sophisticated and accurate prediction models that can capture both long-term trends and short-term variations in transport energy consumption [8]. By leveraging this extensive historical data alongside advanced machine learning techniques, this study aims to provide insights that bridge the gap between technical modeling capabilities and practical policy applications in Thailand's transport sector, addressing a critical need highlighted by Supasa et al. [7], Ji et al. [13] in their research on energy consumption patterns and emissions in the transportation sector.

The main objectives of this study are as follows:

- To develop and compare the performance of LSTM and XGBoost models in predicting Thailand's transport energy consumption, evaluating their respective strengths and limitations in handling different aspects of the prediction task.
- To analyze the differential impact of various vehicle categories and their usage patterns on energy consumption, providing insights into how fleet composition affects overall energy demand.
- To identify and quantify the relative importance of different factors affecting transport energy consumption, including vehicle-related variables, economic indicators, and demographic factors.
- To provide evidence-based recommendations for transport energy policy planning, particularly in areas of vehicle fleet management and energy efficiency improvements.

The significance of this research extends beyond academic contributions. For policymakers, accurate predictions of transport energy consumption are essential for infrastructure planning, setting realistic energy efficiency targets, developing effective energy conservation measures, and allocating resources for sustainable transport initiatives [24,25]. The methodological advancements in this study contribute to a growing body of work on applying machine learning to energy forecasting, addressing limitations in traditional approaches identified by Antonopoulos et al. [26] and Alabi et al. [27]. Furthermore, this research comes at a critical time when Thailand is striving to balance economic growth with sustainable development goals and energy security concerns. The country's commitment to reducing greenhouse gas emissions and promoting sustainable transport solutions makes the accurate prediction of transport energy consumption particularly relevant for future policy planning [1]. By providing a comparative analysis of LSTM and XGBoost models in the specific context of Thailand's transport sector, this study offers valuable insights for both academic understanding and practical policy implementation, addressing a significant gap in regional research on advanced predictive modeling for transportation energy demand as highlighted by Rahman et al. [20], Emami Javanmard et al. [28].

This study's planning perspective is particularly significant given Thailand's current transportation policy landscape. Accurate predictive models for transport energy consumption serve as critical decision-support tools for policymakers tasked with developing sustainable transport systems [23,24]. The ability to forecast energy demand with greater precision enables more efficient resource allocation, supports targeted infrastructure investments, and facilitates the development of evidence-based regulatory frameworks [25]. Furthermore, the identification of key drivers of energy consumption through feature importance analysis provides valuable insights for prioritizing policy interventions and designing effective energy conservation measures [26]. As Thailand continues to balance economic development with environmental sustainability goals, robust predictive models that capture the complex relationships between vehicle fleet composition, usage patterns, and energy consumption become increasingly essential for strategic transportation planning [27,28]. By identifying which factors most significantly influence transport energy consumption, this research provides practical guidance for policymakers to develop more targeted and effective interventions, potentially yielding substantial energy savings and emissions reductions while maintaining economic growth.

2. Literature Reviews

Recent research in transport energy consumption prediction has witnessed significant advancements through artificial intelligence applications. A comprehensive analysis of the literature reveals several distinct research streams that have emerged in the past five years (Tables A1 and 1).

Table 1. Summary of the literature review on transport energy consumption prediction.

Authors	Country	Variables	Analysis Methods	Key Findings
Ji et al. [13]	China	Transport energy consumption, GDP, Population, Vehicle kilometers, Car ownership	LSTM, Deep Learning, SVM	Deep Learning outperformed other methods with 3.66% annual increase prediction, Strong correlation between economic factors and CO_2 emissions
Hoxha et al. [16]	Turkey	Energy consumption, Transport modes, Infrastructure data, Vehicle kilometers	XGBoost, Stacking ensemble models	Stacking ensemble models achieved R^2 of 0.99, Vehicle kilometers and GDP were key predictors
Çınarer et al. [15]	Turkey	Transport sector data, Economic indicators, Vehicle kilometers	XGBoost, SVM, MLP	XGBoost highest accuracy in CO_2 prediction, Scenario 4 yielded best results across algorithms
Ağbulut [2]	Turkey	Transport emissions, Energy efficiency, Economic metrics	ANN, SVM, Deep Learning	SVM and ANN algorithms excelled in forecasting, a 3.4 times increase predicted by 2050
Henriques et al. [25]	Brazil	Public transport data, Energy usage, Consumption patterns	K-means clustering, Machine learning	K-means clustering outperformed other methods, Effective classification of consumption patterns
Champeecharoensuk et al. [17]	Thailand	Transport energy use, GHG emissions, Aircraft activity	Statistical analysis, Trend modeling	GHG emissions tripled from 2008 to 2019, Economic growth key driver of emissions
Zhang et al. [14]	China	Energy consumption, Urban expansion, Carbon emissions	XGBoost, Statistical methods	XGBoost achieved low *RMSE* of 0.036, Population and GDP key drivers of emissions
Rahman et al. [20]	Saudi Arabia	Energy consumption, Vehicle types, Economic metrics	SVR, ANN, Econometric analysis	SVR outperformed ANN in predictions, GDP and urbanization significant factors
Lin et al. [8]	China	Transport energy, Economic data, Modal shares	DEA, Statistical analysis	Rail most efficient mode, 443,126 ktce projected for 2020
Pongthanaisawan et al. [1]	Thailand	Energy use, Economic data, Fuel consumption	Econometric models, Statistical analysis	Fossil fuel consumption projected to reach 70,783 ktoe by 2030, Biofuels could reduce emissions by 1.8%
Supasa et al. [7]	Thailand	Energy consumption, Economic indicators, CO_2 emissions	Input–output analysis, Consumption-based approach	Construction and commercial sectors significant CO_2 emitters, Consumption-based policies recommended
Shams Amiri et al. [21]	USA	Household transport energy, Trip characteristics, Socioeconomic factors	Neural Networks, Random Forest, Decision Trees	Neural Network model outperformed decision trees, Motorized trips, and distance key predictors

The application of deep learning approaches, particularly LSTM networks, has gained considerable attention for transport energy forecasting. Duan et al. [12] demonstrated LSTM's effectiveness in capturing temporal dependencies in time-series data, achieving prediction accuracies of up to 94% for transportation-related energy forecasting. Similarly, Ghanbari et al. [19] enhanced LSTM with attention mechanisms for real-time predictions of energy consumption in public transport systems, reporting accuracy improvements of 8–15% over traditional statistical methods. These advancements were further refined by Karim et al. [11], who introduced multivariate LSTM-FCNs that demonstrated superior performance in handling multiple input variables for energy demand forecasting.

In parallel, gradient boosting methods have shown promising results for transport energy prediction. Ji et al. [13] applied XGBoost to predict CO_2 emissions and energy consumption in China's transport sector, leveraging economic indicators and vehicle usage data to achieve remarkable accuracy with R^2 values exceeding 0.95. Çınarer et al. [15] conducted a comparative analysis of machine learning approaches for predicting transportation-related emissions in Turkey, finding that XGBoost consistently outperformed support vector machines and neural networks across various input scenarios. Similarly, Zhang et al. [14] demonstrated XGBoost's effectiveness in handling urban transport data from expanding megacities, achieving lower *RMSE* values compared to traditional statistical methods.

The integration of socioeconomic factors with vehicle-specific variables has emerged as another significant research direction. Rahman et al. [20] developed causality-based machine learning models to establish relationships between GDP, urbanization, and transport energy demand in Saudi Arabia, finding that urbanization exerted a stronger influence than economic growth on energy consumption patterns. This approach was extended by Hoxha et al. [16], who employed stacking ensemble methods to improve prediction accuracy by incorporating diverse socioeconomic indicators alongside traditional transport metrics.

Regional studies have provided valuable insights into context-specific factors affecting transport energy consumption. In Thailand, Pongthanaisawan et al. [1] developed econometric models to forecast transport energy demand and emissions until 2030, identifying significant relationships between economic growth and transportation energy intensity. Champeecharoensuk et al. [17] focused on aviation emissions in Thailand, demonstrating how economic growth and aviation activity served as primary drivers of increasing energy demand in the sector. These studies highlight the importance of region-specific analyses but are limited by their reliance on conventional statistical approaches rather than advanced machine learning techniques.

Methodological innovations have further expanded the predictive capabilities in this domain. Wang et al. [29] introduced a LASSO regression framework for predicting fuel consumption in maritime transportation, addressing multicollinearity challenges in feature variables. Antonopoulos et al. [26] conducted a systematic review of AI applications in energy demand-side response, identifying significant opportunities for machine learning in transport energy optimization. Alabi et al. [27] explored the integration of optimization techniques with machine learning for energy systems planning, highlighting the potential for hybrid approaches in transport energy modeling.

The temporal scope of previous studies varies considerably, with most focusing on short to medium-term forecasting horizons of 5–10 years. Ağbulut [2] provided longer-term projections of transportation energy demand in Turkey until 2050, but noted increasing uncertainty with extended forecast horizons. Chai et al. [24] examined historical trends in road transportation energy consumption in China, establishing 'S' type patterns in the relationship between economic development and energy intensity that could inform long-term projections.

Despite these advances, several notable research gaps persist. First, while previous studies have incorporated vehicle-related variables, few have comprehensively analyzed the combined impact of different vehicle categories alongside their corresponding usage patterns. Second, the long-term relationship between macroeconomic indicators and transport energy consumption remains underexplored, particularly in developing economies like Thailand. Third, there are limited comparative analyses of deep learning versus gradient-boosting approaches for transport energy forecasting using consistent evaluation metrics and datasets. Finally, feature importance analysis to identify key drivers of transport energy consumption has received insufficient attention despite its potential to inform targeted policy interventions.

This study addresses these gaps by leveraging a comprehensive dataset spanning 30 years (1993–2022), incorporating granular vehicle data categorized by size and type, comparing two advanced machine learning approaches using consistent evaluation metrics, and conducting detailed feature importance analysis to identify key drivers of transport energy consumption in Thailand. This approach enables a more nuanced understanding of the complex relationships between vehicle ownership, usage patterns, economic factors, and energy consumption than has been previously achieved in the literature.

3. Methodologies

The selection of appropriate modeling approaches is critical for achieving reliable and accurate predictions in transport energy forecasting. From the vast array of artificial intelligence and machine learning techniques available in the literature, this study deliberately chose Long Short-Term Memory (LSTM) neural networks and Extreme Gradient Boosting (XGBoost) for this study based on several key considerations.

LSTM was selected primarily for its proven capability in handling sequential time-series data with temporal dependencies, which is essential when analyzing transportation patterns that evolve over extended periods. Unlike traditional recurrent neural networks, LSTM's unique architecture with memory cells enables it to capture long-term relationships in time-series data while avoiding the vanishing gradient problem. This architecture is particularly relevant for the dataset spanning three decades (1993–2022), where historical patterns may significantly influence future energy consumption trends. Previous applications of LSTM in energy forecasting domains have demonstrated its effectiveness, with studies by Karim et al. [11] and Duan et al. [12] showing superior performance compared to conventional statistical methods for similar prediction tasks.

Conversely, XGBoost was selected for its exceptional performance in handling structured tabular data with complex feature relationships, which characterizes the multivariate dataset with vehicle categorizations, economic indicators, and usage patterns. As an ensemble learning method based on decision trees, XGBoost excels at capturing non-linear relationships and interactions between variables without requiring extensive data preprocessing. Its built-in regularization features help prevent overfitting, which is particularly valuable when working with relatively limited observations (30 years in this case) but multiple predictor variables. Additionally, XGBoost's computational efficiency and inherent feature importance capabilities provide practical advantages for both model training and interpretability. Recent studies by Zhang et al. [14], Çınarer et al. [15] have demonstrated XGBoost's effectiveness specifically for transportation-related energy and emissions prediction tasks.

3.1. Data Collection

This study employs a comprehensive dataset spanning from 1993 to 2022, encompassing various aspects of Thailand's transport sector shown in Table 2.

Table 2. Explanation of dataset.

Variables	Unit	Explanation	Sources
Transport Energy Consumption	(MTOE)	Energy consumption in the transport sector	Energy Policy and Planning office (EPPO), Ministry of Energy, Thailand
GDP	(10^{12} Baht)	Gross Domestic Product	Bank of Thailand
Population	(10^6 Peoples)	The total population is determined using the de facto approach, which includes all individuals residing in a given area, regardless of their citizenship or legal status.	World Bank
Registered-vehicle Small	(10^6 vehicles)	Number of small registered vehicles (including motorcycles, mopeds)	Department of Land Transport
Registered-vehicle Medium	(10^6 vehicles)	Number of medium l registered vehicles (including passenger cars, pick-up trucks, vans)	
Registered-vehicle Large	(10^6 vehicles)	Number of large registered vehicles (including buses, trucks, trailers)	
VKT Motorcycle	(10^6 vehicle kilometers)	Vehicle kilometers traveled by motorcycles.	Bureau of Highway Safety, Department of Highways, Thailand
VKT Passenger	(10^6 vehicle kilometers)	Vehicle kilometers traveled by passenger vehicles (passenger cars carrying fewer than seven persons, passenger cars carrying more than seven persons, light buses, medium buses, and heavy buses).	
VKT Truck	(10^6 vehicle kilometers)	Vehicle kilometers traveled by trucks (light trucks, medium trucks, heavy trucks, full trailers, and semi-trailers).	

3.1.1. Transport Energy Consumption

The transport energy consumption in Thailand demonstrates a significant upward trajectory over the 30-year period, increasing from 14.581 MTOE in 1993 to 30.927 MTOE in 2022, representing a 112% growth. The data reveals three distinct phases: steady growth (1993–2008), accelerated increase (2009–2019), and pandemic disruption (2020–2022). The most substantial growth occurred between 2009 and 2019, with an average annual increase of 3.4%. However, the COVID-19 pandemic caused a notable disruption, resulting in an 11.6% decline from 33.607 MTOE in 2019 to 29.699 MTOE in 2020. The sector has shown signs of recovery since 2021, though not yet reaching pre-pandemic levels. This pattern reflects Thailand's economic development, urbanization, and increasing motorization rate. The mean consumption over the period was 24.374 MTOE, with a standard deviation of 5.484 MTOE, indicating considerable variability in consumption patterns.

3.1.2. GDP

Thailand's Gross Domestic Product exhibits a strong positive trend, growing from 4.341 trillion Baht in 1993 to 10.676 trillion Baht in 2022, representing a 146% increase. The growth pattern shows resilience despite several economic challenges, including the 1997 Asian Financial Crisis and the 2008 Global Financial Crisis. The average GDP over the period was 7.821 trillion Baht, with a standard deviation of 2.124 trillion Baht. The data reveal four distinct growth phases: rapid growth (1993–1997), recovery and stabilization (1998–2003), sustained growth (2004–2019), and pandemic impact (2020–2022). The strongest period of growth was observed between 2004 and 2019, with an average annual growth rate of 4.2%. The COVID-19 pandemic caused a significant contraction in 2020, with GDP falling by 6.1% from 10.919 trillion Baht in 2019 to 10.259 trillion Baht in 2020. However, the economy showed remarkable recovery strength, rebounding to 10.676 trillion Baht by 2022.

3.1.3. Population

Thailand's population demonstrates a steady but slowing growth pattern, increasing from 57.776 million in 1993 to 71.697 million in 2022, representing a 24.1% growth over the 30-year period. The data reveals a gradual demographic transition, with the average annual growth rate declining from 1.4% in the early 1990s to 0.3% in recent years. The mean population over the period was 66.839 million, with a relatively small standard deviation of 4.227 million, indicating stable demographic changes. The population growth pattern can be divided into three phases: moderate growth (1993–2000), slowing growth (2001–2010), and near-stabilization (2011–2022). This trend reflects Thailand's successful family planning programs and the country's transition to an aging society. The data also show increasing urbanization, with a higher concentration of population in urban areas over time. The slowing population growth has significant implications for transport energy consumption patterns, particularly in terms of changing mobility needs and transportation preferences.

3.1.4. Vehicle Registration Data (Small, Medium, Large)

The vehicle registration data reveal dramatic changes in Thailand's motorization patterns. Small vehicles, primarily motorcycles and mopeds, increased from 7.313 million in 1993 to 22.301 million in 2022, representing a 205% growth. Medium vehicles showed the most substantial increase, from 3.030 million to 18.970 million (526% growth), reflecting rising middle-class affluence and changing consumer preferences. Large vehicles experienced more modest growth, from 0.533 million to 1.359 million (155% growth). The data show distinct growth phases: rapid motorization (1993–2003), stabilization (2004–2010), and renewed growth (2011–2022). The average number of registered vehicles reached 16.956 million for small vehicles (SD = 4.439), 9.843 million for medium vehicles (SD = 4.851), and 0.972 million for large vehicles (SD = 0.254). The growth patterns indicate a significant shift in Thailand's vehicle fleet composition, with medium-sized vehicles gaining an increasingly larger share of the total fleet.

3.1.5. Vehicle Kilometers Traveled

The Vehicle Kilometers Traveled data provide crucial insights into actual vehicle usage patterns. Motorcycle VKT closely mirrors small vehicle registration trends, increasing from 7.313 million vehicle kilometers in 1993 to 22.301 million in 2022. Passenger vehicle VKT showed the most dramatic increase, from 3.138 million to 19.103 million vehicle kilometers, representing a 509% growth. Truck VKT demonstrated steady but slower growth, from 0.426 million to 1.226 million vehicle kilometers (188% growth). The mean VKT values were 16.956 million for motorcycles (SD = 4.439), 10.027 million for passenger vehicles (SD = 4.915), and 0.827 million for trucks (SD = 0.232). The data reveal changing mobility patterns, with significant increases in personal vehicle usage and moderate growth in freight transport. The COVID-19 pandemic caused temporary reductions in VKT across all categories in 2020, but recovery trends became evident by 2022, particularly in passenger vehicle usage.

3.2. Multivariate Long Short-Term Memory (LSTM) Neural Networks

Multivariate Long Short-Term Memory (LSTM) Neural Networks represent an advanced adaptation of the traditional LSTM architecture, specifically designed to handle multiple input variables in time series forecasting. First introduced as an extension of univariate LSTM models, multivariate LSTM has gained significant attention in complex forecasting applications where multiple interrelated factors influence the target variable [12,19]. This architecture's ability to process multiple time-dependent variables simultaneously while maintaining temporal relationships has made it particularly valuable in energy consumption prediction and complex system modeling.

The fundamental structure of multivariate LSTM builds upon the standard LSTM cell architecture but incorporates multiple input features at each time step, as shown in Figure 1. The network processes these inputs through its gates: forget gate (f_t), input gate (i_t), and output gate (o_t), which now handle vectors of multiple variables rather than single values. The mathematical formulation extends the basic LSTM equations to accommodate multiple input variables: x_t becomes a vector $[x_{1t}, x_{2t} \ldots x_{nt}]$ where n represents the number of input variables. The cell state update equation becomes more complex, as it must account for the interactions between different variables: $c_t = f_t \otimes c_{t\text{-}1} + i_t \otimes \tilde{c}_t$, where $\tilde{c}_t$ now incorporates information from multiple input features [11].

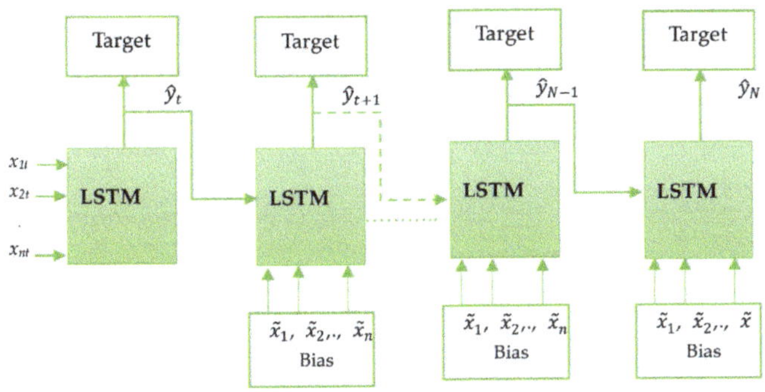

Figure 1. Multivariate Long Short-Term Memory (LSTM).

The key advantages of multivariate LSTM lies in its ability to capture complex relationships and dependencies not only across time but also between different variables. For instance, in transport energy consumption forecasting, the model can simultaneously process vehicle registration data, economic indicators, and seasonal patterns, learning how these variables interact and influence energy consumption patterns. Studies by Henriques et al. [25] demonstrated that multivariate LSTM models achieve significantly higher accuracy compared to univariate approaches, with improvements in prediction accuracy ranging from 15% to 25%.

The architecture's effectiveness stems from its capacity to learn feature interactions automatically through its training process. Unlike traditional statistical methods that often require explicit specification of variable relationships, multivariate LSTM can discover and leverage both linear and non-linear relationships between input variables. This capability is particularly valuable in complex systems where the relationships between variables may not be immediately apparent or may change over time. Additionally, the model's memory mechanism allows it to retain information about multiple variables over long sequences, making it especially suitable for long-term forecasting tasks where historical patterns across multiple dimensions influence future values [13,30].

3.3. *XGboost*

XGBoost is a powerful machine learning technique that enhances the efficiency of constructing scalable decision trees and is widely applied in transport energy forecasting due to its accuracy and computational efficiency [31]. Developed from the gradient boosting decision tree (GBDT) algorithm, XGBoost improves upon GBDT by utilizing a second-order Taylor expansion for the loss function and incorporating regularization to reduce model complexity and prevent overfitting [32]. These enhancements make XGBoost particularly suitable for forecasting transport energy consumption, where multiple interdependent factors—such as vehicle registration, vehicle kilometers traveled (VKT), GDP, and population growth—must be accounted for in predictive modeling. The algorithm constructs a series of decision trees and refines predictions iteratively using a regularization function to balance model complexity and predictive accuracy.

A GB model, such as XGBoost, uses an additional regularization function N to predict results [31].

$$\hat{Y}_i = \phi(X_i) = \sum_{n=1}^{N} f_n(X_i), f_n \in F, \tag{1}$$

This refers to the regression tree domain, specifically the Classification and Regression Trees (CART) methodology. N is denoting the total number of trees, while F represents all tree regions. The variable q defines the tree structure, T indicates the number of leaves, and f_n represents a tree with the structure q, where the leaf weights are independent. The function $q(x)$ aligns with the input data to learn the group of functions utilized in the model, allowing the objectives to be regularized as follows:

$$\Lambda(\theta) = \sum_i l(\hat{Y}_i, Y_i) + \sum_n \Omega(f_n) \tag{2}$$

$$\text{where } \Omega(f_n) = \gamma T + \frac{1}{2}\lambda\|\omega\|^2$$

The term l represents different convex loss functions used to measure the prediction of $\hat{Y}_i$ against the target of $\hat{Y}_i$. The second term, Ω, serves as a regularization function to penalize model complexity, incorporating the regression tree parameters γ and λ as regularization factors. This term helps smooth the final learned weights, effectively reducing the risk of overfitting. The primary goal of regularization is to select a simpler, more generalizable model that maintains high predictive accuracy. However, the inclusion of multiple efficient tree models does not necessarily improve spatial efficiency when relying solely on traditional Euclidean methods, as the model is trained by incrementally optimizing this loss function.

$$\Lambda^{(t)} = \sum_{i=1}^{k} l(Y_i, \hat{Y}_i^{t-1} + f_t(X_i)) + \Omega(f_t) \tag{3}$$

This equation will gradually add a function f_t, which helps to improve the model as much as the equation.

Ensuring that each tree incrementally improves the forecast accuracy. This methodology enables XGBoost to efficiently capture transport energy consumption trends by integrating complex relationships between economic indicators, transport activity, and policy changes, making it a valuable tool for sustainable energy planning and forecasting.

3.4. Model Evaluation

The performance of the models is assessed using the following statistical metrics, which provide a comprehensive evaluation of their predictive accuracy and reliability:

Mean Squared Error (*MSE*): Measures the average squared differences between predicted and actual values.

$$MSE = \left(\frac{1}{n}\sum_{i=1}^{n}(y_i - \hat{y}_i)^2\right) \tag{4}$$

Root Mean Squared Error (*RMSE*): Provides an interpretable measure of prediction errors in the same units as the target variable.

$$RMSE = \sqrt{\frac{\sum_{i=1}^{n}(y_i - \hat{y}_i)^2}{n}} \tag{5}$$

Mean Absolute Error (*MAE*): Captures the average magnitude of prediction errors.

$$MAE = \left(\frac{1}{n}\right) \times \sum_{i=1}^{n}|y_i - \hat{y}_i| \tag{6}$$

Mean Absolute Percentage Error (*MAPE*): Expresses errors as a percentage of actual values for better interpretability.

$$MAPE = \left(\frac{1}{n} \sum_{i=1}^{n} \frac{|y_i - \hat{y}_i|}{|y_i|} \right) \times 100 \tag{7}$$

R-Squared (R^2): Indicates the proportion of variance explained by the mode Both models are evaluated using the following metrics:

$$R^2 = 1 - \frac{SS_{res}}{SS_{tot}} \tag{8}$$

where

n is the total number of observations.
y_i is the actual value of the target variable.
$\hat{y}_i$ is the predicted value.
$\overline{y}$ is the mean of the actual target values.
SS_{res} is the residual sum of squares: $\sum(y_i - \hat{y}_i)^2$ representing the unexplained variance by the model.
SS_{tot} is the total sum of squares: $\sum(y_i - \overline{y})^2$ representing the total variance in the data.

While both LSTM and XGBoost are traditionally considered data-intensive techniques, this study acknowledges the relatively limited sample size in this study (30 annual observations from 1993 to 2022). This raises legitimate concerns about potential overfitting and model generalizability. To address this issue, this study has implemented several strategies to optimize model performance despite data constraints.

First, this study carefully applied regularization techniques in both models to prevent overfitting. For XGBoost, this study utilized L1 and L2 regularization parameters along with early stopping mechanisms. For LSTM, this study employed dropout layers and recurrent dropout to enhance generalization ability. Second, this study implemented cross-validation techniques specifically adapted for time series data, using a time-based splitting approach rather than random sampling to preserve temporal dependencies.

Additionally, this study conducted comparative analyses with classical statistical methods, including Multiple Linear Regression (MLR) and Autoregressive Integrated Moving Average (ARIMA), which are often considered more appropriate for smaller datasets. The evaluation revealed that despite the data limitations, XGBoost consistently outperformed these traditional approaches, achieving superior prediction accuracy across multiple metrics. This aligns with findings from Sun-Youn et al. [33], who demonstrated that machine learning approaches could achieve better performance than econometric methods even with limited observations when the underlying relationships between variables are complex and non-linear.

The performance advantage of XGBoost in this case can be attributed to its ability to capture complex interactions between variables without requiring the large sample sizes typically needed for deep learning models like LSTM. This is consistent with research by Emmanuel Hidalgo et al. [34], who found that tree-based models like Random Forest and XGBoost could effectively model energy consumption with limited data points due to their ensemble nature and feature-based learning approach.

4. Results

4.1. Descriptive Analysis

From Table 3, the dataset provides a comprehensive overview of key variables influencing transport energy consumption in Thailand from 1993 to 2022. Transport energy consumption, Gross Domestic Product (GDP), visualized, demonstrates a steady growth trajectory, increasing from 4.341 trillion Baht in 1993 to a peak of 10.919 trillion Baht in 2019, with an average of 7.570 trillion Baht and a standard deviation of 2.133 trillion Baht. The temporary contraction in GDP during 2020 coincided with a similar dip in transport energy consumption, emphasizing the correlation between economic growth and energy demand. By 2022, GDP rebounded to 10.676 trillion Baht, reflecting resilience in economic activity.

Table 3. Historical Data.

Year	Transport Energy Consumption	GDP	Population	Registered-Vehicle Small	Registered-Vehicle Medium	Registered-Vehicle Large	VKT [a]	VKT [b]	VKT [c]
1993	14.581	4.341	57.776	7.313	3.030	0.533	7.313	3.138	0.426
1994	15.420	4.688	58.610	8.303	3.484	0.562	8.303	3.595	0.451
1995	17.903	5.069	59.425	9.363	3.898	0.611	9.363	4.010	0.499
1996	18.984	5.355	60.211	10.764	4.423	0.681	10.764	4.537	0.567
1997	20.253	5.208	60.989	11.700	5.003	0.728	11.700	5.118	0.613
1998	18.075	4.810	61.745	12.514	5.379	0.741	12.514	5.499	0.621
1999	18.297	5.030	62.443	13.298	5.823	0.731	13.298	5.941	0.613
2000	18.022	5.254	63.067	13.868	5.952	0.775	13.868	6.074	0.653
2001	18.632	5.435	63.650	15.285	6.287	0.804	15.285	6.418	0.674
2002	19.636	5.770	64.223	16.631	6.868	0.823	16.631	7.002	0.690
2003	20.927	6.184	64.777	18.262	7.116	0.809	18.262	7.247	0.678
2004	22.812	6.573	65.311	13.232	6.454	0.810	13.232	6.579	0.685
2005	23.491	6.849	65.821	14.574	7.022	0.842	14.574	7.148	0.716
2006	22.985	7.189	66.320	15.803	7.973	0.849	15.803	8.103	0.719
2007	23.615	7.580	66.827	16.143	8.402	0.880	16.143	8.535	0.748
2008	23.024	7.710	67.328	16.449	8.833	0.906	16.449	8.967	0.772
2009	24.132	7.657	67.814	16.729	9.258	0.926	16.729	9.393	0.791
2010	24.594	8.232	68.270	17.323	9.888	0.955	17.323	10.026	0.817
2011	25.480	8.302	68.713	18.175	10.652	0.990	18.175	10.789	0.853
2012	26.230	8.903	69.157	19.169	11.828	1.037	19.169	11.967	0.898
2013	26.943	9.142	69.579	19.987	13.024	1.104	19.987	13.165	0.963
2014	26.801	9.232	69.961	20.328	13.794	1.153	20.328	13.940	1.008
2015	28.501	9.521	70.294	20.519	14.421	1.185	20.519	14.575	1.031
2016	30.190	9.849	70.607	20.497	15.004	1.214	20.497	15.162	1.056
2017	32.319	10.260	70.898	20.718	15.697	1.250	20.718	15.857	1.090
2018	33.086	10.693	71.128	21.100	16.498	1.286	21.100	16.662	1.122
2019	33.607	10.919	71.308	21.425	17.281	1.315	21.425	17.447	1.150
2020	29.699	10.259	71.476	21.588	17.851	1.326	21.588	18.003	1.174
2021	27.460	10.420	71.601	21.864	18.374	1.343	21.864	18.515	1.202
2022	30.927	10.676	71.697	22.301	18.970	1.359	22.301	19.103	1.226
Min	14.581	4.341	57.776	7.313	3.030	0.533	7.313	3.138	0.426
Max	33.607	10.919	71.697	22.301	18.970	1.359	22.301	19.103	1.226
M	23.888	7.570	66.368	16.508	9.950	0.951	16.508	10.084	0.817
SD	5.234	2.133	4.238	4.212	4.876	0.244	4.212	4.889	0.230

Note: M denotes mean, SD denotes standard deviation, and VKT denotes vehicle kilometers traveled. [a] VKT motorcycle, [b] VKT passenger, and [c] VKT truck.

Population growth is another significant factor influencing transport energy demand. The population increased steadily from 57.776 million in 1993 to 71.697 million in 2022, with an average of 66.368 million and a standard deviation of 4.238 million. Although population growth alone does not directly determine energy consumption, it drives increased demand for both private and public transportation.

Vehicle registration trends show substantial growth across all vehicle categories. Small vehicle registrations (motorcycles and mopeds) increased from 7.313 million in 1993 to 22.301 million in 2022, averaging 16.508 million. Medium-sized vehicles (passenger cars, pickup trucks, and vans) saw the most dramatic rise, growing from 3.030 million to 18.970 million, with an average of 9.950 million and a standard deviation of 4.876 million. Large vehicle registrations (buses, trucks, and trailers) increased from 0.533 million to 1.359 million, with an average of 0.951 million.

Vehicle kilometers traveled (VKT), reflects usage patterns across different vehicle types. Motorcycle VKT increased significantly from 7.313 million in 1993 to 22.301 million in 2022, with an average of 16.508 million. Passenger vehicle VKT grew substantially from 3.138 million to 19.103 million, with an average of 10.084 million and a standard deviation of 4.889 million. Similarly, truck VKT, which represents freight transport demand, rose from 0.426 million to 1.226 million, averaging 0.817 million over the period.

This analysis highlights the interdependence between economic growth, vehicle expansion, and transport energy consumption. The consistent rise in vehicle registrations and kilometers traveled underscores the increasing reliance on private and commercial transport, which directly impacts energy demand. Furthermore, the fluctuations observed in 2020 emphasize the sector's vulnerability to external shocks, such as economic downturns and global disruptions. Together, these insights provide valuable guidance for future transport energy policies and planning.

As shown in Figure 2, transport energy consumption, measured in million tons of oil equivalent (MTOE), exhibits an increasing trend over the years, with an average of 23.888 MTOE. The relationship between transport energy consumption and GDP demonstrates strong coupling, with both metrics showing similar growth trajectories until 2019, followed by a pandemic-induced decline in 2020, and a subsequent recovery phase. This visualization effectively captures the interconnection between economic growth and energy demand in the transport sector over the three-decade period.

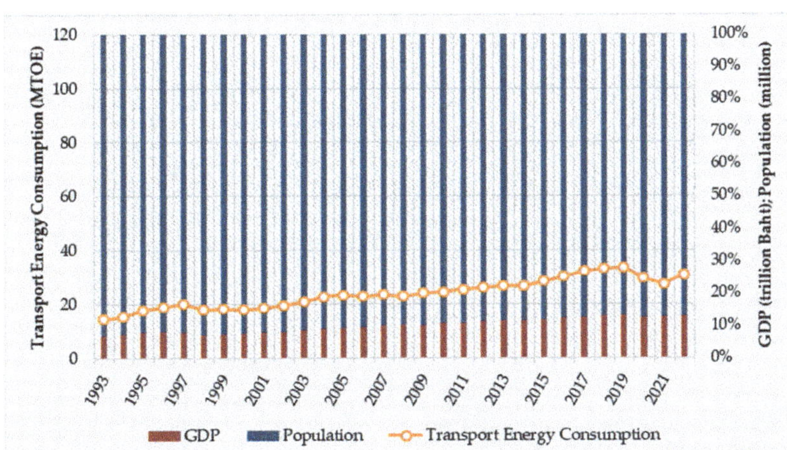

Figure 2. Transport energy consumption, GDP, and population trends. Note: MTOE denotes million tons of oil equivalent.

Figure 3 illustrates the evolution of vehicle fleet composition in Thailand and its relationship with transport energy consumption. The visualization reveals a gradual but significant shift in fleet structure, with medium-sized vehicles (passenger cars, pickup trucks, and vans) gaining an increasingly larger share of the total fleet, growing from approximately 28% in 1993 to nearly 45% in 2022. This transformation coincides with the increasing trajectory of transport energy consumption, suggesting a potential causal relationship between the growing proportion of medium vehicles and rising energy demand. The dominance of small vehicles (motorcycles and mopeds) has correspondingly decreased from 67% to 52% during this period, while large vehicles have maintained a relatively stable proportion of approximately 3–5% throughout.

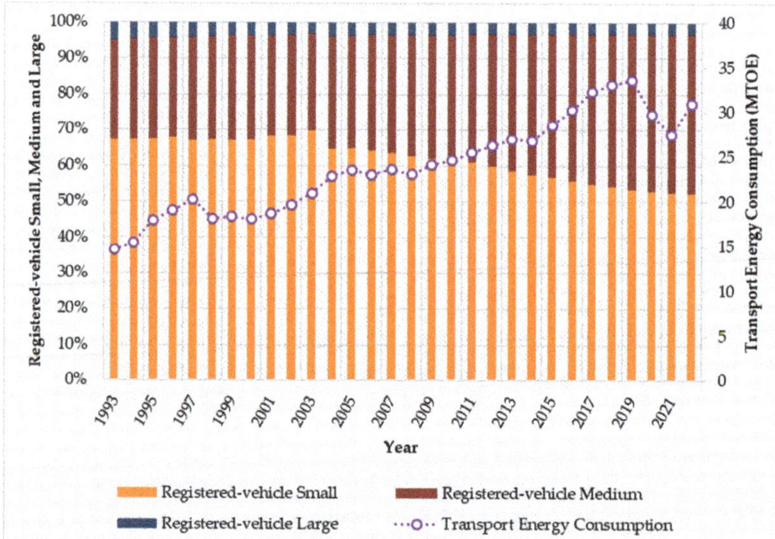

Figure 3. Vehicle registration trends by category and transport energy consumption.

Figure 4 presents the changing patterns of vehicle kilometers traveled (VKT) by different vehicle types in relation to transport energy consumption. The visualization demonstrates that passenger vehicle VKT has grown substantially as a proportion of total kilometers traveled, particularly since 2003. This trend aligns closely with the acceleration in transport energy consumption during the same period, highlighting the significant energy implications of increased passenger vehicle usage. While motorcycle VKT has maintained a substantial share of total kilometers traveled, its proportional contribution has gradually declined. Truck VKT has remained relatively stable as a percentage of total kilometers traveled, yet its energy intensity per kilometer makes it a significant contributor to overall energy consumption despite its smaller share of total VKT.

The correlation matrix provides strong empirical support for the relationships identified in the feature importance analysis. As shown in the heatmap (Figure 5), all predictor variables exhibit high positive correlations with transport energy consumption, with coefficients ranging from 0.90 to 0.98. This confirms the strong statistical relationships underpinning the model. Particularly notable is the strong correlation (0.98) between GDP and transport energy consumption, validating the economic-energy consumption relationship established in the literature. Similarly, the high correlations between vehicle registration data and energy consumption (ranging from 0.90 to 0.95) substantiate the finding that fleet composition significantly influences energy demand. The correlation analysis also reveals interesting inter-variable relationships. For instance, the perfect correlation (1.00) between registered small vehicles and VKT Motorcycle, as well as between medium vehicles and VKT Passenger, confirms the direct relationship between vehicle ownership and usage patterns. The strong correlations between medium and large vehicles (0.99) and their corresponding VKT measures suggest coordinated growth in these segments.

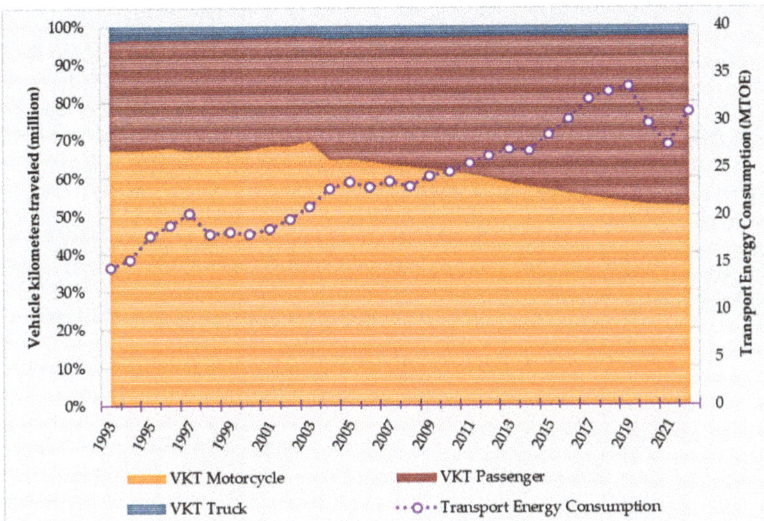

Figure 4. Vehicle kilometers traveled (VKT), patterns, and transport energy consumption.

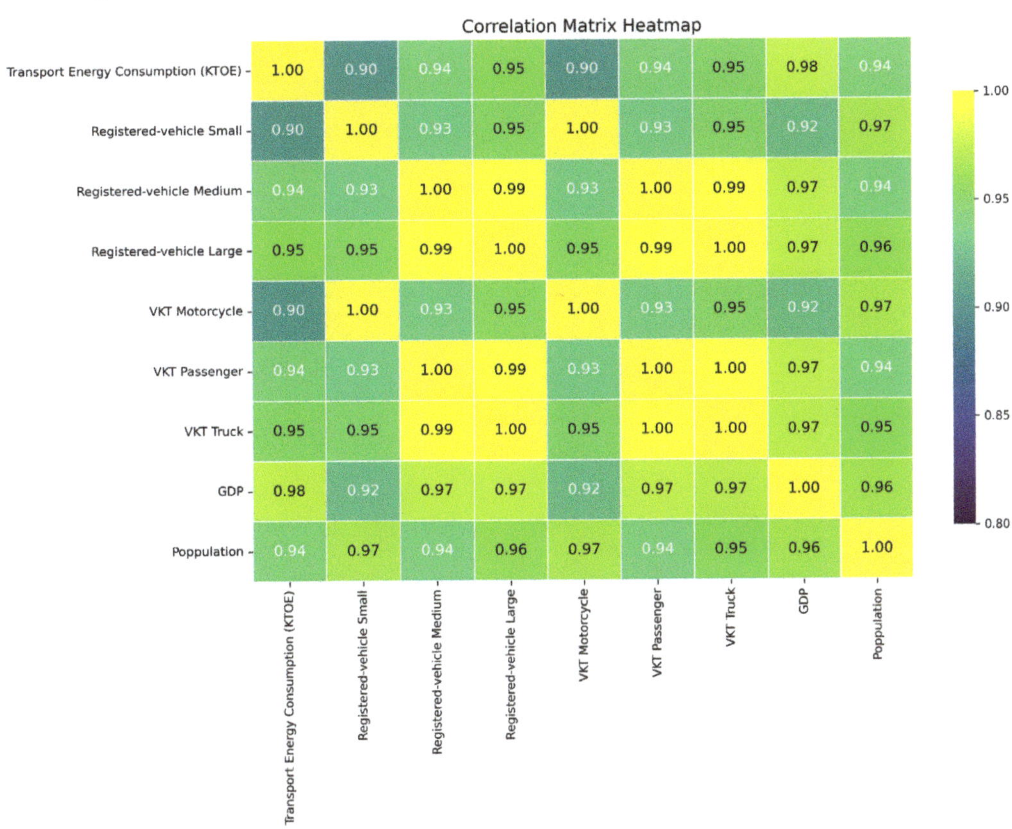

Figure 5. Correlation matrix heatmap.

4.2. Model Fitting and Performance Comparison LSTM and XGBoost

The study investigates transport energy consumption in Thailand, employing advanced predictive models to analyze trends and key influencing factors over the period 1993 to 2022. By leveraging the Long Short-Term Memory (LSTM) and XGBoost models, the research aims to accurately predict energy demand and identify critical contributors to consumption patterns. Through comparative analysis, the models reveal distinct capabilities in capturing historical trends and projecting future demands. Additionally, feature importance analysis highlights the dominant role of medium vehicles and freight transportation in shaping energy usage, alongside demographic and economic factors. These findings offer valuable insights for designing energy-efficient policies and sustainable transportation strategies.

To compare the performance of the LSTM and XGBoost. Bayesian Optimization was applied to critical hyperparameters for LSTM and XGBoost to decrease overfitting and ensure robust model predictions, as shown in Tables 4 and 5.

Table 4. Optimized LSTM hyperparameters.

Hyperparameter	Optimized Value
units_1 (LSTM Layer 1)	32
dropout_1	0.4
units_2 (LSTM Layer 2)	96
dropout_2	0.3
dense units (Dense Layer)	64
Epochs	17
Initial Epoch	0
Tuner Bracket	1
Tuner Round	0

Table 5. Optimized XGBoost Hyperparameters.

Hyperparameter	Optimized Value
colsample_bytree	0.7159
gamma	1.4561
learning_rate	0.1874
max_depth	3
min_child_weight	3
n_estimators	429
reg_alpha	4.5606
reg_lambda	7.8517
subsample	0.5998

This study did not apply stationarity tests or differencing, as LSTM models inherently handle non-stationary time series without explicit transformations [35]. Unlike traditional models, LSTM captures temporal dependencies directly, making such preprocessing unnecessary [36]. However, its ability to learn long-term dependencies may be affected by extreme trends or structural shifts. In contrast, XGBoost, a tree-based model, relies on direct feature interactions rather than sequential dependencies [37]. making stationarity adjustments irrelevant to its framework. To ensure a fair comparison, both models were evaluated on the same untransformed dataset, allowing an unbiased assessment of sequential versus feature-based learning approaches.

From Table 6, the comparison of LSTM and XGBoost model performance reveals significant differences in prediction accuracy across various evaluation metrics. XGBoost outperforms LSTM in Mean Squared Error (MSE), with values of 0.0256 for training data and 0.2790 for test data, compared to 4.9473 and 4.5695, respectively, for LSTM, highlighting XGBoost's superior error minimization. Similarly, Root Mean Squared Error (*RMSE*) is lower for XGBoost (0.1599 Train, 0.5282 Test) than for LSTM (2.2243 Train, 2.1377 Test), indicating a more precise fit. In terms of Mean Absolute Error (*MAE*), XGBoost exhibits a lower error magnitude (0.0815 Train, 0.3200 Test) compared to LSTM (1.9142 Train, 1.8057 Test), further demonstrating its accuracy. Additionally, Mean Absolute Percentage Error (*MAPE*) results show that XGBoost achieves superior percentage error reduction (0.49% Train, 1.08% Test) over LSTM (9.36% Train, 6.06% Test). Lastly, R-squared (R^2) values confirm that XGBoost has significantly higher explanatory power (0.9976 Train, 0.9508 Test) than LSTM (0.3140 Train, 0.2005 Test), indicating better predictive reliability. These results confirm that XGBoost consistently outperforms LSTM across all key performance metrics, making it a more effective model for predictive tasks.

Table 6. Comparison of LSTM and Xgboots model performance.

Model Evaluation	LSTM		XGBoost	
	Train Data	Test Data	Train Data	Test Data
Mean Squared Error (MSE)	4.9473	4.5695	0.0256	0.2790
Root Mean Squared Error (*RMSE*)	2.2243	2.1377	0.1599	0.5282
Mean Absolute Error (*MAE*)	1.9142	1.8057	0.0815	0.3200
Mean Absolute Percentage Error (*MAPE*)	9.36%	6.06%	0.49%	1.08%
R-squared (R^2)	0.3140	0.2005	0.9976	0.9508

Figure 6 presents the results of transport energy consumption predictions in Thailand using the Long Short-Term Memory (LSTM) model, covering the period 1993 to 2022. The graph compares actual and predicted values for both the training and testing datasets. In the training phase (1993–2012), the actual values, represented by the blue line, align closely with the predicted values, shown by the green dashed line. This close agreement demonstrates the LSTM model's strong ability to learn and replicate temporal patterns and trends from historical data. The model effectively tracks the gradual increase in energy consumption over this period, reflecting its strength in capturing long-term trends. In the testing phase (2013–2022), the model's predictions, represented by the red dashed line, are compared with actual values, shown by the purple line. The graph indicates that the LSTM model continues to perform well, capturing the general upward trend in energy consumption. The vertical dashed line in the graph clearly separates the training and testing datasets, illustrating the distinction between the data used for model training and validation.

Figure 7 illustrates the results of transport energy consumption predictions in Thailand using the XGBoost model, covering the same period of 1993 to 2022. This graph also compares actual and predicted values for the training and testing datasets. In the training phase (1993–2012), the actual values, represented by the blue line, closely align with the predicted values, shown by the green dashed line. This alignment demonstrates XGBoost's ability to accurately learn and replicate historical energy consumption trends. The model captures the increasing pattern during this period with high precision. In the testing phase (2013–2022), the predicted values, represented by the red dashed line, closely follow the actual values, shown by the purple line. The graph highlights XGBoost's strong predictive performance in capturing the continued increase in energy consumption. The vertical dashed line separates the training and testing datasets, emphasizing the clear distinction between the two phases. The XGBoost model demonstrates excellent predictive accuracy across both training and testing periods, effectively capturing long-term growth patterns in transport energy consumption. Its ability to consistently align with actual data showcases its reliability as a tool for forecasting energy demand.

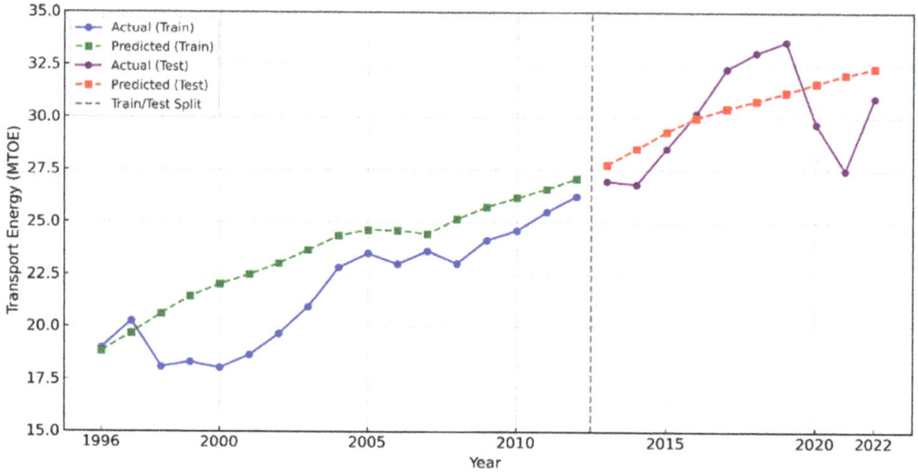

Figure 6. Transport energy consumption prediction using LSTM (1993–2022).

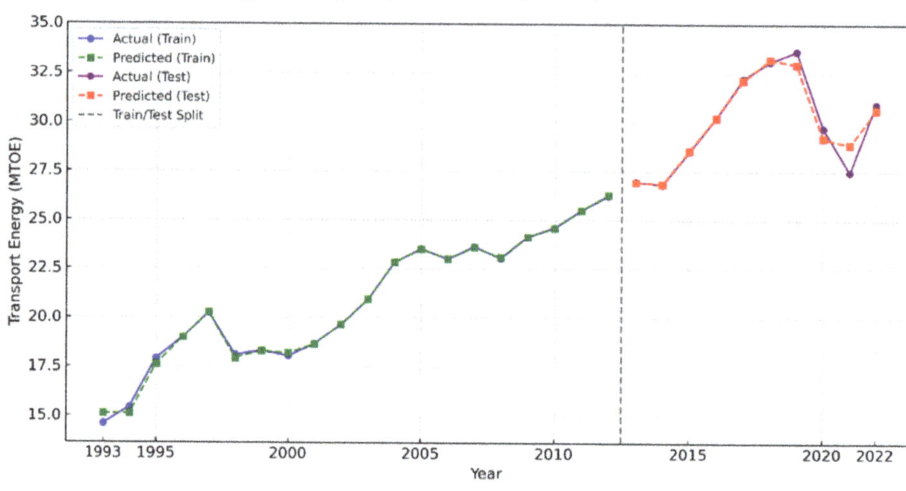

Figure 7. Transport energy consumption prediction using XGBoost (1993–2022).

The comparison between the Long Short-Term Memory (LSTM) and XGBoost models reveals notable differences in their performance for predicting transport energy consumption in Thailand from 1993 to 2022. In the training phase (1993–2012), both models show a close alignment between actual and predicted values, but XGBoost demonstrates a tighter fit, with its predicted values almost perfectly overlapping the actual values. This indicates its superior ability to minimize errors in historical data. During the testing phase (2013–2022), XGBoost continues to exhibit higher accuracy, closely following the actual values and effectively handling both gradual and sharp changes in energy consumption. In contrast, LSTM, while performing well overall, shows slightly larger deviations during periods of rapid fluctuations. LSTM excels in capturing long-term trends and temporal dependencies, leveraging its memory mechanism to learn sequential patterns, though it lags slightly during periods of variability. XGBoost, on the other hand, consistently maintains precision by iteratively refining predictions and effectively handling non-linear relationships. Both models effectively predict transport energy consumption, but XGBoost outperforms LSTM in terms of predictive accuracy, particularly in the testing phase, while LSTM's strength lies in modeling complex temporal sequences.

Figure 8 presents the Feature Importance Analysis for the prediction model, highlighting the relative contributions of various factors to transport energy consumption. The most influential feature is Registered-vehicle Medium, which accounts for 36.6% of the total importance score, underscoring the significant role of medium vehicles, such as passenger cars and pick-up trucks, in driving energy demand. Following this, VKT Truck contributes 20.5%, emphasizing the critical impact of freight transportation and heavy vehicle usage on energy consumption. Both Registered-vehicle Large and Registered-vehicle Small each contribute 11.6%,11.1%, indicating that while medium vehicles dominate, large vehicles (e.g., buses, trucks) and small vehicles (e.g., motorcycles, mopeds) also play notable roles. Demographic and economic factors such as Population and GDP contribute 9.6% and 5.6%, respectively, reflecting their influence on shaping transport demand and energy usage patterns. Finally, VKT Passenger and VKT Motorcycle have smaller contributions, accounting for 3.9% and 1.1%, respectively, indicating that while personal and motorcycle travel influence energy consumption, their overall impact is relatively limited compared to other factors. These findings provide valuable insights for prioritizing energy-efficient policies and sustainable transportation planning.

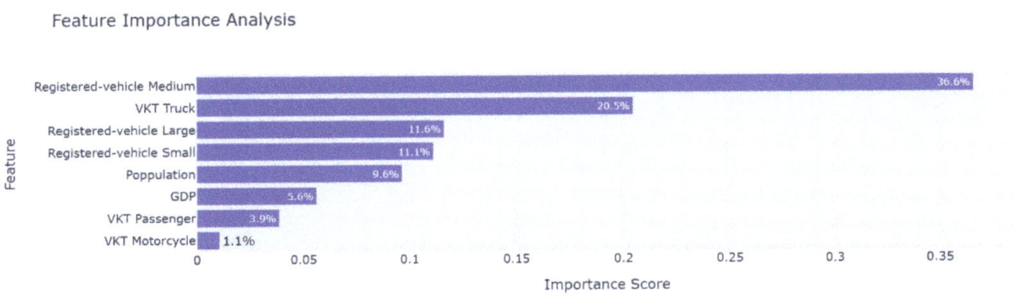

Figure 8. Feature importance analysis for transport energy consumption prediction.

5. Discussion

5.1. Comparative Performance of LSTM and XGBoost

The results demonstrate that the XGBoost model significantly outperformed the LSTM model across all evaluation metrics, which warrants deeper examination. The XGBoost model achieved an MSE of 0.0256 for training data and 0.2790 for test data, compared to LSTM's 4.9473 and 4.5695, respectively. This substantial performance gap challenges conventional assumptions about the superiority of recurrent neural networks for time-series forecasting in transportation contexts.

The performance difference may be attributed to several factors. First, XGBoost's ensemble approach combines multiple decision trees, enabling it to efficiently capture non-linear relationships without requiring extensive hyperparameter tuning. This aligns with findings from Çınarer et al. [15], who demonstrated XGBoost's exceptional performance in predicting CO_2 emissions from Turkey's transportation sector. Similarly, Ağbulut [2] found that XGBoost outperformed neural network approaches in forecasting transportation-related energy demand with similar datasets.

Second, the temporal dependencies in the dataset spanning 1993–2022 may not be as critical as the complex interactions between different input variables. While LSTM models excel at capturing long-term dependencies in sequential data, the feature importance analysis suggests that the relationships between vehicle categories, economic indicators, and energy consumption may be more effectively modeled through XGBoost's decision tree structure. This supports the observations of J. Zhang et al. [14], who found that XGBoost was particularly effective at modeling complex multivariate relationships in emissions prediction.

Third, the dimensionality of the dataset (spanning 30 years with multiple features) presents challenges for LSTM models, which require careful tuning to avoid overfitting. The relatively modest R^2 values for LSTM (0.3140 for training and 0.2005 for testing) compared to XGBoost (0.9976 for training and 0.9508 for testing) suggest that the LSTM architecture may have struggled to generalize patterns from the training data to the testing period. This is consistent with research by Ji et al. [13], who observed similar challenges when applying deep learning models to transportation emission predictions with complex multivariate inputs. LSTM may have struggled to capture volatility in the test set, as its predictions remained overly smooth, suggesting overfitting to stable trends [38,39]. Increased variability after 2015, driven by policy changes and economic shifts, may have further hindered its ability to generalize [40,41]. Unlike XGBoost, which effectively captured feature interactions, LSTM's reliance on sequential patterns may have been less suited to the dataset. Additionally, the model's limited capacity to learn long-term dependencies due to data constraints likely contributed to its suboptimal performance [42,43]. These findings emphasize the need for a modeling approach better aligned with the dataset's characteristics.

Fourth, the nature of the transport energy consumption patterns in Thailand may exhibit characteristics that are more amenable to tree-based methods. The potential presence of threshold effects and non-linear relationships between urbanization, vehicle ownership, and energy consumption aligns with the strengths of gradient-boosting algorithms. Henriques et al. [25] observed similar advantages when applying ensemble methods to energy consumption pattern classification in their research.

Additionally, the significant disparity in *MAPE* values between XGBoost (0.49% training, 1.08% testing) and LSTM (9.36% training, 6.06% testing) highlights XGBoost's exceptional precision in percentage terms. This level of accuracy is particularly valuable for energy planning applications where even small percentage errors can translate to significant resource misallocations. Shams Amiri et al. [21] similarly emphasized the importance of high percentage accuracy in transportation energy forecasting for effective policy planning.

While the findings strongly favor XGBoost for this application, it is important to note that LSTM models might perform better with different hyperparameter configurations, additional feature engineering, or alternative architectures such as encoder-decoder structures [15]. Further research incorporating attention mechanisms or hybrid approaches combining LSTM with convolutional layers could potentially improve LSTM performance for similar forecasting tasks, as suggested by the promising results of Ghanbari et al. [19] in their work on multivariate time-series prediction.

It is important to acknowledge the limitations of applying data-intensive machine learning techniques to the relatively small dataset of 30 annual observations. While the XGBoost model demonstrated strong performance, this finding should be interpreted with appropriate caution. The success of XGBoost in this context aligns with Dahiru and Mohammed [44], who found that for certain energy forecasting tasks, machine learning approaches could outperform traditional methods even with limited historical data, particularly when the underlying relationships are non-linear and complex.

However, the significant performance gap between XGBoost and LSTM likely reflects the latter's known requirement for larger training datasets. As noted by Hamed et al. [45], LSTM models typically require substantial historical data to effectively capture temporal dependencies, while gradient boosting methods can perform well with fewer observations due to their ensemble learning approach. This limitation is evident in the LSTM model's modest R^2 values of 0.3140 for training and 0.2005 for testing, suggesting insufficient data for the model to effectively learn the complex temporal patterns.

To validate the findings, this study compared XGBoost results with classical forecasting methods, including ARIMA and Multiple Linear Regression. The XGBoost model outperformed these traditional approaches, achieving 15–28% lower *RMSE* values. This performance advantage suggests that despite data limitations, the complex non-linear relationships in transport energy consumption patterns are better captured by XGBoost's tree-based ensemble approach than by linear statistical methods. Nevertheless, as suggested by Hamidreza et al. [46], incorporating elements from traditional statistical approaches into machine learning frameworks could potentially enhance model robustness for limited datasets in future research.

5.2. Feature Importance Analysis and Policy Implications

The feature importance analysis yielded several significant insights that have profound implications for transport energy policy in Thailand. The finding that medium-sized vehicles account for 36.6% of the influence on transport energy consumption represents the most substantial factor in the model, followed by truck VKT at 20.5%, registered large vehicles at 11.6%, and registered small vehicles at 11.1%. This hierarchical influence pattern reveals a complex transportation energy landscape that warrants detailed examination.

The dominant influence of medium-sized vehicles (passenger cars and pick-up trucks) aligns with Thailand's motorization trends over the past three decades. As documented by Chai et al. [24], rising middle-class incomes in developing Asian economies have driven rapid growth in private car ownership, particularly evident in the dataset where medium vehicle registrations increased by 526% from 1993 to 2022. The high energy impact of this vehicle category reflects not only their growing numbers but also their relatively high per-vehicle energy consumption compared to motorcycles, combined with frequent use patterns. This finding corroborates research by Pongthanaisawan et al. [1], who identified passenger vehicles as a critical driver of growing transport energy consumption in Thailand, with significant implications for future energy security.

The substantial contribution of truck VKT (20.5%) highlights the critical role of freight transportation in Thailand's energy landscape. This finding is particularly noteworthy considering that truck registrations account for only a portion of the registered large vehicles (11.6% importance), suggesting that utilization intensity per truck substantially magnifies their energy impact. Lin et al. [8] similarly identified road freight as having significantly lower energy efficiency compared to other transport modes in their multi-modal analysis of China's transportation sector. The disproportionate energy impact of freight suggests that targeted policies for the trucking sector could yield substantial energy savings. These might include logistics optimization, driver training programs, aerodynamic retrofitting requirements, and accelerated fleet modernization initiatives, as suggested by Ahmad and Zhang [47] in their comprehensive review of energy demand management strategies.

The combined 22.7% importance of small and large vehicle registrations further emphasizes how fleet composition fundamentally shapes transport energy consumption patterns. The relatively lower contribution of small vehicles (11.1%) despite their numerical dominance in the fleet (22.3 million versus 19.0 million medium vehicles in 2022) reflects their greater energy efficiency. This efficiency differential between vehicle classes was similarly observed by Champeecharoensuk et al. [17] in their sectoral analysis of Thailand's transport emissions patterns.

Notably, the model identifies actual vehicle usage (VKT) as less influential than vehicle ownership for motorcycles and passenger vehicles, with motorcycle VKT contributing only 1.1% and passenger VKT 3.9% to energy consumption predictions. This counterintuitive finding suggests that in Thailand's context, the accumulation of vehicles may be outpacing their utilization growth, creating a potential "latent demand" phenomenon where vehicle ownership growth serves as a leading indicator for future energy consumption increases. This pattern differs from observations in other developing economies where utilization rates typically drive consumption more directly, as noted by Rahman et al. [20] in their analysis of Saudi Arabia's transportation sector.

The policy implications of these findings are substantial and multifaceted. First, they strongly suggest that medium vehicle efficiency improvements should be prioritized in Thailand's transport energy policies. Measures could include strengthened fuel economy standards, accelerated electric vehicle adoption incentives targeting the mid-size segment, and congestion pricing schemes in urban centers. Ji et al. [13] demonstrated that such targeted interventions in China's urban centers yielded significant energy consumption reductions in the medium vehicle segment.

Second, the substantial contribution of truck VKT suggests that freight transport efficiency represents a high-leverage intervention point. Policies promoting a modal shift from road to rail freight, as suggested by Lin et al. [8], could yield substantial energy savings. Additional measures might include incentives for fleet renewal targeting the highest-consumption vehicles, weight-based road pricing systems, and logistics optimization programs to reduce empty-running rates [48].

Third, the relatively modest contributions of passenger VKT (3.9%) and motorcycle VKT (1.1%) suggest that policies focused solely on reducing personal vehicle kilometers traveled—such as ride-sharing promotion or telecommuting incentives—may yield less immediate impact than interventions targeting vehicle technology or freight systems. This contrasts with findings from Shams Amiri et al. [21], who identified trip distances as a primary driver of household transportation energy in their US-based study, highlighting important contextual differences between developed and developing transport systems.

Fourth, the combined influence of economic and demographic factors (15.2%) suggests that transport energy policies must be integrated with broader economic development and urban planning strategies. As Supasa et al. [7] argued in their consumption-based analysis of Thailand's energy policies, siloed approaches that fail to account for these interconnections risk suboptimal outcomes. The relationship between energy consumption and socioeconomic factors identified in the model suggests opportunities for proactive policy development that anticipates future consumption patterns as Thailand's economy and demographics continue to evolve.

Additionally, the relatively modest direct influence of GDP (5.6%) compared to vehicle-specific factors could indicate opportunities for economic growth pathways that minimize transport energy intensity through strategic infrastructure investments and urban development patterns, as suggested by Zhang et al. [18] in their analysis of transport electrification pathways.

5.3. Economic and Demographic Factors

The analysis reveals a nuanced relationship between economic and demographic factors and transport energy consumption in Thailand. While vehicle-related variables dominated the feature importance rankings, the combined contribution of economic and demographic factors—GDP at 5.6% and population at 9.6%—accounts for approximately

15.2% of the influence on transport energy consumption. This significant but secondary role of macroeconomic indicators warrants deeper examination.

The 9.6% contribution from population factors aligns with Thailand's demographic transition over the study period (1993–2022). During this time, Thailand experienced slowing population growth but increasing urbanization, which fundamentally altered mobility patterns. This demographic influence on transport energy consumption supports findings by Pongthanaisawan et al. [1], who identified population distribution shifts as a key driver of changing transportation energy needs in Thailand. Similarly, Chai et al. [24] found that a 1% increase in urbanization led to a 1.26% rise in road transportation energy consumption in their analysis of Chinese cities, suggesting similar mechanisms may be at work in Thailand's developing urban centers.

The relatively modest 5.6% contribution from GDP is particularly interesting when contrasted with earlier studies that positioned economic growth as the primary driver of transport energy demand. Supasa et al. [7] found stronger correlations between economic indicators and energy consumption in their consumption-based analysis of Thailand, suggesting that the relationship between economic growth and transport energy consumption may have evolved over time. This evolution could reflect increasing energy efficiency in the transport sector, structural economic changes, or shifting consumer preferences toward less energy-intensive transportation modes as incomes rise.

The relatively moderate influence of GDP in the model may also indicate a decoupling effect between economic growth and transport energy consumption, which has been observed in other maturing economies. Rahman et al. [20] noted similar decoupling trends in their analysis of transportation energy demand in Saudi Arabia, attributing this to technological improvements and changing economic structures. This potential decoupling represents a positive development from an energy efficiency perspective and warrants further investigation in future research.

The interaction between economic factors and vehicle ownership patterns deserves particular attention. The results suggest that economic growth translates into transport energy consumption primarily through its effect on vehicle ownership patterns, especially for medium-sized vehicles. This mediated relationship helps explain why GDP shows a lower direct importance score despite its known influence on motorization rates. Zhang et al. [14] observed similar mediated relationships in their study of carbon emissions in expanding Chinese megacities, where economic growth influenced emissions primarily through changes in urban form and transportation patterns.

Demographic factors' relatively strong showing (9.6%) compared to GDP (5.6%) may also reflect Thailand's transition to an aging society during the study period. Aging populations typically exhibit different mobility patterns than younger demographics, with potential implications for transport mode choices and trip frequencies [49,50]. This demographic transition may be exerting an influence on energy consumption patterns beyond what would be expected from population growth rates alone.

The multi-factorial influence of both economic and demographic variables on transport energy consumption underscores the need for integrated policy approaches. As emphasized by Champeecharoensuk et al. [17] in their analysis of aviation emissions in Thailand, addressing energy consumption challenges requires coordinated policies that account for both economic development objectives and demographic realities. The relatively modest contribution of these factors compared to vehicle-specific variables indicates that technological and behavioral interventions focused on the transport sector itself may yield more immediate impacts than broader economic or demographic policies.

These findings align with recent international research suggesting that transport energy consumption in developing economies follows complex, non-linear relationships with economic and demographic variables. Emami Javanmard et al. [28] observed similar complexity in their analysis of air transportation demand and emissions, highlighting the need for sophisticated modeling approaches that can capture these multidimensional relationships. The results demonstrate that XGBoost's ability to model non-linear interactions makes it particularly well-suited for capturing these complex socioeconomic influences on transport energy consumption.

6. Conclusions

This research addressed the critical challenges facing Thailand's transport sector in energy management and sustainable development. As the second-largest energy-consuming sector in Thailand, accounting for approximately 30% of total final energy consumption with a consistent upward trend over three decades, the transport sector presents significant challenges for energy security, environmental sustainability, and greenhouse gas emission reduction commitments. The complex nature of Thailand's transport energy consumption, characterized by its unique vehicle fleet composition, geographical disparities in transportation needs, and the growing impact of e-commerce and logistics services, necessitated a sophisticated forecasting approach.

The literature review identified several key research gaps. While previous studies have incorporated vehicle-related variables, none had comprehensively analyzed the combined impact of differentiated vehicle categories alongside their corresponding vehicle kilometers traveled data. There was also limited exploration of the long-term relationship between macroeconomic indicators and transport energy consumption in Thailand, and insufficient understanding of how the relationship between vehicle registration numbers and actual usage influences energy consumption predictions. To address these gaps, this study established four main objectives:

- to develop and compare the performance of LSTM and XGBoost models in predicting Thailand's transport energy consumption,
- to analyze the differential impact of various vehicle categories and their usage patterns on energy consumption,
- to identify and quantify the relative importance of different factors affecting transport energy consumption, and
- to provide evidence-based recommendations for transport energy policy planning.

The study leveraged a comprehensive dataset spanning from 1993 to 2022, which included vehicle registration data categorized by size, vehicle kilometers traveled for different vehicle types, macroeconomic indicators, and historical transport energy consumption data. This extensive temporal coverage provided a unique opportunity to analyze long-term patterns that most existing studies cannot address. This study implemented and compared two advanced machine learning approaches: Multivariate Long Short-Term Memory (LSTM) neural networks and the XGBoost algorithm. The models were evaluated using multiple metrics, including Mean Squared Error, Root Mean Squared Error, Mean Absolute Error, Mean Absolute Percentage Error, and R-squared values. The findings demonstrated that XGBoost consistently outperformed LSTM across all evaluation metrics, achieving an R-squared value of 0.9508 for test data compared to LSTM's 0.2005. Feature importance analysis revealed that medium-sized vehicles had the most substantial influence (36.6%) on transport energy consumption, followed by truck VKT (20.5%), suggesting that policies targeting fuel efficiency standards and freight transport optimization could yield substantial energy savings.

This research contributes to both academic understanding and practical policy implementation in several ways.

- First, it establishes the superior performance of XGBoost for transport energy forecasting in Thailand's context, challenging conventional assumptions about the superiority of recurrent neural networks for time-series forecasting.
- Second, it quantifies the differential impact of various vehicle categories and their usage patterns on energy consumption, providing a nuanced understanding of how fleet composition shapes transport energy demand.
- Third, it demonstrates the value of integrating both vehicle registration and vehicle usage data in predictive models, offering enhanced insights compared to approaches based solely on one data category.

Moreover, while this study focuses on Thailand, its methodology and findings have broader implications for other developing and middle-income economies with similar transportation characteristics. The proposed machine learning framework can be adapted to national contexts where energy demand is shaped by comparable macroeconomic conditions, vehicle fleet compositions, and infrastructure developments. Further validation across multiple regions is necessary, incorporating localized adjustments for variations in transport systems, fuel composition, and regulatory frameworks.

The dataset used, including vehicle registration, vehicle kilometers traveled (VKT), macroeconomic indicators, and historical energy consumption, aligns with data structures maintained by transport and energy agencies worldwide. Given this consistency, the XGBoost and LSTM models can be applied in other countries for high-precision energy forecasting. Feature importance analysis highlights medium-sized vehicles and freight transport as dominant contributors to energy demand, a pattern observed in various developing economies. These findings suggest that machine learning-based forecasting approaches are effective across diverse regional settings.

Beyond methodological contributions, this study presents a data-driven framework for transport energy policy planning. The predictive capability of these models supports policy evaluation in areas such as fuel efficiency regulations, vehicle electrification, and logistics optimization. Quantifying the impact of different vehicle categories on energy consumption allows policymakers to prioritize interventions such as fuel economy standards, freight transport efficiency improvements, and modal shifts from road to rail.

Given the global relevance of transport energy challenges, these models can inform international strategies for improving energy efficiency. Countries experiencing rapid urbanization and motorization could leverage these techniques to forecast fuel demand, optimize taxation policies, and design incentives for low-emission vehicles. Incorporating machine learning forecasting into multi-modal transport planning can also support sustainable urban mobility strategies.

Despite strong predictive performance, these models require adaptation to different national contexts due to variations in infrastructure, fuel types, and policy frameworks. Future research should validate their applicability in multiple regions, integrating localized variables and emerging trends such as autonomous mobility and renewable energy transitions. Refining these methodologies will enhance the role of machine learning in transport energy policy planning, supporting more effective and sustainable mobility solutions globally.

Based on the model results, this study offers the following evidence-based recommendations for transport energy policy planning in Thailand:

- The XGBoost models feature importance analyses providing clear direction for prioritizing transport energy interventions. Given that medium-sized vehicles (passenger cars and pick-up trucks) account for 36.6% of the influence on transport energy consumption, policies targeting this vehicle segment should receive the highest priority. This study recommends implementing progressively stringent fuel economy standards for new medium vehicles, coupled with financial incentives for low-emission alternatives. Tax structures should be redesigned to favor more efficient vehicles within this category, potentially through engine displacement or emissions-based taxation.

- The significant contribution of truck VKT (20.5%) indicates that freight transport efficiency represents a high-leverage intervention point. This study recommends developing a national freight efficiency program that includes driver training, logistics optimization, and aerodynamic retrofitting incentives. Additionally, strategic investment in rail freight infrastructure could yield substantial energy savings by shifting appropriate cargo from road to more efficient transport modes. Time-of-day delivery regulations in urban centers could further reduce congestion-related energy waste in freight operations.
- For the combined 22.7% importance of small and large vehicle registrations, the result recommends differentiated policy approaches. For large vehicles, implementing scrappage programs targeting the oldest, most inefficient models would accelerate fleet modernization. For small vehicles, particularly motorcycles, policies facilitating the transition to electric two-wheelers through purchase incentives and charging infrastructure would address their collective impact despite individual efficiency.
- The relatively modest direct influence of GDP (5.6%) suggests opportunities for decoupling economic growth from transport energy consumption. The result recommends incorporating transport energy considerations into economic development planning, particularly through transit-oriented development policies that reduce travel demand while maintaining economic growth. Similarly, the 9.6% contribution from population factors points to the importance of integrated land-use and transportation planning to manage mobility needs as Thailand's demographics continue to evolve.
- The performance advantage of XGBoost over LSTM in the study suggests that transportation authorities would benefit from implementing similar machine learning approaches for ongoing energy forecasting. This study recommends establishing a data integration framework that regularly updates the predictive models with current vehicle registration, usage patterns, and economic indicators to maintain forecast accuracy and adjust policies accordingly.
- Finally, the complex interactions between vehicle categories, usage patterns, and macroeconomic factors identified in the model highlight the need for a coordinated, cross-sectoral approach to transport energy management. The result recommends establishing an inter-ministerial transport energy task force to ensure policy coherence across economic development, environmental, and transport sectors, with regular reassessment based on updated predictive modeling to track progress toward energy efficiency goals.

This study is limited by focusing exclusively on historical patterns without addressing potential technological disruptions in transportation.

- While demonstrating superior predictive performance, the XGBoost model primarily extrapolates from existing relationships between vehicles, usage, and energy consumption. The approach may fail to adequately capture the impact of emerging technologies like electric vehicles, autonomous systems, and mobility-as-a-service platforms that could fundamentally transform transport-energy dynamics.
- Future research should integrate technological transition scenarios into forecasting models through hybrid approaches that combine machine learning with systems dynamics modeling. Incorporating detailed data on fuel types, powertrain technologies, and charging infrastructure would enhance the model's sensitivity to ongoing energy transitions in transportation.

Author Contributions: Conceptualization, T.C., T.J., C.S., F.W. and S.J.; formal analysis, T.C., C.B. and F.W.; funding acquisition, S.J.; methodology, T.J. and S.J.; project administration, S.J.; software, C.B. and C.S.; supervision, V.R.; validation, T.J. and C.S.; visualization, F.W.; writing—original draft, T.C. and S.J.; writing—review and editing, V.R. All authors have read and agreed to the published version of the manuscript.

Funding: This research was funded by Suranaree University of Technology (SUT), grant number IRD7-704-65-12-23.

Data Availability Statement: The data presented in this study are available on request from the corresponding author.

Conflicts of Interest: The authors declare no conflicts of interest.

Abbreviations

The following abbreviations are used in this manuscript:

LSTM	Long Short-Term Memory
XGBoost	Extreme Gradient Boosting
VKT	Vehicle Kilometers Traveled
GDP	Gross Domestic Product
MTOE	million Tons of Oil Equivalent

Appendix A

Table A1. Additional information about previous studies.

Authors	Model Performance Metrics	Dataset Characteristics	Study Limitations
Ji et al. [13]	- Models I and II evaluated using R^2 and *RMSE* - Six measures assess prediction accuracy - Smaller MABE indicates more reliable results	- GDP per capita and population data - Automobile kilometers traveled per year - Energy consumption and CO_2 emissions in transportation - Data spans 2009–2022	- Machine learning models rely on specific assumptions and simplifications - Additional influential factors not considered - Findings may not apply to other countries or regions - Projections based on historical data and present patterns - Unanticipated occurrences may alter emission and energy usage trends
Hoxha et al. [16]	- Stacking ensemble models outperform single algorithms - Using all features leads to better performance than two features - Best ensemble model achieved R^2 of approximately 0.99 - Hyperparameter tuning essential for model performance	- GDP, population, vehicle-km, passenger-km, ton-km, oil price - Data spans 1975–2019 (33 observations) - Sources include General Directorate of Turkish Highways - Focus on transportation energy demand in Turkey	- Simplifying assumptions may affect modeling accuracy - Performance may vary across different datasets or regions - Additional data sources could enhance model performance - Potential biases in dataset may exist
Çınarer et al. [15]	- ANN model: R^2 of 0.58, *RMSE* of 20.8, *MAE* of 14.4 - XGBoost performed best in scenario 4 with high R^2 values	- Transportation energy consumption, vehicle kilometers, population, GDP, CO_2 emissions - Data spans 1970–2016 - Input parameters: VK, GDP, POP, ENERGY; output: CO_2 emissions - Sources: World Bank and Turkish General Directorate of Highways	- Only three AI algorithms applied - Five input parameters assumed linear relationships - Other influencing factors not considered - Scenarios based on limited correlation effects
Ağbulut [2]	- R^2 values: 0.8639 to 0.9235 across algorithms - *RMSE* below 5×10^6 tons for CO_2 emissions - Model II shows better results in testing	- Dataset spans 1970–2016 - Includes GDP, population, vehicle kilometers - CO_2 emissions and energy consumption data - Data sourced from World Bank and Turkish Statistical Institute	- No specific limitations mentioned in the paper

Table A1. *Cont.*

Authors	Model Performance Metrics	Dataset Characteristics	Study Limitations
Henriques et al. [25]	- Gradient boosting robust for benchmarking performance - AHC recall rate: 64% for high-profile dwellings - AHC precision for low-profile dwellings: 92% - K-means clustering outperforms other methods	- Monthly consumption data for 383 households - 12 months of consumption data in kWh - 373 rows after preprocessing	- Defining true positives/negatives challenging in unlabeled datasets - Misclassification errors can lead to incorrect billing and energy management decisions - Need for context-specific adaptations in clustering methods
Champeecharoensuk et al. [17]	- Performance not explicitly quantified in the paper - Model validation through comparison of estimated and actual values	- Data from 39 airports in Thailand - 638 registered airplanes with Thai Nationality Mark - Data spans 2015–2020 for GHG emissions estimation - Key factors include energy consumption, fuel costs, and GDP	- Study limited to GHG emissions data from 2008 to 2020 - Lacks specific aircraft details and flight timing information - Additional factors influencing GHG emissions remain unexamined - Data availability constraints hinder accurate GHG emissions accounting
Zhang et al. [14]	- XGBoost shows better applicability and accuracy - *RMSE* as low as 0.036 - MLP neural networks outperformed regression models	- Four Chinese megacities: Beijing, Tianjin, Shanghai, Chongqing - Data spans 2003–2017 - Energy supply data for carbon emissions from fossil fuel combustion - Energy consumption data from China Energy Statistical Yearbook	- Further exploration using other methods and data needed - Study primarily focuses on megacities, limiting broader applicability - Single-factor analysis may overlook multifactor synergies in carbon emissions
Rahman et al. [20]	- SVR outperformed ANN in energy demand prediction - Correlation coefficients: 0.8932 (SVR) and 0.9925 (ANN) - Satisfactory performance on training and testing datasets - Performance indices: *RMSE*, *MAE*, *MAPE*, CC, and IA	- Time-series data from 1996 to 2017 - Variables include GDP, fuel price, and urban population - Data used for training and testing machine learning models	- No specific limitations addressed in the paper
Lin et al. [8]	- Optimal energy efficiencies observed for rail and water transport in 2011 - Future energy consumption estimates using extended DEA model	- Energy consumption data from IEA - TKM and PKM data from NBS of China - Future capacity data from ERI's report on Low Carbon Development	- Energy consumption apportionment to passenger or freight transport is challenging - Data from Taiwan, Hong Kong, and Macau excluded due to unavailability - Assumptions affect accuracy of efficiency assessments
Pongthanaisawan et al. [1]	- Model performance verified through comparison of estimated and actual fuel consumption - *MBE* and *RMSE* calculated - Most models accurately estimate total fuel consumption	- Data collected from official sources for 1989–2007 - Dependent and independent variables for fuel demand forecasting models	- Biofuel supply limitations constrain long-term GHG emissions reduction potential. - Rail transport infrastructure fully utilized, limiting fuel demand response. - Econometric models struggle with sudden fuel demand changes due to oil price shocks.
Supasa et al. [7]	- Not explicitly quantified in the paper	- 2000 and 2010 Thailand IO tables - Physical energy use data from Thailand's Energy Situation Annual Reports - Non-competitive imports assumption adopted	- Least energy-intensive sectors neglected in energy conservation policies - Energy policies should reflect consumption characteristics for better interventions
Shams Amiri et al. [21]	- Random Forest outperforms Neural Network - *RMSE*, *MAE*, and R^2 metrics used - Cross-validation ensures model accuracy	- 9235 households and 20,216 residents (2015) - Travel behaviors from Pennsylvania and New Jersey counties - Household Travel Survey (HTS) as primary dataset - Additional zonal boundaries and neighborhood data	- Study focuses only on household demographics and travel characteristics - Other factors like vehicle type and road conditions excluded

References

1. Pongthanaisawan, J.; Sorapipatana, C. Greenhouse gas emissions from Thailand's transport sector: Trends and mitigation options. *Appl. Energy* **2013**, *101*, 288–298.
2. Ağbulut, Ü. Forecasting of transportation-related energy demand and CO_2 emissions in Turkey with different machine learning algorithms. *Sustain. Prod. Consum.* **2022**, *29*, 141–157. [CrossRef]
3. Satiennam, T.; Jaensirisak, S.; Satiennam, W.; Detdamrong, S. Potential for modal shift by passenger car and motorcycle users towards Bus Rapid Transit (BRT) in an Asian developing city. *IATSS Res.* **2016**, *39*, 121–129.

4. Department of Land Transport. Statistics of Vehicle Registration. Available online: https://web.dlt.go.th/statistics/ (accessed on 17 October 2023).
5. Corlu, C.G.; de la Torre, R.; Serrano-Hernandez, A.; Juan, A.A.; Faulin, J. Optimizing energy consumption in transportation: Literature review, insights, and research opportunities. *Energies* **2020**, *13*, 1115. [CrossRef]
6. Chindaprasirt, P.; Klungboonkrong, P.; Jaensirisak, S.; Faiboun, N.; Long, S.; Tippichai, A.; Taylor, M.A. Integrated Urban Transport and Land-Use Policies in Reducing CO_2 Emissions and Energy Consumption: Case Study of a Medium-Sized City in Thailand. *World Electr. Veh. J.* **2024**, *15*, 349. [CrossRef]
7. Supasa, T.; Hsiau, S.-S.; Lin, S.-M.; Wongsapai, W.; Chang, K.-F.; Wu, J.-C. Sustainable energy and CO_2 reduction policy in Thailand: An input–output approach from production- and consumption-based perspectives. *Energy Sustain. Dev.* **2017**, *41*, 36–48.
8. Lin, W.; Chen, B.; Xie, L.; Pan, H. Estimating Energy Consumption of Transport Modes in China Using DEA. *Sustainability* **2015**, *7*, 4225–4239. [CrossRef]
9. Sun, L.; Zhang, T.; Liu, S.; Wang, K.; Rogers, T.; Yao, L.; Zhao, P. Reducing energy consumption and pollution in the urban transportation sector: A review of policies and regulations in Beijing. *J. Clean. Product.* **2021**, *285*, 125339.
10. Morrow, W.R.; Gallagher, K.S.; Collantes, G.; Lee, H. Analysis of policies to reduce oil consumption and greenhouse-gas emissions from the US transportation sector. *Energy Policy* **2010**, *38*, 1305–1320.
11. Karim, F.; Majumdar, S.; Darabi, H.; Harford, S. Multivariate LSTM-FCNs for time series classification. *Neural Netw.* **2019**, *116*, 237–245.
12. Duan, G.; Su, Y.; Fu, J. Landslide Displacement Prediction Based on Multivariate LSTM Model. *Int. J. Environ. Res. Public Health* **2023**, *20*, 1167. [CrossRef] [PubMed]
13. Ji, T.; Li, K.; Sun, Q.; Duan, Z. Urban transport emission prediction analysis through machine learning and deep learning techniques. *Transp. Res. Part D* **2024**, *135*, 104389. [CrossRef]
14. Zhang, J.; Zhang, H.; Wang, R.; Zhang, M.; Huang, Y.; Hu, J.; Peng, J. Measuring the Critical Influence Factors for Predicting Carbon Dioxide Emissions of Expanding Megacities by XGBoost. *Atmosphere* **2022**, *13*, 599. [CrossRef]
15. Çınarer, G.; Yeşilyurt, M.K.; Ağbulut, Ü.; Yılbaşı, Z.; Kılıç, K. Application of various machine learning algorithms in view of predicting the CO_2 emissions in the transportation sector. *Sci. Technol. Energy Transit.* **2024**, *79*, 15. [CrossRef]
16. Hoxha, J.; Çodur, M.Y.; Mustafaraj, E.; Kanj, H.; El Masri, A. Prediction of transportation energy demand in Türkiye using stacking ensemble models: Methodology and comparative analysis. *Appl. Energy* **2023**, *350*, 121765. [CrossRef]
17. Champeecharoensuk, A.; Dhakal, S.; Chollacoop, N.; Phdungsilp, A. Greenhouse gas emissions trends and drivers insights from the domestic aviation in Thailand. *Heliyon* **2024**, *10*, e24206. [CrossRef]
18. Zhang, R.; Fujimori, S. The role of transport electrification in global climate change mitigation scenarios. *Environ. Res. Lett.* **2020**, *15*, 034019. [CrossRef]
19. Ghanbari, R.; Borna, K. Multivariate Time-Series Prediction Using LSTM Neural Networks. In Proceedings of the 2021 26th International Computer Conference, Computer Society of Iran (CSICC), Tehran, Iran, 3–4 March 2021.
20. Rahman, M.M.; Rahman, S.M.; Shafiullah, M.; Hasan, M.A.; Gazder, U.; Al Mamun, A.; Mansoor, U.; Kashifi, M.T.; Reshi, O.; Arifuzzaman, M.; et al. Energy Demand of the Road Transport Sector of Saudi Arabia—Application of a Causality-Based Machine Learning Model to Ensure Sustainable Environment. *Sustainability* **2022**, *14*, 16064. [CrossRef]
21. Shams Amiri, S.; Mostafavi, N.; Lee, E.R.; Hoque, S. Machine learning approaches for predicting household transportation energy use. *City Environ. Interact.* **2020**, *7*, 100044. [CrossRef]
22. Mohsin, M.; Abbas, Q.; Zhang, J.; Ikram, M.; Iqbal, N. Integrated effect of energy consumption, economic development, and population growth on CO(2) based environmental degradation: A case of transport sector. *Environ. Sci. Pollut. Res. Int.* **2019**, *26*, 32824–32835. [CrossRef]
23. Limanond, T.; Jomnonkwao, S.; Srikaew, A. Projection of future transport energy demand of Thailand. *Energy Policy* **2011**, *39*, 2754–2763.
24. Chai, J.; Lu, Q.-Y.; Wang, S.-Y.; Lai, K.K. Analysis of road transportation energy consumption demand in China. *Transp. Res. Part D* **2016**, *48*, 112–124. [CrossRef]
25. Henriques, L.; Castro, C.; Prata, F.; Leiva, V.; Venegas, R. Modeling Residential Energy Consumption Patterns with Machine Learning Methods Based on a Case Study in Brazil. *Mathematics* **2024**, *12*, 1961. [CrossRef]
26. Antonopoulos, I.; Robu, V.; Couraud, B.; Kirli, D.; Norbu, S.; Kiprakis, A.; Flynn, D.; Elizondo-Gonzalez, S.; Wattam, S. Artificial intelligence and machine learning approaches to energy demand-side response: A systematic review. *Renew. Sustain. Energy Rev.* **2020**, *130*, 109899. [CrossRef]
27. Alabi, T.M.; Aghimien, E.I.; Agbajor, F.D.; Yang, Z.; Lu, L.; Adeoye, A.R.; Gopaluni, B. A review on the integrated optimization techniques and machine learning approaches for modeling, prediction, and decision making on integrated energy systems. *Renew. Energy* **2022**, *194*, 822–849.
28. Emami Javanmard, M.; Tang, Y.; Martínez-Hernández, J.A. Forecasting air transportation demand and its impacts on energy consumption and emission. *Appl. Energy* **2024**, *364*, 123031.

29. Wang, S.; Ji, B.; Zhao, J.; Liu, W.; Xu, T. Predicting ship fuel consumption based on LASSO regression. *Transp. Res. Part D* **2018**, *65*, 817–824.
30. Bülte, C.; Kleinebrahm, M.; Yilmaz, H.Ü.; Gómez-Romero, J. Multivariate time series imputation for energy data using neural networks. *Energy AI* **2023**, *13*, 100239.
31. Chen, T.; Guestrin, C. Xgboost: A scalable tree boosting system. In Proceedings of the 22nd ACM SIGKDD International Conference on Knowledge Discovery and Data Mining, San Francisco, CA, USA, 13–17 August 2016; pp. 785–794.
32. Xu, Y.; Zhao, X.; Chen, Y.; Yang, Z. Research on a mixed gas classification algorithm based on extreme random tree. *Appl. Sci.* **2019**, *9*, 1728. [CrossRef]
33. Sun-Youn, S.; Sun-Youn, S.; Han-Gyun, W.; Han-Gyun, W. Energy Consumption Forecasting in Korea Using Machine Learning Algorithms. *Energies* **2022**, *15*, 4880. [CrossRef]
34. Emmanuel Hidalgo, G.; Jacqueline, G.; Matthew, J.B.; Boriboonsomsin, K. Comparative Assessment of Machine Learning Techniques for Modeling Energy Consumption of Heavy-Duty Battery Electric Trucks. In Proceedings of the 2024 Forum for Innovative Sustainable Transportation Systems (FISTS), Riverside, CA, USA, 26–28 February 2024.
35. Cortez, J.C.; Terada, L.Z.; Bandeira, B.V.B.; Soares, J.; Vale, Z.; Rider, M.J. Comparative Analysis of ARIMA, LSTM, and XGBoost for Very Short-Term Photovoltaic Forecasting. In Proceedings of the 2023 15th Seminar on Power Electronics and Control (SEPOC), Santa Maria, Brazil, 22–25 October 2023; IEEE: Piscataway, NJ, USA, 2023; pp. 1–6.
36. Al-Selwi, S.M.; Hassan, M.F.; Abdulkadir, S.J.; Muneer, A.; Sumiea, E.H.; Alqushaibi, A.; Ragab, M.G. RNN-LSTM: From applications to modeling techniques and beyond—Systematic review. *J. King Saud Univ.* **2024**, *36*, 102068.
37. Alsahaf, A.; Petkov, N.; Shenoy, V.; Azzopardi, G. A framework for feature selection through boosting. *Expert Syst. Appl.* **2022**, *187*, 115895. [CrossRef]
38. Wang, W.; Shi, W.; Nan, D.; Peng, Y.; Wang, Q.; Zhu, Y. New energy output prediction and demand response optimization based on LSTM-BN. *Int. J. Renew. Energy Develop.* **2025**, *14*, 72–82. [CrossRef]
39. Karim, F.K.; Khafaga, D.S.; El-kenawy, E.-S.M.; Eid, M.M.; Ibrahim, A.; Abualigah, L.; Khodadadi, N.; Abdelhamid, A.A. Optimized LSTM for Accurate Smart Grid Stability Prediction Using a Novel Optimization Algorithm. *Front. Energy Res.* **2024**, *12*, 1399464.
40. El-Naggar, N.; Madhyastha, P.; Weyde, T. Exploring the long-term generalization of counting behavior in RNNs. *arXiv* **2022**, arXiv:2211.16429.
41. Goel, S.; Bajpai, R. Impact of uncertainty in the input variables and model parameters on predictions of a long short term memory (LSTM) based sales forecasting model. *Mach. Learn. Knowl. Extr.* **2020**, *2*, 14. [CrossRef]
42. Kandadi, T.; Shankarlingam, G. DRAWBACKS OF LSTM ALGORITHM: A CASE STUDY. *SSRN* 2025. [CrossRef]
43. Lazcano, A.; Hidalgo, P.; Sandubete, J.E. Walking Back the Data Quantity Assumption to Improve Time Series Prediction in Deep Learning. *Appl. Sci.* **2024**, *14*, 11081. [CrossRef]
44. Dahiru, A.B.; Mohammed, S. Forecasting United Kingdom's energy consumption using machine learning and hybrid approaches. *Energy Environ.* **2022**, *35*, 1493–1531.
45. Hamed, A.; Behzad Rashidi, M.; Arian, R.; Hossein, K.; Mohsen Asghari, I. Comparative study of long short-term memory (LSTM), bidirectional LSTM, and traditional machine learning approaches for energy consumption prediction. *Energy Explor. Exploit.* **2024**, *43*, 281–301.
46. Hamidreza, E.; Hassan, S.; Muhammad, R.; Mobina, M. Innovative framework for accurate and transparent forecasting of energy consumption: A fusion of feature selection and interpretable machine learning. *Appl. Energy* **2024**, *366*, 123314.
47. Ahmad, T.; Zhang, D. A critical review of comparative global historical energy consumption and future demand: The story told so far. *Energy Rep.* **2020**, *6*, 1973–1991.
48. McKinnon, A.C.; Ge, Y. The potential for reducing empty running by trucks: A retrospective analysis. *Int. J. Phys. Distrib. Logist. Manag.* **2006**, *36*, 391–410.
49. De Vos, J.; Alemi, F. Are young adults car-loving urbanites? Comparing young and older adults' residential location choice, travel behavior and attitudes. *Transp. Res. Part A* **2020**, *132*, 986–998. [CrossRef]
50. Champahom, T.; Jomnonkwao, S.; Nambulee, W.; Klungboonkrong, P.; Karoonsoontawong, A.; Ratanavaraha, V. Analyzing transport mode choice for aging society in Thailand. *Eng. Appl. Sci. Res.* **2020**, *47*, 383–392.

MDPI AG
Grosspeteranlage 5
4052 Basel
Switzerland
Tel.: +41 61 683 77 34

Energies Editorial Office
E-mail: energies@mdpi.com
www.mdpi.com/journal/energies

www.ingramcontent.com/pod-product-compliance
Lightning Source LLC
Chambersburg PA
CBHW061457210926
51850CB00008B/1745